Encyclopaedia of Mathematics

Encyclopaedia of Mathematics
(Set of 3 Vol.)

Prof.Rajiv Kumar Gupta

CYBER TECH PUBLICATIONS
4264/3, Ansari Road, Daryaganj New Delhi - 110002 (India)
Phone : 011-23244078, 011-43559448, Fax : 011-23280028
Email : cybertechpublications@gmail.com

CYBER TECH PUBLICATION
4264/3, Ansari Road, Daryaganj
New Delhi - 110002 (India)
Phone : 011-23244078, 011-43559448,
Fax : 011-23280028
E-mail : cybertechpublications@gmail.com
Website : www.cybertechpublications.com

Encyclopaedia of Mathematics

First Edition : 2012

Rs. 2950.00

Printed at :

New Green Print Solutions, New Delhi 110078

PREFACE

The present book deals with in general properties of matter and sound. Reason fore this change is that paper division in the examination is different Universities. One of salient course syllabi features of the present edition of the book is the inclusion of a large number of typical worked out problems which will elucidate various abstract principles and theories discussed in the text. It covers the complete syllabus of physics prescribed by Technical Universities.

The treatment given is simple, lucid and comprehensive. The language used is simple. The classification of the questions, specially the short-answer type questions, will, I hope, help the students to prepare for the final examination. I have spared no pains in applying my long experience of degree class teaching to make the book useful to the students. I shall, therefore, consider, my labours amply rewarded if the book receives warm reception and appreciation from the teachers and the taught. Though nothing can be claimed as original but the subject matter has been arranged in my own style. Suggestion for improvement of the book will be gratefully acknowledged.

—Authors

Contents

1

INTIGRATION OF VECTORS

INTRODUCTION

Line Integral : Let dl be an element of length at a point on a smooth curve AB drawn in a vector field and F, a continuous vector point function, or vector; inclined at an angle θ to dl. Such that it continuously varies in magnitude as well as direction as we proceed along the curve. Then, the integral

$$\int_A^B F.dl = \int_A^B F\cos\theta dl$$

is referred to as the line integral of vector F along the curve AB.

In terms of the components of F along the three Cartesian coordinates, we have

$$\int_A^B F\ dl = \int_A^B \left(F_x\ dx + F_y dy + F_z dz\right)$$

Thus, if F represents the force acting on a particle moving along the curve from A to B, the line integral $\int_A^B F.dl$ represents the total work done by the force during the motion of the particle over its entire path from A to B.

If the value of the line integral depends only upon the location of the two points in the vector field and not upon the actual path taken between them, the vector field is referred to as a conservative field. Familiar examples of such fields are the electrostatic, magnetic and gravitational fields.

If, therefore, F represents the value of the electric (or magnetic) field intensity at the point P, the line integral $\int_A^B F.'dl$ represents the work done on unit charge (or unit pole) during its motion from A to B (*i.e.*, the potential difference between A and B) *irrespective of the path taken.*

In Hydrodynamics, the line integral of a continuous vector point function or vector (F) along a closed curve is called the circulation of F along the curve.

And, if the circulation of a vector point function along every closed curve in a region be zero, it is said to be irrotational in that region.

Relationship Between Line Integral and Curl

There is a definite relationship between the curl of a vector field F at a point and its line integral along the boundary C of an infinitesimal plane area around and including that point. To clearly bring this out, let us, for convenience, calculate the line integral of F around an infinitesimal plane rectangular area ABCD, of sides AB = δx and BC = δy in the x = y plane surrounding and including a point P.

Let the components of F along the axes of x and y be F_x and F_y respectively. Then, since the sides of the rectangle are small, the average values of these components along them may respectively be taken to be the same as those at their mid-points. So that, we have

average value of x-component along the path AB $= F_x + \frac{\partial F_x}{\partial x}.\frac{\delta x}{2}$

and average value of x-component along the path CD

$$= F_x + \frac{\partial F_x}{\partial x}.\frac{\delta x}{2} + \frac{\partial F_s}{\partial y}\delta y.$$

Remembering that the path CD is oppositely directed to the x-axis, we have *line integral of* F *along* AB + *line integral of* F *along* BC given by

$$\int_A^B F\ dl + \int_C^D F.dl = \left(F_x + \frac{\partial F}{\partial x}\partial x\right)\delta x - \left(F_x + \frac{\partial F_x}{\partial x}\frac{\delta x}{2} + \frac{\partial F_x}{\partial y}\partial y\right)\delta x$$

$$= \frac{F_x}{\partial y}\partial x\delta y \qquad \text{...(i)}$$

Exactly, in the same manner and remembering that DA is oppositely directed to the y-axis, we have

line integral of F *along* BC + *line integral of* F *along* DA *given by*

$$\int_B^C F.dl + \int_D^A F.dl = \frac{\partial F_y}{\partial y}\partial x\ \delta x\ \partial y \qquad \text{...(ii)}$$

Thus, adding relation (i) and (ii), we have line integral over the entire boundary C the rectangle ABCD given by

$$\oint_{ABCD} F.dl = \oint_C F.dl = \left(\frac{\partial F_x}{\partial x} - \frac{\partial F_x}{\partial y}\right)\delta x\delta y \qquad \text{...(iii)}$$

Now, the z-component of curl F at the point being the above line integral per unit area, we have

$$\left(\frac{\partial F_y}{\partial x} - \frac{\partial F_x}{\partial y}\right) = \left[(\text{curl}\,F).k\right] = (\text{curl}\,F)_z$$

So that, if the small area δxδy be regarded as the vector area $\delta_{zs}k$ (where k is the unit vector along the axis of z), we have

$$\oint_{ABCD} F.dl = \oint_C F.dl = (\text{curl } F)_z.\delta_{s_z}k.$$

It follows, therefore, that if we have an infinitesimal plane area at the point in question given by $\partial s = \delta_{s_x}i + \delta_{s_z}j + \delta_{s_z}k$, with a boundary C and ted in any direction, we have

$$\oint_C F.dl = \text{curl } F.\delta s, \text{ the maximum of } \oint F.dl \text{ being } |\text{curl } F|\ \delta s.$$

We thus see that *the magnitude of curl* F *at a point is the maximum value of the line integral of* F *per unit area along the boundary* C *of an infinitesimal plane area* (δs) *at that point.*

And the direction of the curl is perpendicular to the plane of the infinitesimal area (when the value of the line integral is the line integral is the maximum) in accordance with the right-handed screw rule.

It has been mentioned earlier that the curl of a conservative vector field is zero at all points in space. This may easily be seen from the following:

The line integral between any two points P and Q in the case of a conservative vector field is, as we know, a constant property of those points, irrespective of the path taken between them.

It follows, therefore that if we choose two different paths I and II between the two points, we shall have

$$\underset{\text{Path I}}{\int_P^Q F.dl} = \underset{\text{Path II}}{\int_P^Q F.dl}$$

If Q be brought nearer and nearer to P, path I becomes shorter and in the limit when Q is made to coincide with P, path I is reduced to zero and hence the integral along this, or any other path too, is reduced to zero. Thus the line integral of F along any closed path (also called *circulation of field* F *around the closed path*), beginning and ending at the same point, is zero in the case of a conservative field, *i.e.*,

$$\oint_{\text{Closed path}} F.dl = 0,$$

Since this relation holds good for all conservative fields irrespective of the closed path chosen these field must have zero curl at all points in space. That is why they are referred to as curl-free or non-curl fields.

Surface Integral

Imagine a smooth surface S drawn in a vector field and a continuously varying vector point function or vector F at a point P in a small element dS of the surface, at an angle θ with the normal to the surface, at the point (drawn outwards if the surface be closed and always towards the same side otherwise). Then, the integral $\iint_S \mathrm{F.SdS} = \iint_S \mathrm{F}\cos\theta\, \mathrm{dS}$ over the entire surface is called the *normal surface integral,* or generally, simply the *surface integral* of vector F over the surface. In terms of the Cartesian components of F, we have

$$\iint_S \mathrm{F.dS} = \iint\left(F_x dS_x + F_y dS_y + F_z dS_z\right)$$

The surface integral of the normal component of a continuous vector point function F over a closed surface S is called the *flux of* F across the surface. As we have seen before (ii), if the flux of a vector point function across every closed surface in a region be zero, it is said to be *solenoidal* in that region.

Example:

(i) if F represents the electric or magnetic induction at the point P the surface integral $\iint_S \mathrm{F.dS}$ *represents the total normal induction over the surface.*

(ii) If the surface S be drawn over a region of a moving or a flowing fluid such that its velocity v varies from point the surface integral of v. viz., $\iint_S \mathrm{v.dS}$ *gives the volumes rate of the fluid cross the surface.*

(iii) Volume integral. Suppose we have a surface enclosing a region of volume V and that F is a vector point function at a point in a small element dV of the region. Then, the integral $\iiint_V \mathrm{F\, dV}$, *covering the entire region, is called the volume integral of vector F over the surface.*

In terms of the Cartesian components, we have

volume integral $\iiint_V \mathrm{F\, dV} = i\iiint \hat{V}\, F_x dxdydz + j\iiint \hat{V}\, F_y dxdydz$

$$+k\iiint \hat{V}\, F_z dxdydz..$$

LINE INTEGRAL OF A VECTOR FIELD AROUND AN INFINITESIMAL RECTANGLE

The curl of a vector field $\vec{F}$ at a point is related to the line integral of $\vec{F}$ along the boundary C of an infinitesimal plane area around that point which encloses it. To establish the relation between these two quantities (*i.e.*, curl and line integral), let us evaluate the line integral of the vector field $\vec{F}$ around an infinitesimal rectangle of sides PQ = δx and QR = δy in the X – Y plane. Let the x and y-components of $\vec{F}$ at the point P be F_x and F_y respectively. Since, we are dealing with an infinitesimal rectangle it means the sides are very small, so that, the component of the vector at the middle point of any side may be taken as the average value for all points on it. Thus we have

The average value of x-component along the path $PQ = F_x + \frac{\delta F_x}{\delta x}\frac{\delta x}{2}$ and along the path

$$RS = F_x + \frac{\delta F_x}{\delta x}.\frac{\delta x}{2} + \frac{\delta F_x}{\delta y}\delta y$$

Since the direction of the path RS is towards the negative direction of x, therefore, the line integral of $\vec{F}$ along PQ plus line integral of $\vec{F}$ along RS is given by,

$$\int_P^Q \vec{F}.d\vec{l} + \int_R^S \vec{F}.d\vec{l} = \left(F_x + \frac{\delta F_x}{\delta x}.\frac{\delta x}{2}\right)\delta x - \left(F_x + \frac{\delta F_x}{\delta x}.\frac{\delta x}{2} + \frac{\delta F_x}{\delta y}\right)\delta x$$

$$= -\frac{\delta F_x}{\delta y}\delta x\, \delta y$$

Again remembering that SP is towards the negative direction of y and proceeding in the same way, we get,

$$\int_Q^R \vec{F}.d\vec{l} + \int_S^P \vec{F}.d\vec{l} = \frac{\delta F_y}{\delta x}\delta x\, \delta y$$

We get the line integral,

$$\oint_{PQRS} \vec{F}.d\vec{l} = \left(\frac{\delta F_y}{\delta x} - \frac{\delta F_x}{\delta y}\right)\delta x\, \delta y$$

That $\left(\frac{\delta F_y}{\delta x} - \frac{\delta F_x}{\delta y}\right)$ is the z-component of curl $\vec{F}$ or $\vec{\nabla} \times \vec{F}$. If the small area δxδy may be regarded as the magnitude of the vector area $\delta S_z \hat{k}$, thus,

$$\oint_{PQRS} \vec{F}.d\vec{l} = (\text{curl } \vec{F})_z dS_z = (\text{curl } \vec{F})_z.\delta S_z \hat{k}$$

This relation is valid for any infinitesimal area at P, if

$\delta S = \delta S_x \hat{i} + \delta S_y \hat{j} + \delta S_z \hat{k}$ with boundary C than we have

$$\oint_C \vec{F}.d\vec{l} = \text{curl } \vec{F}.\delta\vec{S}$$

The maximum value of $\oint_C \vec{F}.d\vec{l} = \left|\text{curl } \vec{F}\right| \delta S$

$$\left|\text{curl } \vec{F}\right| = \frac{\left[\oint_C \vec{F}.d\vec{l}\right]_{max}}{\delta S}$$

Thus, *the magnitude of curl* $\vec{F}$ *at a point is the maximum value of the line integral of* $\vec{F}$ *per unit area along the boundary C of an infinitesimal plane area at that point.* And the direction of the curl is perpendicular to the plane to the infinitesimal area, corresponding to the maximum value of integral, there will be an orientation of the area for which the line integral is maximum *i.e.*, the value of line integral depends upon the orientation of the given small vector area at the point.

LAPLACIAN TRANSFORM

The div grad ϕ gives a new operator as given below :

$$\vec{\nabla}\cdot(\vec{\nabla}\,\phi) = \vec{\nabla}\cdot\vec{\nabla}\phi = (\vec{\nabla}\cdot\vec{\nabla})\,\phi = \nabla^2\phi$$

It is read as del squared (or nabla squared) ϕ. It appears oflen in physics, it has been given a special name – the Laplacian operator

$$= \vec{\nabla}\,.\,\vec{\nabla}$$

$$\text{Laplacian} = \nabla^2 = \frac{\delta^2}{\delta x^2} + \frac{\delta^2}{\delta y^2} + \frac{\delta^2}{\delta z^2}$$

As it is the dot or scalar product of two vectors ($\vec{\nabla}\cdot\vec{\nabla}$), it is a scalar operator

Curl of the Curl

Let us evaluate $\vec{\nabla}\times\vec{\nabla}\times\vec{F}$ *i.e.*, the curl of the curl, which can be written by the use of vector equality,

$$\vec{A}\times(\vec{B}\times\vec{C}) = \vec{B}(\vec{A}.\vec{C}) - (\vec{A}.\vec{B})\vec{C}$$

here $\quad \vec{\nabla}\times(\vec{\nabla}\times\vec{F}) = \vec{\nabla}(\vec{\nabla}\cdot\vec{F}) - (\vec{\nabla}\cdot\vec{\nabla})\vec{F}$

or $\quad \vec{\nabla}\times(\vec{\nabla}\times\vec{F}) = \vec{\nabla}(\vec{\nabla}\cdot\vec{F}) - \nabla^2\vec{F}$

Finally $\vec{\nabla}(\vec{\nabla}\cdot\vec{F})$. **It is a vector field,**

So we can write as our conclusion,

(a) $\vec{\nabla}.(\vec{\nabla}.\phi) = \nabla^2\phi$ = a scalar field
(b) $\vec{\nabla}\times(\vec{\nabla}\phi) = 0$
(c) $\vec{\nabla}(\vec{\nabla}.\vec{F})$ = a vector field
(d) $\vec{\nabla}.(\vec{\nabla}\times\vec{F}) = 0$
(e) $\vec{\nabla}\times(\vec{\nabla}\times\vec{F}) = \vec{\nabla}(\vec{\nabla}.\vec{F}) - \nabla^2\vec{F}$
(f) $(\vec{\nabla}.\vec{\nabla})\vec{F} = \nabla^2\vec{F}$ = a vector field

FLUX THROUGH A SURFACE OR SURFACE INTEGRAL

Let us imagine a smooth surface in the vector field then, consider a continuously varying $\vec{F}$ at a point P inside a small surface element of area dS. Draw the positive normal of unit length n on the element and let θ be the angle between $\hat{n}$ and $\vec{F}$.

The component of $\vec{F}$ perpendicular to dS is F Cos θ = $\vec{F}.\hat{n}$. Let us consider the flow of fluid, then the *net amount of fluid going out through the surface is the "flux of velocity"* through surface. For any arbitrary closed surface, the net outward flow or flux is *the average outward normal component of the velocity times the area or the surface.* So that, we can write,

Flux = (average normal component) (Surface area)

Here, the flux of $\vec{F}$ through the element $d\vec{S}$ is

$$\vec{F}.\,\hat{n}\;dS = F\,\text{Cos}\,\theta\;dS$$

The integral of this taken over the entire surface is called the *total flux or surface integral of* $\vec{F}$ through the whole surface S *i.e.*,

$$\iint_S \vec{F}.\hat{n}\;dS = \iint_S F\,\text{Cos}\,\theta\;dS$$

= Total fluid flowing outward through S

In erms of Cartesian components, we have,

$$\iint_S \vec{F}.\hat{n}\;dS = \iint_S (F_x\,dS_x + F_y\,dy + F_z dz)$$

$$\left[(\text{as } \hat{n}dS = d\vec{S} = \hat{i}dS_x + \hat{j}dS_y + \hat{k}dS_z)\right]$$

In the case *where the vector does not represent the flow of anything,* for example, *the electric field* $\vec{E}$, *the integral of the normal component of electric field over an area is still representing the flux.* In other words, it is *not necessary that the vector should represent the flow of something we write,*

Flux of $\vec{E}$ through the surface $= \iint_S \vec{E}.\hat{n}\, dS$

or Flux of $\vec{E}$through the surface $= \iint_S \vec{E}.\hat{n}\, dS$

(the sign $\iint_S$ is simply taken to distinguish from the line integral $\int_Q^P$, we shall represent flux by $\int_S$).

So that, we generalize the word "flux" to mean the "surface integral of the normal component" of a vector.

Volume Integral

Let us consider a surface enclosing volume v in a vector field $\vec{F}$ is a vector point function at a point in a small volume element dv. Then the $\iiint_v \vec{F}.dv$ is known as the volume integral of the vector field $\vec{F}$ for entire volume v over the surface. In cartesian components,

$$\iiint_v \vec{F}.dv = \int_x\int_y\int_z (F_x\hat{i} + F_y\hat{j} + F_z\hat{k})\, dx\, dy\, dz \text{ where } dv = dx\, dy\, dz.$$

THE CLASSIFICATION OF VECTOR FIELDS

So far, we have discussed only two kinds of vector fields, viz.

(1) *Lamellar or Scalar Potential Field* : It is characterised by zero curl and can be expressed as the gradient of a scalar potential and it is, therefore, called a lamellar or scalar potential field.

(2) *Solenoidal Field* : It is characterised by zero divergence which means there are no sources of vector flux in the field. The flux lines are closed curves and the field is called solenoidal.

Since the div, curl of a vector $\{\vec{\nabla}.(\vec{\nabla} \times \vec{F}) = 0\}$ is zero, therefore, *a solenoidal vector can always be expressed as the curl of another vector as given below.*

On the basis of different *dissociations of curl and divergence the vector fields are of four kinds given below* :

(a) The Laplace Equation

Curl $\vec{F} = 0$ and div $\vec{F} = 0$. According to first condition curl $\vec{F} = 0$ the field is lamellar or irrotational. The second condition makes the field solenoidal or incompressible.

So from first condition $\vec{F}$ = grad ϕ and from second condition div grad $\phi = 0$ or $\vec{\nabla}\cdot\vec{\nabla}\,\phi = \nabla^2\,\phi = 0$

which is *Laplace equation.*

The examples of this type of field are electric intensity in free space due to static charges on boundary of conductors, or the irrotational motion of an incompressible fluid.

(b) The Poisson's Equation

Curl $\vec{F} = 0$ but div $\vec{F} \neq 0$. Here the field is irrotational but not solenoidal. From first condition, we have,

$$\vec{F} = \text{grad}\ \phi$$

Applying second condition, we have,

$$\text{div grad}\ \phi \neq 0$$

or $$\nabla^2\,\phi \neq 0$$

which is Poisson's equation. The field must contain a spatial distribution of sources of flux. The *examples* are the *electric field of a volume distribution of charge, e.g. electrons in a thermionic tube,* or the gravitational force inside mass, or the irrotational motion of compressible fluid.

(c) The Idea of Vector Potential

Curl $\vec{F} \neq 0$ but div $\vec{F} = 0$. The field is rotational and cannot have a scalar potential, but it is solenoidal. Let $\vec{A}$ be a vector such that $\vec{F}$ = curl $\vec{A}$ then div $\vec{F}$ = div curl $\vec{A} = 0$, as required by second condition and curl $\vec{F}$ = curl $\vec{A}$ = grad div $\vec{A} - \nabla^2\,\vec{A}$. The nature of the vector $\vec{A}$ is a matter of choice, let $\vec{A}$ to be solenoidal, then div $\vec{A} = 0$ so that

$$\text{curl}\ \vec{F} = -\nabla^2\,\vec{A}$$

This relation is identical to Poisson's equation but contains a vector instead of a scalar operand. That is why, $\vec{A}$ is called the vector potential of F.

(d) The Electro-magnetic Field

Curl $\vec{F} \neq 0$ and div $\vec{F} = 0$. This is the most general type of vector field as rotational motion of compressible fluid and in electromagnetic theory. *Here the field can be expressed as the sum of two fields.* The first is a *lamellar vector,* which has no curl but can have divergence, and the second is a *solenoidal vector,* which has no divergence but can have curl.

Lamellar vector can be taken as grad ϕ and solenoidal vector as curl $\vec{A}$. Since the curl of a gradient of scalar function is zero and divergence of a curl of a vector is also zero, hence the required condition is satisfied. So,

$$\vec{F} = \text{grad } \phi + \text{curl } \vec{A}$$

Hence $\quad \text{div } \vec{F} = \text{div grad } \phi + \text{div curl } \vec{A}$

or $\text{div } \vec{F} = \vec{\nabla}\cdot\vec{\nabla}\ \phi$, since $\text{div curl } \vec{A} = 0$

$$= \nabla^2 \phi = 0$$

This is Poisson's equation and solution of it gives ϕ, and

$\text{curl } \vec{F} = \text{curl grad } \phi + \text{curl curl } \vec{A}$

or $\text{curl } \vec{F} = \text{curl curl A}$. (Since curl grad f = 0)
Since,

$\text{curl curl } \vec{A} = \text{grad div } \vec{A} - \vec{\nabla}^2\ \text{A}.$

But for solenoidal field $\vec{A}$ = curl A, so first term on right hand side is zero, hence,

$$\text{curl } \vec{F} = -\nabla^2 \vec{A}$$

The solution of this equation gives $\vec{A}$. *Thus, the general vector field maybe decomposed into two fields, one being lamellar, with scalar potential* ϕ *and the other solenoidal, with vector potential* $\vec{A}$. This decomposition is known as *Helmholtz's Theorem.*

SOLVED PROBLEMS

1. *Find the Laplace operator (transform) of the waveform (saw-tooth wave)* $f(t) = \frac{2}{3}t,\ 0 \le t \le 3.$

Solution : Here f(t) is a periodic function with period 3

$$\therefore \text{L [f(t)]} = \int_0^3 e^{-st}\ f(t)\,dt$$

$$= \frac{\int_0^3 e^{-st}\ \frac{2}{3}t\ dt}{1 - e^{-35}}$$

$$= \frac{1}{1 - e^{-35}}\left[\frac{2}{3}\int_0^3 e^{-st}\ t\,dt\right]$$

$$= \frac{2}{3\left(1 - e^{-35}\right)}\left[\frac{-e^{st}}{s^2}(ts + t)\right]_0^3$$

$$= \frac{2}{3\left(1 - e^{-3s}\right)}\left[\frac{-e^{3s}}{s^2}(3s + 1) + \frac{1}{s^2}\right]$$

$$= \frac{2}{3\left(1 - e^{-3s}\right)}\left[\frac{\left(1 - e^{-3s}\right) - 3se^{-3s}}{s^2}\right]$$

$$= \frac{2}{3}\left[\frac{1}{s^2} - \frac{3e^{-3s}}{s\,(1 - e^{-3s})}\right]$$

2. *Find the Laplace transform of the function f(t) of period 2π which on the internal [0, 2π] is given by*

$$f(t) = \begin{cases} \sin t, & 0 \le t \le \pi \\ 0, & \pi \le t \le 2\pi \end{cases}.$$

Solution : $L\,[f(t)] = \dfrac{\int_0^T e^{-st}\, f(t)\, dt}{1 - e^{st}}$

$$= \frac{\int_0^\pi e^{-st}\sin t\, dt + \int_\pi^{2\pi} e^{-st}(0)\, dt}{1 - e^{-2\pi s}}$$

$$= \frac{1}{1 - e^{-2\pi s}}\int_0^\pi e^{-st}\sin t\, dt$$

$$= \frac{1}{1 - e^{-2\pi s}}\left[\frac{e^{-st}}{s^2 + 1}(i - s\sin t - \cos t)\right]_0^\pi$$

$$= \left[\text{since } \int e^{ax}\sin t \times dx = \frac{e^{ax}}{a^2 + b^2}(a\sin bx - b\cos bx)\right]$$

$$= \frac{1}{1 - e^{-\pi s}}\left(\frac{1 + e^{-\pi s}}{s^2 + 1}\right)$$

$$= \frac{1}{1 - e^{-\pi s}}\left(\frac{1 + e^{-\pi s}}{s^2 + 1}\right).$$

3. *Evaluate* $\int_c A \,.\, dr$, *where* $A = (2x + y^2)\, i + (3y - 4x)\, j$ *and C is the axis from (0, 0) to (2, 0) and along the line x = 2 from (2, 0) to (2, 1).*

Solution : Here C consists of two parts OD and DB. Let OD and DB be denoted by C_1 and C_2.

Then $\int_C A . dr$

$= \int_{C1} A . dr + \int_{C2} A . dr$

Now A . dr

$= [(2x + y^2)\, i + (3y - 4x)\, j] . [i\, dx + j\, dy]$

$= (2x + y^2)\, dx + (3y = 4x)\, dy$

Also along C_1, y = 0 i.e. y remains constant and along C_2 from it is evident that x remains constant.

∴ From (i) we have

$\int_C A . dr = \int_{C1} (2x + y^2)\, dx + \int_{C2} (3y - 4x).$ [∵ dy = 0 on C_1] and dx = 0 on C_2

$$= \int_{x=0,\, y=0}^{x=2,\, y=0} (2x + y^2)\, dx + \int_{x=2,\, y=0}^{x=2,\, y=1} (3x - 4x)\, dx$$

$$= \int_{x=0}^{2} 2x\, dx + \int_{y=0}^{1} (3y - 8)\, dy,$$

[∵ y = 0 for the first integral and x = 2 for the second integral.]

$$\int_C A . dr = \left(x^2\right)_0^2 + \left(\frac{3}{2}y^2 - 8y\right)_0^1 = 4 + \frac{3}{2} - 8 = \frac{5}{2} \quad \textbf{Ans.}$$

4. *Express the function f(t)* $= \begin{cases} 0, & t < 2 \\ 6, & t \geq 2 \end{cases}$ *in terms of the unit step function and hence obtain its Laplace transform.*

Solution : $f(t) = \begin{cases} 0, & t < 2 \\ 6, & t \geq 2 \end{cases} = 6U(t-2)$

$$\therefore \quad L[f(t)] = 6L[U(t - 2)] = 6\left(\frac{e^{2s}}{s}\right).$$

5. *Express the function f(t)* $= \begin{cases} -1, & 0 \leq t \leq 4 \\ 1, & t \geq 4 \end{cases}$ *in terms of the unit step function and hence find its Laplace transform.*

Solution : $f(t) = \begin{cases} -1, & 0 \leq t < 4 \\ 1, & t \geq 4 \end{cases} = -1 + \begin{cases} 0, & 0 \leq t < 4 \\ 2, & t \geq 4 \end{cases}$

$$= -1 + 2U(t - 4)$$

$$L[f(t)] = L[-1] + 2L[V(t-4)]$$

$$= -\frac{1}{s} + 2\frac{s^{-4s}}{s} = \frac{2e^{-4s} - 1}{s}.$$

6. *Find L[f(t)], where f(t) =* $\begin{cases} t^2, 0, t < 1 \\ 4t, t > 1 \phi \end{cases}$ *by expressing it in terms of Heaviside unit step functions.*

Solution : $f(t) = t^2\, U(t) + (4t - t^2)\, U(t-1)$ ...(1)

$$L[t^2U(t)] = L[t^2] = \frac{2!}{s^3} = \frac{2}{s^3}$$

$$4t - t^2 = -(t^2 - 4t)$$

$$= -[(t-1)^2 - 2t - 1]$$

$$= -[(t-1)^2 - 2(t-1) - 3]$$

$$= 3 + 2(t-1) - (t-1)^2$$

$$= f(t-1)$$

$\therefore$ $F(t) = 3 + 2t - t^2$

$\therefore$ $\phi(s) = L^{-1}[F(t)]$

$$= 3.\frac{1}{s} + 2.\frac{1}{s^2} - \frac{2}{s^3}$$

$$= \frac{3}{s} + \frac{2}{s^2} - \frac{2}{s^3}$$

$\therefore$ $L[(4t - t^2)\, V(t-1)] = e^{-as}\, \phi(s)$ [where a = 1]

$$= \left(\frac{3}{s} + \frac{2}{s^2} - \frac{2}{s^3}\right) e^{-5}$$

Hence from (i) $L[f(t)] = \frac{2}{s^2} + \left(\frac{3}{s} + \frac{2}{s^2} - \frac{2}{s^3}\right) e^{-5}$

7. *Find the Laplace transform of*

$$f(t) = \begin{cases} \cos\left(t - \frac{2\pi}{3}\right), & t \geq \frac{2\pi}{3} \\ 0 & , t < \frac{2\pi}{3} \end{cases}.$$

Solution : $f(t) = V\left(t-\frac{2\pi}{3}\right)\cos\left(t-\frac{2\pi}{3}\right)$

By second shifting theorem, we have

$$L[f(t)] = L\left[V\left(t-\frac{2\pi}{3}\right)\cos\left(t-\frac{2\pi}{3}\right)\right]$$

$$= e^{-\frac{2\pi}{3}s} L[\cos t]$$

$$= e^{-\frac{2\pi}{3}s} \frac{s}{s^2+1}.$$

8. *Evaluate* $\int_c F \,.\, dr$, *where* $F = yz\, i + zx\, j + xy\, k$ *and C is the portion of the curve* $r = (a \cos t)\, i + (b \sin t)\, j + (c\, t)\, k$, *frm* $t = 0$ *t,* $t = \pi/2$.

Solution : The curve is

$$r = (a \cos t)\, i + (b \sin t)\, j + (c\, t)\, k \qquad \ldots (i)$$

$\therefore$ Its parametric equations are

$$x = a \cos t,\ y = b \sin t,\ z = ct \qquad \ldots (ii)$$

From (i) we get $dr/dt = (-a \sin t)\, i + (b \cos t)\, j + (c)\, k]$

$$\therefore\ F.\left(\frac{dr}{dt}\right) = (yzi + zxj + xyk)\,.\,[(-a \sin t)\, i + (b \cos t)\, j + ck]$$

$$= -ayz \sin t + b\, zx \cos t + cxy, \quad [\because i\,.\,i = 1,\ i\,.\,j = 0 \text{ etc.}]$$

$$= -abc\ t \sin^2 t + abc\ t \cos^2 t + abc \sin t \cos t, \text{ from (ii)}$$

$$= -abc\ [(\cos^2 t - \sin^2 t)\, t + 1/2\, (2 \sin t \cos t)]$$

or $\quad F.\, dr = abc\ [t \cos 2t + 1/2 \sin 2t]\, dt$

$$\therefore\ \int_C F\,.\,dr = abc \int_{t=0}^{\pi/2}\left[t \cos 2t + \frac{1}{2}\sin 2t\right] dt$$

$$= abc\left[\int_0^{\pi/2} t \cos 2t\, dt + \frac{1}{2}\int_0^{\pi/2} \sin 2t\, dt\right]$$

$$= abc\left[\left\{\frac{1}{2} t \sin 2t\right\}_0^{\pi/2} - \int_0^{\pi/2} 1\left(\frac{1}{2}\sin 2t\, dt\right) + \frac{1}{2}\int_0^{\pi/2} \sin 2t\, dt\right]$$

$$= abc\left[\frac{1}{2}\left(\frac{1}{2}\pi \sin \pi\right)\right] = 0 \ \textbf{Ans.}$$

9. *Evaluate $\int_c F \, . \, dr$, where $F = xy\, i + (x^2 + y^2)\, j$ and C is the rectangle in the xy-plane bounded by the lines y = 2, x = 4, y = 10 and x = 1.*

Solution : AB is the first part C_1 of the curve C and for AB we have y = 2 and ∴ dy = 0.

Similarly, for the second part C_2 (*i.e.* BD) of the curve we have x = 4, dx = 0, for the third third part C_3 (*i.e.* DE) of the curve C we have y = 10, dy = 0 and for the last part C_4 (*i.e.* EA) of the curve we have x = 1, dx = 0.

Also F . dr $[xyi + (x^2 + y^2)\, j] \, . \, [idx + jdy]$

$$= xy\, dx + (x^2 + y)\, dy$$

Now $\int_C F \, . \, dr = \int_{C1} F \, . \, dr + \int_{C2} F \, . \, dr + \int_{C3} F \, . \, dr + \int_{C4} F \, . \, dr$

$= \int_{C1} xy\, dx + \int_{C2} (x^2 + y^2)\, dy + \int_{C3} xy\, dx + \int_{C4} (x^2 + y^2)\, dy.$

[∵ dx = 0 for C_2 and C_4; dy = 0 for C_1 and C_2]

$$= \int_{x=1}^{1} 2x\, dx + = \int_{y=2}^{10} (16 + y^2)\, dy + \int_{x=4}^{1} 10\, x\, dx + \int_{y=10}^{2} (1 + y^2)\, dy,$$

Since for C_1, y = 2 and x varies from 1 to 4; for C_2, x = 4 and varies from 2 to 10; for C_3, y = 10 and x varies form 4 to 1 and for C_1, x = 1 and x varies from 10 to 2.

$$\therefore \int_C F \, . \, dr = \left(x^2\right)_1^1 + \left(16y + \frac{1}{3}y^3\right)_2^{10} \left(5x^2\right)_2^1 + \left(y + \frac{1}{3}y^3\right)_{10}^{2}$$

= 15 + 128 + 1/3 (992) – 75 – 8 – 1/3 (992) = 60 **Ans.**

10. *If in the above C is around the triangle from O to D, D to B and B to O, evaluate $\int_c A. \, dr$.*

Solution : Here let C_3 denote the line from B to O whose equation is y = 1/2x.

Then $\int_{C2} A \, . \, dr = \int_{C2} [(2 + y^2)dx + (3y - 4x)\, dy]$,

$$= \int_{C2} \left[\left(2x + \frac{1}{4}x^2\right) dx + \left(\frac{3}{2}x - 4x\right)\frac{1}{2}\, dx\right], \text{ putting } y = 1/2\, x.$$

$$= \int_{x=2}^{0} \left(\frac{3}{4}x + \frac{1}{4}x^2\right) dx, \qquad [\because \text{ x varies from 2 to 0 along } C_2]$$

$$= \frac{1}{4}\left[\frac{3}{2}x^2 + \frac{1}{3}x^3\right]_2^0 = -\frac{1}{4}\left[6 + \frac{8}{3}\right] = -\frac{13}{6}$$

∴ Required value

$= \int_C A \,.\, dr = \int_{C1} A \,.\, dr + \int_{C2} A \,.\, dr + \int_{C3} A \,.\, dr$

$= -\frac{5}{2} - \frac{13}{6}, \int_{C1} A \,.\, dr + \int_{C2} A \,.\, dr = -\frac{5}{2}$

$= -\frac{28}{6} = -\frac{14}{3}$ **Ans.**

11. *Find the total work done in moving a particle in a force field given by $F = 2x\,y\,i - 3\,x\,j = 5z\,k$ along the curve C given by $x = t$, $y = t^2 + 1$, $z = 2r^2$ from $t = 0$ to $t = 1$.*

Solution : Total work done $= \int_C F \,.\, dr$

$$= \int_C (2xyi - 3xj - 5zk) \,.\, (idx + jdy + kdz)$$

$$= \int_C = (2xy\, dx = 3x\, dy - 5z\, dz)$$

$$= \int_C [(2t\,(t^2 + 1)\, d\,(t) - 3t\, d\,(t^2 + 1) - 5\,(2t^2)\, d\,(2t^2)]$$

$$= \int_{t=0}^{1} [(2t^2 + 2t)\, dt - 3t\,(2t\, dt) - 10t^2\,(4t\, dt)]$$

$$= \int_0^1 (-38t^3 - 6t^2 + 2t)\, dt$$

$$= \left[-\frac{19}{2}t^4 - 2t^3 + t^2\right]_0^1 = -\left(\frac{19}{2}\right) - 2 + 1$$

$= -21/2$ **Ans.**

12. *Find the circulation of F round the curve C, where $F = (x - y)\, i + (x + y)\, j$ and C is the curve as given.*

Solution : Here F . dr $= [(x - y)\, i\, (x + y)\, j] \,.\, [i\, dx + j\, dy)$

$= (x - y)\, dx + (x + y)\, dy$

∴ The required circulation as in above

$= \int_C F \,.\, dr = \int_{C1} F \,.\, dr + \int_{C2} F \,.\, dr$

Here $\int_C F \,.\, dr$

$= \int_{x=0}^{1} [(x - y)\, dx + (x + y)\, dy]$, for the curve $y = x^2$

$= \int_{x=0}^{1} (x - x^2)\, dx \int_{x=0}^{1} (x + x^2)\, 2x\, dx$, putting y = x2

$$= \left(\frac{1}{2}x^2 - \frac{1}{3}x^3\right)_0^1 + \left(\frac{2}{3}x^3 + \frac{1}{2}x^4\right)_0^1$$

$$= \frac{1}{2} - \frac{1}{3} + \frac{2}{3} + \frac{1}{2} = \frac{4}{3}.$$

And $\int_{C2} F \cdot dr$

$= \int_{y=1}^{0} [(x-y)\,dx + (x+y)\,dy]$, for the curve $y^2 = x$

$= \int_{y=1}^{0} [(y^2-y)\,2y\,dy + (y^2+y)\,dy]$, putting $y^2 = x$

$$= \int_0^0 (2y^3 - y^2 + y) = \left(\frac{1}{2}y^4 - \frac{1}{3}y^3 + \frac{1}{2}y^2\right)_1^0$$

$$= -\frac{1}{2} + \frac{1}{3} - \frac{1}{2} = -\frac{2}{3}$$

$\therefore$ From (i), the required circulation $= \frac{4}{3} - \frac{2}{3} = \frac{2}{3}$ **Ans.**

13. *If $F = 3xy\,i - y^2 j$, evaluate $\int_c F \cdot dr$, where C is the curve in the xy-plane, $y = 2x^2$, from (0, 0) to (1, 2).*

Solution : In the xy-plane, $r = xi + yj$

Also for the given curve $x = t$, $y = 2t^2$

and t varies from t = 0 to t = 1 for the given points (0, 0) to (1, 2)

$$\therefore \int_C F \cdot dr = \int_C (3xy\,i - y^2 j) \cdot (idx + j + dy)$$

$$= \int_C (3xy\,dx - y^2\,dy)$$

$$= \int_{=0}^{1} 3t\,(2t^2)\,d(t) - \int_{t=0}^{1} (2t^2)\,d(2t^2)$$

since $x = t$, $y = 2t^2$

$$= 6\int_0^1 t^3\,dt - 16\int_0^1 t^5\,dt$$

$$= 6\left(\frac{1}{4}t^4\right)_0^1 - 16\left(\frac{1}{6}t^6\right)_0^1 \; \frac{6}{4} - \frac{16}{6}$$

$$= \frac{3}{2} - \frac{8}{3} = \frac{9-16}{6} = -\frac{7}{6}$$ **Ans.**

14. *If $F = (3x^2 + 6y)\,i - 14yzj = 20\,xz^2k$, evaluate $\int F \cdot dr$ from (0, 0, 0) to (1, 1, 1) along the arc of the curve given by $x = t$, $y = t^2$, $z = t^3$.*

Solution : We have $r = x\,i + y\,j + z\,k = t\,i + t^2 j + t^3 k$

$\therefore \qquad \dfrac{dr}{dt} = i + 2rj + 3t^2k$

Also from the equation of the curve viz. $x = t$, $y = t^2$, $z = t^3$, we find that the point (0, 0, 0) and (1, 1, 1) correspond to t = 0 and t = 1.

$\therefore \qquad \int_c F \,.\, dr \int_c F \,.\, \dfrac{dr}{dt}\, dt$

$= \int_c [3x^2 + 6y)\, i - 14yzj + 20xz^2k] \,.\, (i + 2tj + 3t^2k)\, dt$, from (i)

$= \int_{t=0}^{1} [(3x^2 + 6y) - 28yzt + 60xz^2t^2]\, dt$

$= \int_{t=0}^{1} [(3t^2 + 6t^2) - 28t^6 + 60t^2]\, dt$, putting $x = t$, $y = t^2$, $z = t^2$

$= \int_0^1 \left(9t^2 - 28t^6 + 60t^9\right) dt = \left(3t^3 - 4t^7 + 6t^{10}\right)_0^1$

$= 3 - 4 + 6 = 5$ **Ans.**

15. *Evaluate $\int_c F \,.\, dr$, where $F = (2x + y)\, i + (3y - x)\, j + yz\, k$ and C is the curve $x = 2t^2$, $y = t\ z = t^3$, from $t = 0$, to $t = 1$.*

Solution : For the curve C, we have

$$r = x\,i + y\,j + z\,k = 2t^2\, i + t\, j + t^3\, k \qquad \textbf{(Note)}$$

$\therefore \qquad \dfrac{dr}{dt} = 4t\,i + j + 3t^2k$

$\therefore \quad \int_c F \,.\, dr = \int_c F \,.\, \dfrac{dr}{dt}\, dt$

$= \int_c [(2x + y)\, i + (3y - x)\, j + yzk] \,.\, (4ti + j3t^2 k)\, dt$, from (i)

$= \int_c [4(2x + y)t + (3y - x) + 3yzt^2]\, dt$

$[\because\ i \,.\, i = i \,.\, i = 0 \text{ etc.}]$

$= \int_{t=0}^{1} [4(4t^2 + t)\, t + (3t - 2t^2) + 3t \,.\, t^3 \,.\, t^2]\, dt$,

$[\because\ x = 2t^2,\ y = t,\ z = t^3]$

$= \int_0^1 (16\,t^3 + 4t^2\, 3t - 2t^2 + 3t^6)\, dt$,

$= \int_0^1 (3\,t^6 + 16t^3 + 2t^2 + 3t) = \left(\dfrac{3}{7} + 4t^4 + \dfrac{2}{3}t^3 + \dfrac{3}{2}t^2\right)_n^1$

$= \dfrac{3}{7} + 4 + \dfrac{2}{3} + \dfrac{3}{2} = \dfrac{18 + 168 + 28 + 63}{42} = \dfrac{277}{42}$ **Ans.**

EXERCISES

1. What do you understand by the field of any physical quantity? Given examples of scalar and vector fields.

2. What do you understand by gradient of a scalar field ? If $\vec{r}$ be the position vector of any particle find the value of grad (1/r).

   where $\hat{n}$ is unit vector pointing outward at right angle to the surface.

3. What are line and surface integrals ? Clearly explain what they represent by giving examples.

4. If $\vec{A}$ $x^2s\hat{i} - 2y^3s^2\hat{j} + xy^2s\hat{k}$; calculate div $\vec{A}$ (or ∇ $\vec{A}$) at the point (1, –1, 1).

5. If $\vec{r}$ is the position vector, show that

   $\iint_S \vec{r}.\hat{n}dS = 3v$, where v is the volume enclosed by the surface S.

6. In any scalar field, we can draw surfaces corresponding to equal values of the scalar. They are known as

   (a) isothermal, equidensity or equipotential surfaces

   (b) vector line, line of flow or flux line

7. The gradient of scalar field grad f or $\vec{\nabla}$ f is

   (a) a vector (b) a scalar

8. If $\vec{\nabla} \,.\, \vec{D} = 0$

   there is a $\vec{C}$ such that

   (a) $\vec{D} = \vec{\nabla} \cdot \vec{C}$ (b) $\vec{D} = \vec{\nabla} \times \vec{C}$

9. Evaluate $\iint_S \vec{F}.d\vec{S}$ where

   $$\vec{F} = 4xz\hat{i}_1 - y^2\hat{i}_2 + yz\hat{i}_3$$

   S is the surface bounded by x = 0 , x = 1, y = 1, z = 0, z = 1, and $\hat{i}_1, \hat{i}_2, \hat{i}_3$ are unit vectors along x, y, z axes.

   [Hint: $\iint_S \vec{F}.d\vec{S} = \iiint_V \text{div}\,\vec{F}dv = \int_{x=0}^{1}\int_{y=0}^{1}\int_{z=0}^{1} \text{div}\,\vec{F}\,dx\,dy\,dz$]

2

DIFFERENTIATION OF VECTORS

INTRODUCTION

The vector analysis is a very powerful tool visualising the physical meaning of the equations distinctly and exactly. Particularly, in electromagnetic theory change in field and potential can be understood clearly in terms of vector differentiation and integration which is discussed below.

RECIPROCAL VECTORS

If $\vec{a}$, $\vec{b}$ and $\vec{c}$ are the three orthogonal unit vectors along the three axes, then the *Reciprocal Vectors* $\vec{a}^{\times}$, $\vec{b}^{\times}$ and $\vec{c}^{\times}$ are defined as,

$$a^{\times} = \frac{\vec{b}\times\vec{c}}{\vec{a}.(\vec{b}\times\vec{c})},\ b^{\times} = \frac{\vec{c}\times\vec{a}}{\vec{b}.(\vec{c}\times\vec{a})},\ c^{\times} = \frac{\vec{a}\times\vec{b}}{\vec{c}.(\vec{a}\times\vec{b})}$$

the vectors $\vec{a}$, $\vec{b}$ and $\vec{c}$ are assumed to be non-coplanar so that product $\vec{a}.(\vec{b}\times\vec{c})$ is not zero. The vectors $\vec{a}^{\times}$, $\vec{b}^{\times}$ and $\vec{c}^{\times}$ are perpendicular to the planes of $\vec{b}$, $\vec{c}$; $\vec{c}$, $\vec{a}$ and $\vec{a}$, $\vec{b}$ respectively. As the dot product which can be easily seen that,

$$\vec{a}.\vec{a}^{\times} = \vec{b}\vec{b}^{\times} \times \vec{c}\vec{c}^{\times} = 1$$

Hence they are called reciprocal vectors and the dot product of any other pair, one from each system, is zero. Thus,

$$\vec{a}.b^{\times} = 0, \text{because } \vec{a}.b^{\times} = \frac{\vec{a}.(\vec{c}\times\vec{a})}{\vec{a}.(b\times\vec{c})}$$

Here the numerator being the triple product containing two equal vectors *i.e.,* ($\vec{a}$ and $\vec{a}$) is equal to zero.

Similarly,

$$\vec{b}.\vec{a}^{\times} = \vec{b}.\vec{c}^{\times} = \vec{c}.\vec{a}^{\times} = \vec{c}.\vec{b}^{\times} = 0$$

Further, it is true that,

$$\vec{a} = \frac{\vec{b}^x \times \vec{c}^x}{\vec{a}^x.(\vec{b} \times \vec{c}^x)}, \vec{b} = \frac{\vec{c}^x \times \vec{a}^x}{\vec{b}^x \times (\vec{c}^x \times \vec{a}^x)} \; \vec{c} = \frac{\vec{a}^x \times \vec{b}^x}{\vec{c}^x.(\vec{a} \times \vec{b}^x)})$$

Hence the two sets of vectors $\vec{a}$, $\vec{b}$, $\vec{c}$ and $\vec{a}^x$, $\vec{b}^x$, $\vec{c}^x$ are said to be reciprocal to each other.

DIFFERENTIATION OF A VECTOR WITH RESPECT TO TIME

If a vector changes in magnitude as well as direction continuously with respect to come scalar variable, then such a vector is a function of this scalar variable. So that it can be differentiated with respect to that scalar variable and a new vector is obtained. And if the scalar variable is time, the result of differentiation of the vector with time is called time derivative of the vector.

Velocity and Acceleration

Let the position of a particle at any time t be specified by the vector $\vec{r}$ *i.e.,* $\vec{OP}$ which is position vector and vector $\vec{r}$ is a function of variable t. As the time increases, the particle moves and the position vector changes in direction and magnitude, when time changes from t to t + δt, $\vec{r}$ becomes, $\vec{r} + \delta\vec{r}$ at Q.

The average rate of change of the position vector with time is given be,

$$\frac{\vec{r} + \delta\vec{r} - \vec{r}}{t + \delta t - t} = \frac{\delta\vec{r}}{\delta t}$$

This rate of change of $\vec{r}$ with time in the limit when δt → 0 *is called as first time derivative, or the differential coefficient* and is represented by dr/dt. Therefore,

$$\frac{\delta\vec{r}}{\delta t} \quad \lim_{dt \to 0} \frac{\delta\vec{r}}{\delta t}$$

The velocity of a particle is the rate of change of position w.r t. time, thus,

$$\vec{v} = \frac{d\vec{r}}{dt} \qquad ...(1)$$

However, if r is represented in cartesian components, as,

$$\vec{r} = x\hat{i} + y\hat{j} + z\hat{k}$$

Then, $$\vec{v} = \frac{d\vec{r}}{dt} = \frac{dx}{dt}\hat{i} + \frac{dy}{dt}\hat{j} + \frac{dz}{dt}\hat{k}$$

or $$\vec{v} = v_x\hat{i} + v_y\hat{j} + v_z\hat{k} \qquad ...(2)$$

Similarly, the second time derivative of the position vector gives the acceleration $\vec{a}$ of the particle. Thus, we get,

$$\text{Acceleration} = \vec{a} = \frac{d^2r}{dt^2} = \frac{dv}{dt} \qquad ...(3)$$

or $$\vec{a} = \frac{d^2x}{dt^2}\hat{i} + \frac{d^2y}{dt^2}\hat{j} + \frac{d^2z}{dt^2}\hat{k}$$

or $$\vec{a} = \frac{dv_x}{dt}\hat{i} + \frac{dv_y}{dt}\hat{j} + \frac{dv_z}{dt}\hat{k}$$

or $$\vec{a} = a_x\hat{i} + a_y\hat{j} + a_z\hat{k} \qquad ...(4)$$

Force $\overrightarrow{F}$ acting on the particle of mass m is given by Newton's second law of motion,

$$\vec{F} = m\vec{a} = m\frac{d^2\vec{r}}{dt^2} \qquad ...(5)$$

DIFFERENTIATION OF SUMS AND PRODUCTS

(a) *Differentiation of sum of the vectors* : The differential coefficient of sun $\vec{A} + \vec{B}$ both of which are the function of t is equal to the sum of their individual coefficients.

Let $$\vec{R} = \vec{A} + \vec{B}$$
then for a change of time form t to t + δt, we have,

$$\vec{R} + \delta\vec{R} = (\vec{A} + \delta\vec{A}) + (\vec{B} + \delta\vec{B})$$

Hence, $\delta\vec{R} = \delta\vec{A} + \delta\vec{B}$ (Since $\vec{R} = \vec{A} + \vec{B}$)

dividing this by δt and putting the limit, we get,

$$\lim_{\delta t \to 0}\frac{\delta\vec{R}}{\delta t} = \lim_{\delta t \to 0}\frac{\delta\vec{A}}{\delta t} + \lim_{\delta t \to 0}\frac{\delta\vec{B}}{\delta t}$$

Or $$\frac{d\vec{R}}{dt} = \frac{d\vec{A}}{dt} + \frac{d\vec{B}}{dt} \qquad ...(i)$$

so that such *differentiation is distributive* and holds goods for any given number *i.e.,*

$$\frac{d}{dt}(\vec{A} + \vec{B} + \vec{C} + ...) = \frac{d\vec{A}}{dt} + \frac{d\vec{B}}{dt} + \frac{d\vec{C}}{dt} + ... \qquad ...(ii)$$

Similarly we can show that for *difference of vectors*,

$$\frac{d}{dt}(\vec{A}-\vec{B}=\frac{d\vec{A}}{dt}-\frac{d\vec{B}}{dt}$$

(b) *Differentiation of scalar product of two vectors,* $\vec{A}.\vec{B}$

Let $\vec{R}=\vec{A}.\vec{B}$, the increase δt in t given,

$$\vec{R}+\delta\vec{R}=(\vec{A}+\delta\vec{A}).(\vec{B}+\delta\vec{B})$$

or $\vec{R}+\delta\vec{R}=\vec{A}.\vec{B}+\vec{A}.\delta\vec{B}+\vec{B}.\delta\vec{A}+\delta\vec{A}.\delta\vec{B}$

Now, on neglecting the $\delta\vec{A}.\delta\vec{B}$ which is small, we get,

$\delta\vec{R}=\vec{A}.\delta\vec{B}+\vec{B}.\delta\vec{A}$, (Since $\vec{R}=\vec{A}.\vec{B}$)

Dividing by δt and proceeding the limit, we get,

$$\frac{d\vec{R}}{dt}=\frac{d}{dt}(\vec{A}.\vec{B})=\vec{A}.\frac{d\vec{B}}{dt}+\vec{B}.\frac{d\vec{A}}{dt} \quad ...(iii)$$

so that, the operation is *commutative.*

(c) *Differentiation of vector product* $\vec{A}\times\vec{B}$

Let $\vec{R}=\vec{A}\times\vec{B}$

As above, the change in time from t to t + δt gives,

$$\vec{R}+\delta\vec{R}=(\vec{A}+\delta\vec{A})\times(\vec{B}+\delta\vec{B})$$

or $\vec{R}+\delta\vec{R}=\vec{A}\times\vec{B}+\vec{A}\times\delta\vec{B}+\delta\vec{A}\times\vec{B}+\delta\vec{A}\times\delta\vec{B}$

Neglecting $\delta\vec{A}\times\delta\vec{B}$ and dividing by δt whence as before,

$$\frac{d}{dt}(\vec{A}\times\vec{B})=\frac{d\vec{R}}{dt}=\vec{A}\times\frac{d\vec{B}}{dt}+\frac{d\vec{A}}{dt}\times\vec{B} \quad ...(iv)$$

in which the *order of vectors must by strictly maintained.*

(d) *Differentiation of triple products*

In the case of triple product, we proceed as before to obtain the 4–16, 150 differential coefficients, thus,

(i) In scalar triple product, we have,

$$\frac{d}{dt}[\vec{A}.(\vec{B}\times\vec{C})]=\frac{d\vec{A}}{dt}.(\vec{B}\times\vec{C})+\vec{A}.\left(\frac{d\vec{B}}{dt}\times\vec{C}\right)+\vec{A}.\left(\vec{B}\times\frac{d\vec{C}}{dt}\right) \quad ...(v)$$

(ii) In vector triple product we have,

$$\frac{d}{dt}[\vec{A}\times(\vec{B}\times\vec{C})]=\frac{d\vec{A}}{dt}\times(\vec{B}\times\vec{C})+\vec{A}\times\left(\frac{d\vec{B}}{dt}\times\vec{C}\right)+\vec{A}\times\left(\vec{B}\times\frac{d\vec{C}}{dt}\right) \quad ...(IV)$$

SCALAR AND VECTOR FIELDS

(a) *The Field :* If a physical quantity varies from point to point in space it can be expressed as a continuous function of the position of a point in a region of space then such a function is called the *function of position or point function* and *the region in which it specifies the physical quantity is called the field.* There are two main kinds of the fields, corresponding to nature of physical quantities:

(i) Scalar fields corresponding to scalar quantities

(ii) Vector fields corresponding to vector quantities.

(b) *The Scalar Field :* is represented by a continuous *scalar function* ϕ (x, y, z) giving the value of quantity at each point. In all practical cases, *the magnitude of such function does not change abruptly when it passes from any point to another close to it.* For example, distribution of temperature, magnetic and electrostatic potentials, density of any non-directed quantity in a given region of space. In any scalar field, we can draw surfaces corresponding to equal values of the scalar such as *isothermal, equidensity* or *equipotential surfaces.* Such surfaces are called *level surfaces,* each level surface having its own constant value. Two level surfaces cannot cut each other because if they cut, scalar values corresponding to both must hold along their common line, which is contrary to our definition. Thus, scalar point function ϕ (x, y, z) are *single valued* or *uniform* at every point.

(c) *A Vector Field :* A vector field is represented at every point by a continuous vector function $\vec{F}$ (x, y, z). At any given point of field, the function $\vec{F}$ (x, y, z) is specified by a vector of definite magnitude and direction, both of which change continuously from point to point throughout the field region. *Examples* of vector fields : the distribution of velocity in a fluid (liquid or gas), the distribution of electric or magnetic field intensity etc.

Starting from any arbitrary point in the field, proceed an infinitesimal distance in the direction of the vector at that point so as to reach at a closely-neighbouring point. Proceeding in a similar way, we obtain in general, a curved line, called the *vector line, line of flow or flux line.* The tangent at any point of this curved line gives the direction of the vector at that point. For the representation of the magnitude of the vector at any point on a flux line, draw a very small surface perpendicular there to and choose a number of points per unit area upon this surface, which are numerically equal to the magnitude of the vector. Through each of these points flux lines can be drawn. Thus, the field is mapped out by flux lines The direction of flux lines is that of vector function, their density *i.e.,* the number of lines crossing per unit area perpendicular to their direction, is the magnitude of the vector.

SOLVED PROBLEMS

1. *Evaluated $\iint_s$ F. n dS, where F = 18 z i = 12 j + 3y k and S is the surface of the plane 2x + 3y + 6z = 12 in the first octant.*

Solution : The equation of the plane is $\phi \equiv 2x + 3y + 6z = 12 = 0$.

$$\therefore \text{ grad } \phi = i\frac{\partial \phi}{\partial x} + j\frac{\partial \phi}{\partial y} + k\frac{\partial \phi}{\partial z} = 2i + 3j + 6k$$

$\therefore$ n = unit vector in the direction of grad ϕ

$$= \frac{2i + 3j + 6k}{\sqrt{[2^2 + 3^2 + 6^2]}} = \frac{1}{7}(2i + 3j + 6k)$$

$\therefore$ F . n = (18 zi − 12j + 3yk). 1/7 (2i + 3j + 6k)

$$= \frac{1}{7}(36z - 36 + 18y) = \frac{18}{7}(y + 2z - 2) \qquad \text{. . . (i)}$$

$$\text{and} \quad n . k = \frac{1}{7}(2i + 3j + 6k) . k = \frac{6}{7} \qquad \text{. . . (ii)}$$

Also we now that

$\int_S F . n\, dS = \int_{R1} F . n \dfrac{dx\, dy}{m.k}$, where R_1 is the projection of S on the x-y-plane, i.e. R_1 is the region bounded by x-axis, y-axis and the straight line 2x + 3y = 12 or (x/60 + (y/4) = 1.

$$\therefore \iint_s F . n\, dS = \iint_{R1} \frac{18}{7}(y + 2z - 2)\frac{7}{6}\, dx\, dy, \text{ from (i) and (ii)}$$

$$= \iint_{R1} (3y + 6z - 6)\, dx\, dy$$

$$= \iint_{R1} [3y + (12 - 2x - 3y) - 6]\, dx\, dy,$$

from 2x + 3y + 6z = 12

$$= \int_{x=0}^{6} \int_{y=0}^{(12-2x)} (6 - 2x)\, dx\, dy$$

$$= \int_{x=0}^{6} (6 - 2x)\frac{1}{3}(12 - 2x)\, dx$$

$$= \frac{4}{3}\int_0^6 (3 - x)(6 - x)\, dx$$

$$= \frac{4}{3}\int_0^6 (18 - 9x + x^2)\, dx$$

$$= \frac{4}{3}\left[18 - \frac{9}{2}x^2 + \frac{1}{3}x^3\right]_0^6$$

$$= \frac{4}{3}[108 - 162 + 72] = \frac{4}{3}[18] = 24 \text{ Ans.}$$

2. *If $F = yi + (x - 2xz)\, j - x\, y\, k$, evaluate $\iint_s (\nabla \times F) \cdot n\, dS$ where S is the surface of the sphere $x^2 + y^2 + z^2 = a^2$ above the xy-plane.*

Solution : The equation of the sphere is $\phi \equiv x^2 + y^2 = z^2 - a^2 = 0$

$$\therefore \text{ grad } \phi = i\frac{\partial \phi}{\partial x} + j\frac{\partial \phi}{\partial y} + k\frac{\partial \phi}{\partial z} = i\,(2x + j\,(2y) + k(2z)$$

$\therefore$ n = unit vector in the direction of grad ϕ

$$= \frac{(2x)\, i + (2y)\, j + (2z)\, k}{\sqrt{[(2x)^2 + (2y)^2 + (2z)^2]}} = \frac{xi + yj + zk}{\sqrt{(x^2 + y^2 + z^2)}}$$

$= (xi + yi + zk)/a$, since $x^2 + y^2 + z^2 = a^2$.

Now
$$\nabla \times F = \begin{vmatrix} i & j & k \\ \frac{\partial}{\partial x} & \frac{\partial}{\partial y} & \frac{\partial}{\partial z} \\ y & x - 2xz & -xy \end{vmatrix}$$

$$= i\left[\frac{\partial}{\partial y}(-xy) - \frac{\partial}{\partial z}(x - 2xz)\right] + j\left[\frac{\partial}{\partial z}(y) - \frac{\partial}{\partial y}(-xy)\right]$$
$$+ k\left[\frac{\partial}{\partial x}(x - 2xz) - \frac{\partial}{\partial y}(y)\right]$$

$$= i\,[x] + j\,[y] + k\,[-2z]$$

$\therefore (\nabla \times F) \cdot n = (xi + yj - 2zk) \cdot (xi + yj + zk)/a$

$= (x^2 + y^2 - 2z^2)/a$

Also $n \cdot k\, [(xi + yj + zk)/a \cdot k = z/a$

$$\therefore \int_S (\nabla \times F) \cdot n\, dS = \int_{R1} (\nabla \times F) \cdot n \frac{dx\, dy}{n.k}$$

where R_1 is the projection of S on x y-plane

$$= \int_{R1} \frac{(x^2 + y^2 - 2z^2)}{a} \frac{dx\, dy}{(a\, z)}, \text{ where } z^2 = a2 - y^2$$

$$= \int_{x=-a}^{a} \int_{x=-\sqrt{(a^2 - x^2)}}^{\sqrt{(a^2 - x^2)}} \frac{\left(3x^2 + 3y^2 - 2a^2\right) dx\, dy}{\sqrt{\left(a^2 - x^2 - y^2\right)}}$$

(here R_1 is the projection of the sphere on the xy-plane.

$$= 4 \int_{x=-a}^{a} \int_{y=0}^{\sqrt{(a^2-x^2)}} \frac{3x^2 + 3y^2 - 2a^2}{\sqrt{(a^2 - x^2 - y^2)}} \, dx \, dy$$

$= 0$ **Ans.**

3. *Evaluate $\int_s F \cdot n \, dS$ over the surface of the cylinder $x_2 + y_2 = 9$ included in the first octant between $z = 0$ and $z = 4$;*

where $F = zi + xj - y\,z\,k$.

Solution : In such cases of the surface of a cylinder if cylindrical coordinates r, ϕ, z are taken instead of cartesian coordinates x, y, z, the solution becomes easier.

$\therefore$ We take $x = r \cos \phi$, $y = r \sin \phi$ and on the surface of the given cylinder $r = 3$, ϕ varies form 0 to $1/2\,\pi$ and z varies from 0 to 4.

Also here $\phi \equiv x^2 + y^2 = 9 = 0$

$$\therefore \quad \frac{\partial \phi}{\partial x} = 2x, \frac{\partial \phi}{\partial y} = 2y, \frac{\partial \phi}{\partial z} = 0$$

$$\therefore \quad 'n' = \frac{i \frac{\partial \phi}{\partial x} + j \frac{\partial \phi}{\partial y} + k \frac{\partial \phi}{\partial z}}{\sqrt{\left[\left(\frac{\partial \phi}{\partial x}\right)^2 + \left(\frac{\partial \phi}{\partial y}\right)^2 + \left(\frac{\partial \phi}{\partial z}\right)^2\right]}}$$

or $$n = \frac{1}{3}(xi + yj)$$

$$\therefore \quad F \cdot n = (zi + xj = yzk) \cdot 1/3\,(xi + yj)$$
$$= 1/3\,(xz + yx)$$

Also dS = elementary area on the cylindrical surface = $3d\,\phi\,dz$

$$\therefore \int_S F \cdot n \, dS = \int_S 1/3\,(xz + yx)\, 3\, d\phi \, dz$$

$= \int_S x\,(y + z)\, df\, dz$, where $x = 3 \cos \phi$, $y = 3 \sin \phi$

$$= \int_{z=0}^{4} \int_{\phi=0}^{\pi/2} 3 \cos \phi \,(3 \sin \phi + z)\, d\phi \, dz$$

$$= 9 \int_{z=0}^{4} \int_{\phi=0}^{\pi/2} \sin \phi \cos \phi \, d\phi dz + 3 \int_{\phi=0}^{\pi/2} z \cos \phi \, d\phi \, dz$$

$$= 9 \int_0^4 \left(\frac{1}{2} \sin^2 \phi\right)_0^{\pi/2} dz + 3 \int_0^4 (\sin \phi)_0^{\pi/2} z \, dz$$

$$= \frac{9}{2}[z]_0^4 + 3\left[\frac{1}{2}z^2\right]_0^4 = 18 + 24 = 42$$ **Ans.**

4. *Find $L[t^4\ U(t-2)]$.*

Solution : $f(t) = t^4 = \left(\overline{t-2}+2\right)^4$

$= (a+2)^4$ (where $a = t-2$)

$= a^4 + 4a^3.2 + 6a^2.4 + 4a.8 + 16$ (By Binomial Theorem)

$= (t-2)^4 + 8(t-2)^3 + 24(t-2)^2 + 32(t-2) + 16$

$= 16 + 32t + 24t^2 + 8t^3 + t^4$

$$\therefore\ L[f(t)] = 16.\frac{1}{s} + 32.\frac{1}{s^2} + 24.\frac{2'}{s^3} + 8.\frac{3'}{s^4} + \frac{4'}{s^4}$$

$$= \frac{16}{s} + \frac{32}{s^2} + \frac{48}{s^3} + \frac{48}{s^4} + \frac{24}{s^5}$$

$\therefore$ $L[t^4\ V(t-2)] =$

$$\left(\frac{16}{s} + \frac{32}{s^2} + \frac{48}{s^3} + \frac{48}{s^4} + \frac{24}{s^5}\right)e^{-2s}$$

5. *Find $L^{-1}\left[\frac{e^{-3s}}{s^2+8s+25}\right]$.*

Solution : $L^{-1}\left[\frac{e^{-3s}}{s^2+8s+25}\right] = f(t-a)\ U(t-a)$

$= f(t-3)\ U(t-3)$

where $f(t) = L^{-1}\left[\frac{1}{s^2+8s+25}\right]$

$$= L^{-1}\left[\frac{1}{(s+4)^2+9}\right]$$

$$= e^{-4t}\ L^{-1}\frac{1}{s^2+9}$$

$$= e^{-4t}\frac{1}{3}\sin 3t$$

$$\therefore \quad f(t-3) = \frac{1}{3} e^{-4(t-3)} \sin 3(t-3)$$

$$= \frac{1}{3} e^{12-4t} \sin(3t-9)$$

$$\therefore \quad L^{-1}\left[\frac{e^{-3s}}{s^2+8s+25}\right] = \frac{1}{3} e^{12-4t} \sin(3t-9)\, V(t-3).$$

6. *Solve by Laplace transformation method,* $\frac{d^2y}{dt^2} + \frac{dy}{dt} - 2y = 3 \cos$ *3t – 11 sin 3t given that* *y(0) = 0, y'(0) = 6.*

Solution : $\frac{d^2y}{dt^2} + \frac{dy}{dt} - 2y = 3\cos 3t - 11\sin 3t$

$$\Rightarrow \quad y''(t) + y'(t) = 3\cos 3t - 11\sin 3t$$

Taking Laplace transform of both the sides,

$L[y''(t)] + L[y'(t)] - 2L[y(t)] = 3L[\cos 3t] - 11L[\sin 3t]$

$[s^2L[y(t)] - s\,y(0) - y'(0)] + [sL[y(t)] - y(0)] - 2L[y(t)]$

$$= \frac{3s}{s^2+9} - 11\frac{3}{s^2+9}$$

$$\Rightarrow \quad s^2 L[y(t)] - s(0) - 6 + sL[y(t)] - 0 - 2L[y(t)]$$

$$= \frac{3s}{s^2+9} - \frac{33}{s^2+9}$$

$$\Rightarrow \quad (s^2+s-2)\,L[y(t)] = \frac{3s-33}{s^2+9} + 6$$

$$\Rightarrow \quad (s^2+s-2)\,L[y(t)] = \frac{6s^2+3s+21}{s^2+9}$$

$$\Rightarrow \quad L[y(t)] = \frac{3(2s^2+s+7)}{(s^2+9)(s^2+s-2)}$$

$$\Rightarrow \quad L[y(t)] = \frac{3(2s^2+s+7)}{(s^2+9)(s+2)(s-1)}$$

Let $\quad 3(2s^2+s+7) = \frac{As+B}{s^2+9} + \frac{C}{s+2} + \frac{D}{s-1}$

$$\Rightarrow \quad 3(2s^2+s+7) = (As+B)(s+2)(s-1)$$

$$+ C(s^2 + 9)(s - 1) + D(s^2 + 9)(s + 2)$$

Put s = 1 $\quad 30 + D(10)(3)$

$\Rightarrow \quad D = 1$

Put s = – 2

$\Rightarrow \quad 39 = C(13)(-3)$

Equating the coefficients of s^3 on both the sides, we get

$$A + C + D = 0$$

$\Rightarrow \quad A = -C - D$

$\Rightarrow \quad A = 1 - 1 = 0$

Equating the constant terms on both the sides, we get

$$-2B - 9C + 18D = 21$$

$\Rightarrow \quad -2B + 9 + 18 = 21$

$\Rightarrow \quad -2B = -6$

$\Rightarrow \quad B = 3$

$\therefore \quad L[y(t)] = \dfrac{3}{s^2+9} - \dfrac{1}{s+2} + \dfrac{1}{s-1}$

Taking inverse of both the sides, we get

$$y(t) = L^{-1}\left[\frac{3}{s^2+9}\right] - L^{-1}\left[\frac{1}{s+2}\right] + L^{-1}\left[\frac{1}{s-1}\right]$$

$$= \sin 3t - e^{2t} + e^{t}.$$

6. *Solving using, Laplace transformation method, the following differential equation :*

$\dfrac{d^2y}{dt^2} + 2\dfrac{dy}{dt} - 3y = \sin t$, *given that* $y = \dfrac{dy}{dt} = 0$, *when* $t = 0$.

Solution : $\dfrac{d^2y}{dt^2} + 2\dfrac{dy}{dt} - 3y = \sin t$

Taking Laplace Transform of both the sides, we get

$$L\left[\frac{d^2y}{dt^2}\right] + 2L\left[\frac{dy}{dt}\right] - 3L[y(t)] = L[\sin t]$$

$\Rightarrow \quad [s^2L[y(t)] - s\,y(0) - y'(0)] + 2[sL[y(t)] - y(0)] - 3L[y(t)]$

$$= \frac{1}{s^2 + 1}$$

$$\Rightarrow \quad (s^2 + 2s - 3)\, L\,[y(t)] = \frac{1}{s^2 + 1} \quad [\because\ y(0) = y'(0) = 0]$$

$$\Rightarrow \quad L[y(t)] = \frac{1}{(s^2 + 1)(s^2 + 2s - 3)}$$

$$= \frac{1}{(s-1)(s+3)(s^2+1)}$$

Let $\frac{1}{(s-1)(s+3)(s^2+1)} = \frac{A}{s-1} + \frac{B}{s+3} + \frac{Cs+D}{s^2+1}$

$\Rightarrow 1 = A(s + 3)(s^2 + 1) + B(s - 1)(s^2 + 1) + (Cs + D)(s - 1)(s + 3)$

Putting s = 1, we get $A = \frac{1}{8}$

Putting s = – 3, we get $B = -\frac{1}{40}$

Putting s = 0, we get 3A – B – 3D = 1

$$\Rightarrow \qquad D = -\frac{1}{5}$$

Equating the coefficients of s^3 on both the sides, we get

$$A + B + C = 0$$

$$\Rightarrow \qquad C = -\frac{1}{10}$$

$$\therefore \quad L[y(t)] = \frac{1}{8}\cdot\frac{1}{s-1} - \frac{1}{40}\cdot\frac{1}{s+3} - \frac{1}{10}\frac{s}{s^2+1} - \frac{1}{5}\cdot\frac{1}{s^2+1}$$

$$y(t) = \frac{1}{8} L^{-1}\left[\frac{1}{s-1}\right] - \frac{1}{40} L^{-1}\left[\frac{1}{s+3}\right] - \frac{1}{10} L^{-1}\left[\frac{s}{s^2+1}\right] - \frac{1}{5} L^{-1}\left[\frac{1}{s^2+1}\right]$$

$$\Rightarrow \quad y(t) = \frac{1}{8} e^{t} - \frac{1}{40} e^{3t} - \frac{1}{10} \cos t - \frac{1}{5} \sin t\,.$$

8. *Use Laplace transformation to solve the equation :*

$\frac{d^2y}{dt^2} + 4\frac{dy}{dt} + 4y = \sin 2t$, *given that y(0) = y'(0) = 0.*

Solution : Given that :

$$y''(t) + 4\, y'(t) + 4\, y(t) = \sin 2t$$

Taking Laplace Transform of both the sides, we get;

L[y"(t)] + 4L [y'(t)] + 4L [y(t)] = L [sin 2t]

s^2L [y(t)] – s y(0) – y'(0)] + 4 [sL [y(t)] – y(0)] + 4L [y(t)]

$$= \frac{2}{s^2+4}$$

$$\Rightarrow \qquad (s^2 + 4s + 4)\, L\,[y(t)] = \frac{2}{s^2+4}$$

$$\Rightarrow \qquad L[y(t)] = \frac{2}{(s+2)^2(s^2+4)}$$

$$\Rightarrow \qquad y(t) = 2L^{-1}\left[\frac{1}{(s+2)^2\,(s^2+4)}\right]$$

Let $\dfrac{1}{(s+2)^2(s^2+4)} = \dfrac{A}{s+2} + \dfrac{B}{(s+2)^2} + \dfrac{Cs+D}{s^2+4}$

$\Rightarrow \quad 1 = A\,(s + 2)\,(s^2 + 4) + B\,(s^2 + 4)\,(Cs + D)\,(s + 2)^2$

Putting s = – 2, we get;

$$1 = B\,(8)$$

$$\Rightarrow \qquad B = \frac{1}{8}$$

Equating the coefficients of s^3, on both sides, we get;

$$A + C = 0 \qquad ...(i)$$

Equating the coefficients of s^2, we get;

$$2A + B + 4C + D = 0 \qquad ...(ii)$$

Equating the coefficients of s, we get;

$$4A + 4D + 4C = 0 ...(iii)$$

Equating the constants terms on both the sides, we get;

$$8A + 4B + 4D = 1 ...(iv)$$

Solving (i) and (iii), we get; D = 0

Substituting the values of B and D we get from (iv), $A = \dfrac{1}{16}$

Putting the value of A in (i), we get,

$$C = -\frac{1}{16}$$

$$\therefore \quad y(t) = 2\, L^{-1}\left[\frac{1}{16}\cdot\frac{1}{s+2} + \frac{1}{8}\cdot\frac{1}{(s+2)^2} + \frac{-\frac{1}{16}s}{s^2+4}\right]$$

$$\Rightarrow \quad y(t) = \frac{1}{8}L^{-1}\left[\frac{1}{s+2}\right] + \frac{1}{4}L^{-1}\left[\frac{1}{(s+2)^2}\right] - \frac{1}{8}L^{-1}\left[\frac{s}{s^2+4}\right]$$

$$\Rightarrow \quad y(t) = \frac{1}{8}e^{-2t} + \frac{1}{4}te^{-2t} - \frac{1}{8}\cos 2t$$

which is the required general solution.

9. *Solve using Laplace transformation method, the following differential equation :*

$$\frac{d^2y}{dt^2} + 2\frac{dy}{dt} + 2y = e^{-t}\sin 2t, \text{ given } y(0) = 0,\ y'(0) = 1.$$

Solution : The given D.E. is :

$$y''(t) + 2\,y'(t) + 2\,y(t) = e^{-t}\sin 2t$$

Taking Laplace Transform of both the sides, we get

$$L[y''(t)] + 2L[y'(t)] + 2L\,[y(t)] = L\,[e^{-t}\sin 2t]$$

$$s^2L\,[y(t)] - s\,y(0) - y'(0) + 2\,[sL\,[y(t)] - y(0)] = \frac{2}{(s+1)^2+4}$$

$$\Rightarrow s^2L\,[y(t)] - 0 - 1 + 2s\,L\,[y(t)] + 2L\,[y(t)] = \frac{2}{s^2+2s+5}$$

$$\Rightarrow (s^2 + 2s + 2)\,L\,[y(t)] = \frac{2}{s^2+2s+5} + 1$$

$$\Rightarrow L[y(t)] = \frac{2}{(s^2+2s+2)(s^2+2s+5)} + \frac{1}{s^2+2s+2}$$

$$= \frac{2}{3}\left[\frac{3}{(s^2+2s+2)(s^2+2s+5)}\right] + \frac{1}{s^2+2s+2}$$

$$= \frac{2}{3}\left[\frac{(s^2+2s+\ )-(s^2+2s+2)}{(s^2+2s+2)(s^2+2s+5)}\right] + \frac{1}{s^2+2s+2}$$

$$= \frac{2}{3}\,\frac{1}{s^2+2s+2} - \frac{2}{3}\,\frac{1}{s^2+2s+5} + \frac{1}{s^2+2s+2}$$

$$= \frac{5}{3}\frac{1}{s^2+2s+2} - \frac{2}{3}\frac{1}{(s+1)^2+2^2}$$

$$\Rightarrow \quad y(t) = \frac{5}{8}L^{-1}\left[\frac{1}{(s+1)^2+1^2}\right] - \frac{2}{3}L^{-1}\left[\frac{1}{(s+1)^2+2^2}\right]$$

$$\Rightarrow \quad y(t) = \frac{5}{3}e^{-t}\sin t - \frac{1}{3}e^{-t}\sin 2t$$

$$\Rightarrow \quad y(t) = \left(\frac{5}{3}\sin t - \frac{1}{3}\sin 2t\right)e^{-t}$$

is the required general solution.

10. *Find the inverse Laplace transform of log* $\left(\frac{1+s}{s}\right)$.

Solution : Let $F(s) = \log\left(\frac{1+s}{s}\right)$

$$= \log(1+s) - \log s$$

$$\therefore \quad \frac{dF(s)}{ds} = \frac{1}{1+s} - \frac{1}{s}$$

$$\Rightarrow \quad L^{-1}\left[\frac{dF(s)}{ds}\right] = L^{-1}\left[\frac{1}{1+s}\right] - L^{-1}\left[\frac{1}{s}\right]$$

But $$L^{-1}\left[\frac{dF(s)}{ds}\right] = (-1)\,t\,L^{-1}[F(s)]$$

$$\therefore \quad e^{-t} - 1 = (-1)\,t\,L^{-1}[F(s)]$$

$$\Rightarrow \quad L^{-1}[F(s)] = \frac{1-e^{-t}}{t}$$

$$\Rightarrow \quad L^{-1}\left[\log\left(\frac{1+s}{s}\right)\right] = \frac{1-e^{t}}{t}.$$

11. *Find* $L^{-1}\left[\log\left(1+\frac{1}{s^2}\right)\right]$.

Solution : (i) Let $F(s) = \log\left(1+\frac{1}{s^2}\right)$

$$= \log\left(\frac{s^2+1}{s^2}\right)$$

$$= \log(s^2+1) - \log s^2$$

$$= \log(s^2+1) - 2\log s$$

$$\therefore \quad \frac{dF}{ds} = \frac{2s}{s^2+1} - \frac{2}{s} = 2\left(\frac{s}{s^2+1} - \frac{1}{s}\right)$$

$$\Rightarrow \quad L^{-1}\left[\frac{dF(s)}{ds}\right] = 2\,L^{-1}\left[\frac{s}{s^2+1}\right] - 2\,L^{-1}\left[\frac{1}{s}\right]$$

$$\Rightarrow \quad L^{-1}\left[\frac{dF(s)}{ds}\right] = 2\cos t - 2$$

But $$L^{-1}\left[\frac{dF(s)}{ds}\right] = (-1)\,t\,L^{-1}[F(s)]$$

$$\Rightarrow \quad 2(\cos t - 1) = (-1)\,t\,L^{-1}[F(s)]$$

$$\Rightarrow \quad L^{-1}[F(s)] = \frac{2}{t}(1-\cos t)$$

$$\Rightarrow \quad L^{-1}\left[\log\left(1+\frac{1}{s^2}\right)\right] = \frac{2}{t}(1-\cos t)$$

(ii) Here, $F(s) = \dfrac{s}{(s^2-1)^2}$

$$\therefore \quad L^{-1}[F(s)] = L^{-1}\left[\frac{s}{(s^2-1)^2}\right]$$

$$= L^{-1}\left[-\frac{1}{d}\frac{d}{ds}\left(\frac{1}{s^2-1}\right)\right]$$

$$= -\frac{1}{2}L^{-1}\left[\frac{d}{ds}\left(\frac{1}{s^2-1}\right)\right]$$

$$= -\frac{1}{2}(-1)\,t\,L^{-1}\left[\frac{1}{s^2-1}\right]$$

$$= \frac{t}{2} \sin ht.$$

12. *Solve :* $\frac{d^2y}{dt^2} + 9y = \cos 2t,\ y(0) = 1,\ y\left(\frac{\pi}{2}\right) = -1.$

Solution : $\frac{d^2y}{dt^2} + 9y = \cos 2t$

$$y^{11}(t) + 9y(t) = \cos 2t$$

Taking Laplace Transform of both the sides, we get

$$L[y''(t)] = 9L[y(t)] = L[\cos 2t]$$

$$\Rightarrow \quad s^2 L[y(t)] - s\,y(0) + 9L[y(t)] = \frac{s}{s^2+4}$$

$$\Rightarrow \quad s^2 L[y(t)] - s - A + 9L[y(t)] = \frac{s}{s^2+4}$$

where y'(0) = A (say)

$$\Rightarrow \quad (s^2 + 9) L[y(t)] - s - A = \frac{s}{s^2+4}$$

$$\Rightarrow \quad L[y(t)] = \frac{s}{(s^2+4)(s^2+9)} + \frac{s+A}{s^2+9}$$

$$L[y(t)] = \frac{s}{5(s^2+4)} - \frac{s}{5(s^2+9)} + \frac{s+A}{s^2+9}$$

Taking the inverse Laplace transform, we get;

$$y(t) = L^{-1}\left[\frac{s}{5(s^2+4)} - \frac{s}{5(s^2+9)} + \frac{s}{s^2+9} + \frac{A}{s^2+9}\right]$$

$$= \frac{1}{5} L^{-1}\left[\frac{s}{s^2+4}\right] - \frac{1}{5} L^{-1}\left[\frac{s}{s^2+9}\right]$$

$$+ L^{-1}\left[\frac{s}{s^2+9}\right] + A\,L^{-1}\left[\frac{1}{s^2+9}\right]$$

$$\Rightarrow \quad y(t) = \frac{1}{5}\cos 2t - \frac{1}{5}\cos 3t + \cos 3t + \frac{A}{3}\sin 3t$$

$$\Rightarrow \quad y(t) = \frac{1}{5}\cos 2t + \frac{4}{5}\cos 3t + \frac{A}{3}\sin 3t$$

But $y\left(\frac{\pi}{2}\right) = -1$ (given)

$$\therefore \quad -1 = \frac{1}{5}\cos\pi + \frac{4}{5}\cos\frac{3\pi}{2} + \frac{A}{3}\sin\frac{3\pi}{2}$$

$$\Rightarrow \quad -1 = -\frac{1}{5} + 0 - \frac{A}{3}$$

$$\Rightarrow \quad A = \frac{12}{5}$$

$$\Rightarrow \quad y(t) = \frac{1}{5}\cos 2t + \frac{4}{5}\cos 3t + \frac{4}{5}\sin 3t$$

$$\Rightarrow \quad y(t) = \frac{1}{5}(\cos 2t + 4\cos 3t + 4\sin 3t)$$

13. *Solve :* $\frac{d^2y}{dt^2} + 2\frac{dy}{dt} + 2y = 5\sin t,\ y = \frac{dy}{dt} = 0$ *at* $t = 0.$

Solution : $\frac{d^2y}{dt^2} + 2\frac{dy}{dt} + 2y = 5\sin t$

$$y''(t) + 2y'(t) + 2y(t) = 5\sin t$$

Taking the Laplace Transform of both the sides, we get

$$L[y''(t)] + 2L[y'(t)] + 2L[y(t)] = 5L[\sin t]$$

$$\Rightarrow s^2L[y(t)] - s\,y(0) - y'(0) + 2[sL[y(t)] - y(0)] + 2L[y(t)] = \frac{5}{s^2+1}$$

$$\Rightarrow s^2L[y(t)] - s(0) - (0) + 2sL[y(t)] - 2(0) + 2L[y(t)] = \frac{5}{s^2+1}$$

$$\Rightarrow (s^2 + 2s + 2)\,L[y(t)] = \frac{5}{s^2+1}$$

$$\Rightarrow L[y(t)] = \frac{5}{(s^2+1)(s^2+2s+2)}$$

Taking inverse of both the sides :

$$y(t) = L^{-1}\left[\frac{5}{(s^2+1)(s^2+2s+2)}\right]$$

Let $\left[\frac{5}{(s^2+1)(s^2+2s+2)}\right] = \frac{As+B}{s^2+1} + \frac{Cs+D}{s^2+2s+2}$

$$\Rightarrow 5 = (s^2 + 2s + 2)(As + B) + (s^2 + 1)(Cs + D)$$

Equating the coefficients of like powers of s and constant terms on both the sides, we get

A + C = 0

2A + B + D = 0, 2A + B + C = 0, 2B + D = 5

$\Rightarrow$ A = – 2, B = 1, C = 2, D = 3

$$\therefore \frac{5}{(s^2+1)(s^2+2s+2)} = \frac{-2s+1}{s^2+1} + \frac{2s+3}{s^2+2s+2}$$

$$\therefore y(t) = L^{-1}\left[\frac{-2s+1}{s^2+1}\right] + L^{-1}\left[\frac{2s+3}{s^2+2s+2}\right]$$

$$= -2\,L^{-1}\left[\frac{s}{s^2+1}\right] + L^{-1}\left[\frac{1}{s^2+1}\right] + L^{-1}\left[\frac{2(s+1)+1}{s^2+2s+2}\right]$$

$$= -2\cos t + \sin t + L^{-1}\left[\frac{2(s+1)+1}{(s+1)^2+1}\right]$$

$$= -2\cos t + \sin t + L^{-1}\left[\frac{s+1}{(s+1)^2+1}\right] + L^{-1}\left[\frac{1}{(s+1)^2+1}\right]$$

$y(t) = -2\cos t + \sin t + 2e^{-t}\cos t + e^{-t}\sin t.$

14. $L^{-1}\left[\frac{s}{(s^2-1)^2}\right]$.

Solution : $L^{-1}\left[\frac{1}{s+1}\right] = e^{-t}$

$$\therefore \quad L^{-1}\left[\frac{1}{s(s+1)}\right] = \int_0^t e^{-t}dt = \left[\frac{e^{-t}}{-1}\right]_0^t$$

$$= -[e^{-t} - t^0] = -[e^{-t} - 1].$$

15. *Find the inverse Laplace transform of* $\frac{1}{s(s+1)}$.

Solution : $L^{-1}\left[\frac{1}{(s+1)^3}\right] = \frac{t^2e^{-t}}{2}$

$$\therefore \quad L^{-1}\left[\frac{1}{s(s+1)^3}\right] = \frac{1}{2}\int_0^t t^2e^{-t}dt$$

$$= \frac{1}{2}\left[t^2\left(-e^{-t}\right) - 2t\left(e^{-t}\right) + 2\left(-e^{-t}\right)\right]_0^t$$

$$= 1 - e^{-t}\left(1 + t + \frac{t^2}{2}\right).$$

16. *Find the inverse Laplace transform of* $\frac{1}{s^2(s+1)}$.

Solution : $L^{-1}\left[\frac{1}{s^2+1}\right] = \sin t$

$$\therefore \quad L^{-1}\left[\frac{1}{s(s^2+1)}\right] = \int_0^t \sin t \, dt$$

$$= [-\cos t]_0^t = 1 - \cos t$$

$$\therefore \quad L^{-1}\left[\frac{1}{s^2(s^2+1)}\right] = \int_0^t (1 - \cos t)\, dt$$

$$= [t - \sin t]_0^t$$

$$= t - \sin t.$$

17. *Find the inverse Laplace transform of* $\frac{1}{s^2(s^2+4)}$.

Solution : $L^{-1}\left[\frac{1}{s^2+4}\right] = \frac{1}{2}\sin 2t$

$$\therefore \quad L^{-1}\left[\frac{1}{s(s^2+4)}\right] = \frac{1}{2}\int_0^t \sin 2t \, dt$$

$$= \frac{1}{2}\left[\frac{-\cos 2t}{2}\right]_0^t = \frac{1}{2}\left[\frac{-\cos 2t}{2} + \frac{1}{2}\right]$$

$$= \frac{1}{4}[1 - \cos 2t]$$

$$\therefore \quad L^{-1}\left[\frac{1}{s^2(s^2+4)}\right] = \frac{1}{4}\int_0^t (1 - \cos 2t)\, dt$$

$$= \frac{1}{4}\left[t - \frac{\sin 2t}{2}\right]_0^t$$

$$= \frac{1}{4}\left[t - \frac{\sin 2t}{2}\right]$$

$$\therefore \quad L^{-1}\left[\frac{1}{s^3(s^2+4)}\right] = \frac{1}{4}\int_0^t \left(t - \frac{\sin 2t}{2}\right) dt$$

$$= \frac{1}{4}\left[\frac{t^2}{2} + \frac{\cos 2t}{4}\right]_0^t$$

$$= \frac{1}{4}\left[\frac{t^2}{2} + \frac{\cos 2t}{4} - \frac{1}{4}\right]$$

$$= \frac{t^2}{8} - \frac{\cos 2t}{16} - \frac{1}{16}.$$

18. *Find the Laplace transform of* $\frac{\sin 2t}{t}$.

Solution : $L[\sin 2t] = \frac{2}{s^2+4}$

$$L\left[\frac{\sin 2t}{t}\right] = \int_s^t \frac{2}{s^2+4}\, ds$$

$$= 2.\frac{1}{2}\left[\tan^{-1}\frac{s}{2}\right]_s^x = \tan^{-1}\infty - \tan^{-1}\frac{s}{2}$$

$$= \frac{\pi}{2} - \tan^{-1}\frac{s}{2} = \cot^{-1}\frac{s}{2}.$$

19. *Find* $L\left[\frac{1-e^t}{t}\right]$.

Solution : $L[1 - e^t] = L(1) - L(e^t)$

$$= \frac{1}{s} - \frac{1}{s-1}$$

$$\therefore \quad L\left[\frac{(1-e^t)}{t}\right] - \int_s^\infty \left(\frac{1}{s} - \frac{1}{s-1}\right) ds$$

$$= [\log s - \log(s-1)]_s^\infty$$

$$= \left[\log\left(\frac{s}{s-1}\right)\right]_s^\infty = \left[\log\left(\frac{1}{1-\frac{1}{s}}\right)\right]_s^\infty$$

$$= \log 1 - \log\left(\frac{s}{s-1}\right)$$

$$= 0 - \log\left(\frac{s}{s-1}\right) = \log\left(\frac{s}{s-1}\right)^{-1}$$

$$= \log\left(\frac{s}{s-1}\right)^{-1} = \log\left(\frac{s-1}{s}\right).$$

20. *Find L* $\left(\frac{1-\cos t}{t}\right)$.

Solution : $L(1 - \cos t) = L(1) - L(\cos t)$

$$= \frac{1}{s} - \frac{s}{s^2+1}$$

$$L\left(\frac{1-\cos t}{t}\right) = \int_s^\infty \left(\frac{1}{s} - \frac{s}{s^2+1}\right) ds$$

$$= \left[\log s - \frac{1}{2}\log(s^2+1)\right]$$

$$= \frac{1}{2}\left[\log s^2 - \log(s^2+1)\right]_s^\infty$$

$$= \frac{1}{2}\left[\log\frac{s^2}{s^2+1}\right]_s^\infty = \frac{1}{2}\left[\log\frac{1}{1+\frac{1}{s^2}}\right]$$

$$= \frac{1}{2}\left[\log 1 - \log\frac{s^2}{s^2+1}\right] = -\frac{1}{2}\log\frac{s^2}{s^2+1}.$$

21. *Find L* $\left[\frac{e^{-at} - e^{-bt}}{t}\right]$.

Solution : $L[e^{-at} - e^{-bt}] = L[e^{-at}] - L[e^{-bt}]$

$$= \frac{1}{s+a} - \frac{1}{s+b}$$

$$\therefore \quad L\left[\frac{e^{-at}-e^{-bt}}{t}\right]=\int_s^{\infty}\left(\frac{1}{s+a}-\frac{1}{s+b}\right)ds$$

$$=\left[\log(s+a)-\log(s+b)\right]_s^{\infty}$$

$$=\left[\log\left(\frac{s+a}{s+b}\right)\right]_s^{\infty}=\left[\log\left(\frac{1+\frac{a}{s}}{1+\frac{b}{s}}\right)\right]_s^{\infty}$$

$$=\log 1-\log\frac{s+a}{s+b}=\log\frac{s+b}{s+a}.$$

22. *Using Laplace transform method of solve the differntial equation :*

$$\frac{d^2y}{dt^2}+4\frac{dy}{dt}=\sin t,$$

given that $y(0)=y'(0)=0.$

Solution : $\frac{d^2y}{dt^2}+\frac{4dy}{dt}=\sin t$

$\Rightarrow$ $y''(t)+4y'(t)=\sin t$

Taking Laplace transform of both sides

$$\Rightarrow L[y''(t)]-sy(0)-y'(0)+4[sL(y(t)]-y(0)]=\frac{1}{s^2+1}$$

$$\Rightarrow s^2L[y(t)]-s(0)-(0)+4[sL\{y(t)\}=\frac{1}{s^2+1}$$

$$\Rightarrow (s^2+4s)\,L[y(t)=\frac{1}{s^2+1}$$

$$\Rightarrow L[y(t)]=\frac{1}{s(s+4)(s^2+1)}$$

Taking the inverse transform, we get

$$\text{Let } y(t)=L^{-1}\left[\frac{1}{s(s+4)(s^2+1)}\right]$$

$$\frac{1}{s(s+4)(s^2+1)}=\frac{A}{s}+\frac{B}{s+4}+\frac{Cs+D}{s^2+1}$$

$\Rightarrow$ $1+A(s+4)(s^2+1)+Bs(s^2+1)+(Cs+D)\,s(s+4)$

Equating the coefficients of s and constant terms on both the sides we get

$$A + B + C = 0 \quad ...(1)$$

$$4A + 4C + D = 0 \quad ...(2)$$

$$A + B + 4D = 0 \quad ...(3)$$

and $$4A = 1 \quad ...(4)$$

$\Rightarrow$ $$A = \frac{1}{4}$$

Putting the value of A and solving for (1), (2) and (3) we get

$$B = -\frac{65}{68}, C = \frac{41}{34} D = \frac{3}{17}$$

$$y(t) = \frac{1}{4}L^{-1}\left[\frac{1}{s}\right] - \frac{65}{68}L^{-1}\left[\frac{1}{s+4}\right] + \frac{41}{34}L^{-1}\left[\frac{s}{s^2+1}\right] + \frac{3}{17}L^{-1}\left[\frac{1}{s^2+1}\right]$$

23. *Using Laplace transform, solve* $\frac{dx}{dt} = 2x - 3y, \frac{dy}{dt} = y - 2x$ *given that* $x(0) = 8, y(0) = 3$.

Solution : $L[x] = \bar{x}$ and $L[y] = \bar{y}$

$\therefore$ Taking L.T. of both the equation, we get

$$-8 + 5\bar{x} = 2\bar{x} - 3\bar{y}$$

and $$-3 + s\bar{y} = \bar{y} - 2\bar{x}$$

$\therefore$ $$(s-2)\bar{x} + 3\bar{y} = 8$$

and $$2\bar{x} + (s-1)\bar{y} = 3$$

Solving for $\bar{x}$ and $\bar{y}$, we obtain

$$\bar{x} = \frac{85-17}{s^2-35-4} = \frac{3}{s-5} + \frac{5}{s+1} \text{ by P.F.}$$

and $$\bar{y} = \frac{35-22}{5^2-35-4} = \frac{-2}{s-4} + \frac{5}{s+1} \text{ by P.F.}$$

Taking inverse, we get

$\therefore$ $$x = L^{-1}\left[\frac{3}{5-4}\right] + L^{-1}\left[\frac{5}{5+1}\right] = 3e^{4t} + 5e^{-t}$$

and $$y = L^{-1}\left[\frac{-2}{5-4}\right] + L^{-1}\left[\frac{5}{5+1}\right] = -2e^{4t} + 5e^{-t}.$$

24. *Solve* $\frac{dx}{dt} + x + 3\int_0^t y\,dt = \cos t + 3\sin t$

$2\frac{dx}{dt} + 3\frac{dy}{dt} + by = 0$

given $x(0) = -3$ *and* $y(0) = 2$.

Solution : Let $L[x] = \bar{x}$ and $L[y] = \bar{y}$

Taking Laplace Transform of the given equations, we get

$$(3 + 5\bar{x}) + \bar{x} + 3.\frac{1}{5}\bar{y} = \frac{5}{5^2+1} + \frac{3}{5^2+1}$$

and $$2(3 + 5\bar{x}) + 3(-2 + 5\bar{y}) + 6\bar{y} = 0$$

$\Rightarrow$ $$(8 + 1)\,\bar{x} + \frac{3}{5}\bar{y} = \frac{5+3}{8^2+1} - 3$$

$$= \frac{8 - 35^2}{8^2+1}$$

and $$25\bar{x} + (35 + 6)\bar{y} = 0$$

Solving for $\bar{x}$ and $\bar{y}$, we get

$$\bar{x} = -\frac{(5+2)(3s-1)}{(s+3)(s^2+1)} = -\frac{1}{5+3} + \frac{1-25}{s^2+1}$$

and $$\bar{y} = \frac{2}{3}\frac{s(35-1)}{(s+3)(s^2+1)} = \frac{2}{s+3} - \frac{2}{3}.\frac{1}{s^2+1}$$

Taking inverse, wer get

$$x = -e^{-3t} + \sin t - 2\cos t$$

$$y = 2e^{-3t} - \frac{2}{3}\sin t$$

25. *The currents x and y in two coupled circuits are given by*

$$L\frac{dx}{dt} + Rx + R(x - y) = E$$

$$L\frac{dy}{dt} + Ry - R(x - y) = 0,$$

where L, R, E are constants. Find x and y interms of t, if x = 0, y = 0 at t = 0.

Solution : Let $L[x] = \bar{x}$, $L[y] = \bar{y}$

Taking Laplace Transform of both the sides of the given equations, and applying the given initial conditions, we have

$$L\left(-0 + 5\bar{x}\right) + 2R\bar{x} - R\bar{y} = \frac{E}{s}$$

and $$L\left(-0 + s\bar{y}\right) + 2R\bar{y} - R\bar{x} = 0$$

$\Rightarrow$ $$\left(Ls + 2R\right)\bar{x} - R\bar{y} \equiv \frac{E}{s}$$

and $$-R\bar{s} = \left(Ls + 2R\right)\bar{y} = 0$$

Solving for $\bar{x}, \bar{y}$, we have

$$\bar{x} = E.\frac{Ls + 2}{s\left(Ls + R\right)\left(Ls + 3R\right)}$$

$$= E\left[\frac{2}{3R}.\frac{1}{s} - \frac{1}{2R}\frac{1}{2R}\frac{1}{Ls + R} - \frac{1}{6R}\frac{1}{Ls + 3R}\right] \text{ (by P.F.)}$$

and $$\bar{y} = E.R.\frac{1}{s\left(Ls + R\right)\left(Ls + 3R\right)}$$

$$= \frac{E}{6R}\left[4.\frac{1}{s} - 3.\frac{1}{s + \frac{R}{L}} - \frac{1}{s + \frac{3R}{L}}\right]$$

Taking inverse, we have

$$x = \frac{E}{6R}\left[4 - 3e^{-Rt/L} - e^{-3Rt/L}\right]$$

and $$y = \frac{E}{6R}\left[2 - 3e^{-Rt/L} + e^{-3Rt/L}\right]$$

26. *Express the function* $f(t) = \begin{cases} 8, & t < 2 \\ 6, & t \geq 2 \end{cases}$ *in terms of the unit step function and thus obtain the Laplace transform.*

Solution : $f(t)=\begin{cases}8, & t<2\\ 6, & t\geq 2\end{cases}=8+\begin{cases}0, & t<2\\ -2, & t\geq 2\end{cases}$

$$=8-2\begin{cases}0, & t<2\\ 1, & t\geq 2\end{cases}$$

$$= 8 - 2U(t-2)$$

$\therefore$ $L[f(t)] = L[8] - 2L[V(t-2)]$

$$=\frac{8}{5}-2.\frac{e^{-2s}}{s}$$

$$=\frac{8-2e^{-2s}}{s}.$$

27. *Find L[f(t)] where f(t)* $=\begin{cases}e^{t-a}, & t\geq a\\ 0, & t<a\end{cases}$.

Solution : $f(t) = V(t-a)\, e^{t-a}$

$\therefore$ $L[f(t)] = L[V(t-a)\, e^{t-a}]$

$= e^{-as} L[e^t]$ (by second shifting theorem)

$$=e^{-as}\left(\frac{1}{s-1}\right),\ s>1.$$

28. *Find L[t U(t – 3)].*

Solution : $L[t\, V(t-3)]$

$$= L[(t-3+3)\, V(t-3)]$$

$$= L[(t-3)\, V(t-3)] + 3L[V(t-3)]$$

$$= e^{-3s} L[t] + \frac{3e^{-3s}}{s}$$

$$=e^{-3s}\frac{1}{s^2}+\frac{3e^{-3s}}{s}$$

$$=\frac{e^{-3s}}{s^2}(1+3s).$$

29. *Express f(t)* $=\begin{cases}\sin t, & t<\pi\\ t, & t\geq\pi\end{cases}$ *in terms of the unit step function and hence find its Laplace operator.*

Solution : $f(t)=\begin{cases}\sin t, & t<\pi\\ t, & t\geq\pi\end{cases}$

$$=\sin t+\begin{cases}0, & t<\pi\\ t\sin t, & t\geq\pi\end{cases}$$

$$= \sin t + (t = \sin t)\ U(t-\pi)$$

$$= \sin t + t\ U(t-\pi) - \sin t\ V\ (t-\pi)$$

$$= \sin t + \pi\ V(t-\pi) + (t-\pi)\ V\ (t-\pi) + \sin\ (t-\pi)\ V\ (t-\pi)$$

$$\therefore\ L[f(t)] = L[\sin t] + \pi\ L[V(t-\pi)] + L[(t-\pi)\ V\ (t-\pi)] + L[\sin\ (t-\pi)\ V\ (t-\pi)]$$

$$=\frac{1}{s^2+1}+\pi\frac{e^{-\pi s}}{s}+e^{-\pi s}\ L(t)+e^{-\pi s}\ L(\sin t)$$

$$=\frac{1}{s^2+1}+\frac{\pi}{s}e^{-\pi s}+\frac{e^{-\pi i}}{s^2}+\frac{e^{-\pi s}}{s^2+1}$$

$$=\frac{1}{s^2+1}+e^{=\pi s}\left[\frac{\pi}{s}+\frac{1}{s^2}+\frac{1}{s^2+1}\right]$$

30. *Find* $L\left[\sin 2t\ \delta\left(t-\frac{\pi}{4}\right)-t^2\ \delta(t-4)\right]$.

Solution : Let f(t) = sin 2t and $a=\frac{\pi}{4}$

$$\therefore\ L\left[\sin 2t\ \delta\left(t-\frac{\pi}{4}\right)\right]$$

$$= L\ f\ (t)\ \sigma\ (t-A)$$

$$= f(a)\ e^{-\pi s} = f\left(\frac{\pi}{4}\right)e^{-(\pi/4)^s}$$

$$= \sin\frac{2\pi}{4}\ e^{-(\pi/4)^s} = e^{-\pi s/4}$$

Again $L\ [t^2\ \delta\ (t-4)]$

$$= f\ (4)\ e^{-4s} \text{ where } f(t) = t^2 \text{ amd } \pi = 4$$

$$= (4)^2\ e^{-4s} = 16\ e^{-4s}$$

$$\therefore\ L\left[\sin 2t\ \delta\left(f-\frac{\pi}{4}\right)\right]=e^{-\pi s/4}-16e^{-4s}$$

31. *Find L[t U(t – 4) – t^3 δ(t – 2)].*

Solution : Let t = (t – 4) + 4 = f (t – 4)

$$\Rightarrow f(t) = t + 4$$

giving $\neq (s) = \frac{1}{s^2} + \frac{4}{5}$, let a = 4

$$\therefore L[t\,U(t-4)] = e^{-as}\,\phi(s)$$

$$= \left(\frac{1}{s^2} + \frac{4}{5}\right)e^{4s}$$

$L[t^3\,\delta(t-2)] = f(2)\,e^{-2s} = 8e^{-2s}$ where $f(t) = t^3$, a = 2.

32. *Find the Laplace transform of each of the following :*

(i) t^2 U(t – 3) (ii) e^t U(t – 1).

Solution : Let $t^2 = A(t-3)^2 + B(t-3) + C$

$$t^2 = A(t^2 - 6t + 9) + B(t-3) + C$$

Equating the coefficients of t^2, t and constant terms, we get

$$A = 1,\ -6A + B = 0 \text{ and } 9A - 3B + C = 0$$

$\Rightarrow$ $A = 1,\ B = 6$ and $C = 9$

$\therefore$ $t^2 = (t-3)^2 + 6(t-3) + 9$

$\Rightarrow$ $t^2V(t-3) = (t-3)^2\,U(t-3) + 6(t-3)\,V(t-3) + 9U(t-3)$

$$L[t^2\,V(t-3)] = L(t-3)^2\,V(t-3)] + 6L[(t-3)\,V(t-3)]$$

$$= e^{-3s}\,L[t^2] + 6e^{-3s}\,L[t] + 9e^{-3s}$$

$$= e^{-3s}\left(\frac{2'}{s^3}\right) + \frac{6e^{-3s}}{s^2} + \frac{9e^{-3s}}{s}$$

$$= \frac{e^{-3s}\left(9s^2 + 6s + 2\right)}{s^3}$$

(ii) Let $e^tV(t-1) = e^{t-1}\,U(t-1).e$

$\therefore$ $L[e^t\,V(t-1)] = L[e^{t-1}\,(t-1)e]$

$$= e\,L[e^{t-1}\,V(t-1)]$$

$$= e.e^{-5}\,L[e^t]$$

$$= e^{1-5}\left(\frac{1}{s-1}\right) = \frac{e^{-(5-1)}}{(s-1)}.$$

33. *If ϕ (x, y, z) = xy^2z and f = xz i – xy j + yz^2 k, show that* $\frac{\partial^2}{\partial x^2 \partial z}(\phi f)$ *at (2, – 1, 1) is 4i + 2j.*

Solution:

We have $\phi f = xy^2 z\,(xz\ i - xy\ j + yz^2 k)$

$= x^2y^2z^2\ i - x^3y^3z\ j + xy^3z^3\ k.$

$\therefore \quad \frac{\partial}{\partial x}(\phi f) = 2xy^2z^2\ i - 2xy^3z\ j + y^3z^3\ k.$

$$\frac{\partial^2}{\partial x^2}(\phi f) = \frac{\partial}{\partial x}\left[\frac{\partial}{\partial x}(\phi f)\right] = \frac{\partial}{\partial x}[2xy^2z^2\, i - 2xy^3z\, j + y^3z^3\, k]$$

$= 2y^2z^2\ i - 2y^3z\ j + 0\ k = 2y^2z^2\ i - 2y^3z\ j$

and $\frac{\partial^3}{\partial x^2 \partial z}(\phi f) = \frac{\partial}{\partial z}\left[\frac{\partial^2}{\partial x^2}(\phi f)\right]$

$= \frac{\partial}{\partial z}(2y^2z^2\, i - 2y^3 - j) = 4y^2z\ i - 2y^3\ j.$

$\therefore$ at the point (2, – 1, 1), $\frac{\partial^3}{\partial x^2 \partial z}(\phi f)$

$= 4 \,.\, (-1)^2 \,.\, 1\ i - 2 \,.\, (--1)^3\ j = 4\ i + 2\ j.$ Hence Proved.

EXERCISES

1. Which of the following is a vector field?
   (a) distribution of temperature
   (b) magnetic and electrostatics potentials
   (c) density of any non-directed quantity in a given region of space
   (d) the distribution of electric of magnetic field intensity
2. If potential function $\vec{v}$ = – a xy, obtain the value of electric field, where a is constant.
3. What a field is called as conservative field ? Give examples of such field.
4. Why the vector fields are classified on the basis of different association of curl and divergence ? What are these combinations ? Illustrate by giving example.

5. If $F = yi + (x \blacksquare 2xz)\,j \blacksquare xy\,k$, evaluate $\iint_s (\nabla \times F) \cdot n\, dS$ where S is the surface of the sphere $x^2 + y^2 + z^2 = a^2$ above the xy-plane.

6. Show that curl of a gradient of scalar is zero and also that the div of a curl of vector is zero *i.e.,*

   curl grad f = 0, where f is scalar

   div curl $\vec{F} = 0$ where $\vec{F}$ is vector

   What you conclude from these relations? How the idea of vector potential is introduced ?

7. Obtain the expressions for Laplace and Poisson's equations.

8. Explain clearly what do you understand by divergence and curl of a vector field $\vec{F}$ obtain expressions in general orthogonal coordinates for

   (a) div $\vec{F}$ and (b) curl $\vec{F}$.

9. What do you mean by the flux of a vector field? Show that flux of a vector $\vec{F}$ through the surface is given by

   $\iint_S \vec{F} \hat{n}\, dS$ or $\int_S \vec{F} . \hat{n}\, dS$,.

10. If $F = 3xy\,i - y^2\,j$, evaluate $\int_c F \cdot dr$, where C is the curve in the xy-plane, $y = 2x^2$, from (0, 0) to (1, 2).

3

APPLICATION OF THE MULTIPLE VECTORS PRODUCTS

INTRODUCTION

(i) *Deduction of the Cosine law:* Let two vectors A and B be represented by the two sides of a triangle. Then, the third side represents a vector $C = -B + A = A - B$.

Taking the scalar product of each side with itself, we have

$$C.C = (A - B).\ (A - B).$$

Or, $$C^2 = A^2 + B^2 - 2AB, \text{ because } A.B = B.A.$$

Or, $$C^2 = A^2 + B^2 - 2AB \cos (AB)$$

$$= A^2 + B^2 - 2AB \cos \theta.$$

This is the well known trigonometric relation called the *cosine law.*

(ii) *Obtaining the equation of a plane:* Let ABCD be the plane, whose equation is required to be obtained. Drop a normal N on to the plane from an origin O, lying quite outside the plane and let r be the position vector (with respect to O) of any point P in the plane.

Then, since the projecti n normal N must be equal to the magnitude N or N, we have $N = r \cos \theta$ and ∴ $N^2 = Nr \cos \theta = Nr$, which is the required equation of the plano.

Or, if we write N and r in terms of their rectangular components N_x, N_y, N_z and x, y, z respectively, along the three Cartesian coordinate axes, with i, j, k as the orthogonal triad of unit vectors along these, we have

$$N = N_x i + N_y j + N_z k \text{ and } r = xi + yj + zk.$$

i.e., $N^2 = (N_x i + N_y j + N_z k)\ (xi + yj + zk), = xN_x + yN_y + zN_z.$

which is the usual form of the equation met with in Analytical Geometry.

(iii) *Deduction of expressions for work and power*: As we know, work is said to be done by a force acting on a particle when the particle is displaced in a direction other than the one normal to the force. *It is a scalar quantity* proportional to the force and the resolved part of displacement in the direction of the force or to the displacement and resolved part of the force in the direction of displacement. If, therefore, vectors F and d represent force and displacement respectively, inclined to each other at an angle θ we have *work done*, W = (F cos θ) d = F. d, its value being zero only when F and d are at right angles to each other.

Further, if there be more than one force acting on the particle, say, F_1, F_2, F_3 etc., then, work done by them for a displacement d of the particle is respectively F_1 d, F_2. d. F_n. d. We, therefore, have *total work done*,

$$W = \sum_1^n F.d = d.\sum F = d.R,$$

i.e., the effect off all the forces taken together is the same as that of one single force, R *equal to their resultant.*

Now, *power is the rate of doing work.* So that,

$$\text{power } P = \frac{dW}{dt} = \frac{d}{dt}(F.d) = F.\frac{d}{dt}(d).$$

Since $\frac{d}{dt}$ (d) is the velocity v of the particle, we have P = F.v.

EQUALITY OF VECTORS

Although, a vector may refer to a physical quantity defined at a particular point, it does not necessarily have any particular location. So that, we can compare two vectors even though they may measure physical quantities defined at quite different points of space and time.

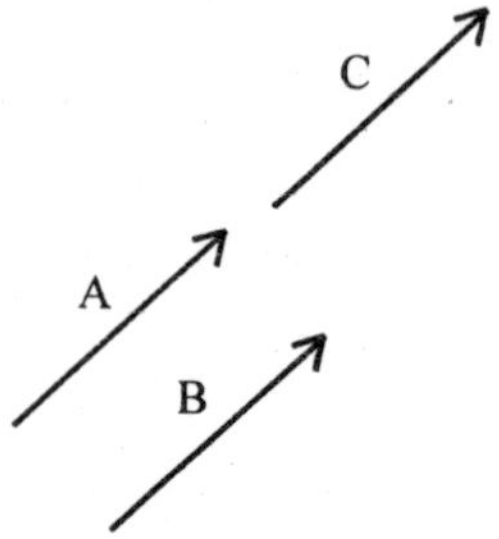

Fig. 1

It follows, therefore, that *all vectors, with the same magnitude and direction, are equal despite their entirely different locations in space and remain so even if moved parallel to themselves*. This, in Fig. 1 vectors A, B and C are all equal, since-they have all the same magnitude and direction, even though they are differently located in space.

Such vectors are called free vectors to distinguish them from a localised vector which must pass through a specified point in space, such as a position vector and is thus a *bound vector*. Their 'invariance' *under parallel translation, i.e.*, when moved parallel to themselves, is a very important property of free vectors.

COLLINEAR VECTORS

The term 'collinear' is used as being synonymous with parallel. So that, vectors having the same (as in the case of A and C) *or parallel lines of action* (as in the case of A and B) *are said to be collinear*. Further, *if the directions of the two parallel vectors be the same, they are referred to as like vectors.* An important property of such (*i.e., like*) vectors is the following:

If A *and* B *be two collinear or like vectors, there exists a scalar k such that* B = kA, *the absolute value of k being the ratio of the lengths of the two collinear or like vectors.*

ADDITION AND SUBTRACTION OF TWO VECTORS

Addition : (i) two vectors may be effected with the help of the *parallelogram law of vector addition*, according to which the sum or *the resultant* R of *two vectors* A *and* B *is the diagonal of the- parallelog am of which* A *and* B *are the adjacent sides.* R = A + B. As will be seen, A, B and R are all concurrent.

(ii) *The triangle law* : The triangle law of vector addition follows from the parallelogram law and states that *if the tail-end of* one vector be placed at the head or the arrow-end of the other, their sum or resultant R is drawn from the tail-end, of the first to the head-end of the other. The resultant R is the same irrespective of the order in which the vectors (A and B) are taken. So that, we have

$$R = A + B = B + A,$$

showing that vector addition too is commutative.

It may be noted that *vectors need not lie in the same plane for the laws of vector addition to be applicable to them.*

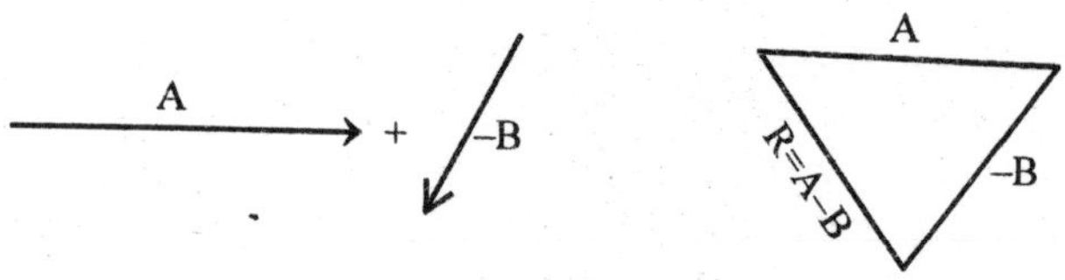

Fig. 2

Subtraction : Clearly, if the sum of two vectors A and B be equal to a vector of zero magnitude, we have A + B = 0, showing that vectors A and B are equal in magnitude but opposite in direction. So that, we have B = –A, indicating that –A *is a vector of the same magnitude as vector A but pointing in a direction opposite to that of* A. And this at once gives us a method of subtracting a vector B from a vector A namely, the addition of the vector A to the vector –B. 50 that, A + (–B) = A – B.

ADDITION OF MORE THAN TWO VECTORS—(COMPOSITION OF VECTORS)

The determination of the sum or the resultant of a system of more-than two vectors (not necessarily in the same plane) is called *composition of vectors*. This may be done with the help of what is called the *polygon law of vector addition* which states that if a vector polygon be drawn, placing the tail-end of each succeeding vector at the head or the arrow-end of the preceding one, their resultant; R is drawn from the tail-end of the first to the head or the arrow-end of the last. This

follows directly from the *Triangle law*. For, what we do, in effect, is to first obtain the resultant (A + B) of the first two vectors (A and B) by the *triangle law* and then the resultant R of (A + B) and C by the, same law. Or, we could fiat obtain the resultant (B + C) of B and C and then that of A and (B + C) which give R, the resultant of A, B and C. We thus have

$$
\begin{aligned}
R &= A + B + C \\
&= (A + B) + C \\
&= A + (B + C),
\end{aligned}
$$

showing that *vector addition too, like scalar addition, is associative* and may be extended to any number of rectors.

VECTORS IN A COORDINATE SYSTEM

Although, it is quite possible to deal with physical law in terms of vectors, it is found most helpful in actual practice to represent them in a coordinate system, and the one most widely favoured is the familiar *Cartesian coordinate system* which may be right-handed or left-handed. The former is shown in Fig. 3 (a) with its three mutually perpendicular axes OX, OY and OZ represented by the extended thumb, fore-finger and middle finger respectively of the right hand. Its mirror image [Fig 3 (b)] forms the left-handed system which is, however, not quite so convenient to me. We shall, therefore, stick to the right-handed system only, in which to an observer at O, right-handed rotations about the axes OX, OF and OZ appear to be from X to Y, Y to Z and Z to X respectively.

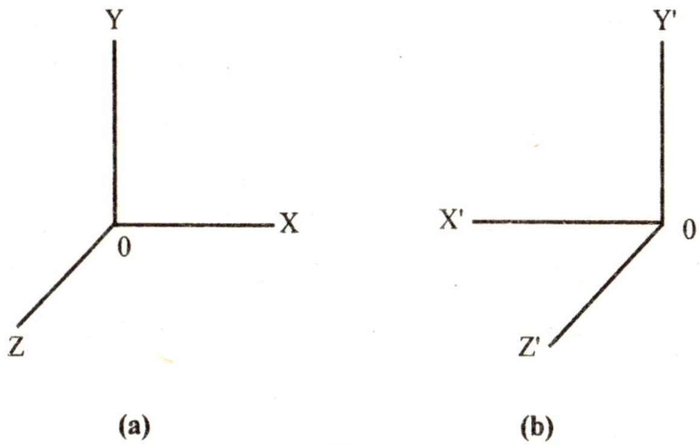

Fig. 3

VECTOR DERIVATIVES-VELOCITY—ACCELERATION

Let r be a *single-valued function* of a scalar variable t such that for every value of t there exists only one value of r. Then, as t varies continuously, r also does so.

We are particularly interested in the case in which t represents the time variable and r stands for the position vector of a moving particle with respect to a fixed origin O. Then, as t varies continuously, the point moves along a continuous curve in space. So that, if r and r + δr be the position vectors of the point in positions P and P' relative to origin O for the values t and t + δt of the scalar variable, we have

change in the value of r = δr.

The quotient $\delta r/\delta t$ (of vector δr by the scalar or a number δt) is also a vector. As $\delta t \to 0$, point P' approaches P and the chord PP' tends to coincide with the tangent of the curve at P.

The limiting value of $\delta r/\delta t$ as δt tends to zero is dr/dt and is a vector *whose direction is that of the tangent at* P *in the sense in which t increases.*

It is called the *time derivative of* r or *the differential coefficient* of r with respect to t. We thus have

$$\frac{dr}{dt} = \underset{\delta t \to 0}{\text{Lt}} \frac{\delta r}{\delta t}.$$

When this limit exists, the function r is said to be differentiable.

The second and third derivatives of r are respectively

$$d^2r/dt^2 \text{ and } d^3r/dt^3.$$

Clearly, δr represents the displacement of the particle in time interval δt and, therefore, δr/δt gives its *average velocity* during interval δt. The limiting value of this average velocity, as $\delta t \to 0$, is the *instantaneous velocity v* of the particle. Thus, we have

$$v = dr/dt$$

along the tangent to the path of the particle.

Proceeding in the same manner if, δv be the increase in the velocity v of the particle during the time-interval δt, the *rate of change of velocity or the average acceleration during the interval* = δv/δt *and therefore, instantaneous acceleration* a of the particle is the limiting value dv/dt of δv/δt as $\delta t \to 0$. Thus,

$$a = \frac{dv}{dt} = \frac{d^2r}{dt^2}$$

Now, since r = xi + yj + zk and since x, y and z are functions of time, we also have

$$v = \frac{dr}{dt} = \frac{dx}{dt}i + \frac{dy}{dt}j + \frac{dz}{dt}k$$

and $$a = \frac{d^2r}{dt^2} = \frac{d^2x}{dt^2}i + \frac{d^2y}{dt^2}j + \frac{d^2z}{dt^2}k.$$

TRIPLE VECTOR PRODUCTS

The cross product of two vectors, one of which is itself the cross product of two vectors, is called a triple vector product (for the simple reason that it is a vector). Thus, if A, B and C be three vectors, we have A × (B × C), B × (C × A) and C × (A × B) as examples of *vector products.*

The triple vector product A × (B × C) is a vector lying in a plane perpendicular to that A and (B × C). But, as we know, vector (B × C) lies in the plane normal to that of B and C. It follows, therefore, that vector A × (B × C) lies in the plane of vectors B and C and is perpendicular.

Similarly, vector B × (A × C) lies in the plane of A and C and is perpendicular of B. And vector C × (A × B) lies in the plane of A and B and is perpendicular to C. So that, in general,

$$A \times (B \times C) = (A \times B) \times C.$$

EVALUATION OF THE VECTOR TRIPLE PRODUCT A × (B × C)

Let, i, j, k be the right handed *orthogonal tracts of unit vectors* with unit vectors i along B, j perpendicular to it in the plane of B and C, and k perpendicular to i and j. Then, we have $B_y = 0$, $B_z = 0$ and $C_z = 0$. So that.

$$A = A_x\, i + A_y\, j + A_z\, k,$$

$$B = B_x\, i + 0 + 0 = B_x\, i, \text{ and}$$

$$C = C_x\, i + C_y\, j + 0 = C_x\, i + C_y\, j.$$

Hence $\quad B \times C = B_x\, i \times (C_x\, i + C_y\, j) = B_x\, C_y\, k$

[∵ $i \times i = 0$ and $i \times j = k$.

∴ $\quad A \times (B \times C) = A_y\, B_x\, C_y\, i - A_x\, B_x\, C_y\, j.$

[∵ $i \times k = -j$, $j \times k = i$ and $k \times k = 0$.

Adding and subtracting $A_x\, B_x\, C_x i$ to the right-hand side, we have

$$A \times (B \times C) = A_x\, B_x\, C_x i + A_y\, B_x\, C_y i - A_z\, B_x\, C_z i - A_x\, B_x\, C_y j$$

$$= (A_x\, C_x + A_y\, C_y)\, B_x i - A_x\, B_x\, (C_x i + C_y j)$$

Now, $\quad A_x\, C_x + A_y\, C_y = A.C,\ A_x\, B_x = A.B,$

[∵ $i.i = j.j = I$ and $i.j = j.k = 0$.

$$B_x i = B \text{ and } C_x i + C_y j = C.$$

we, therefore, have $\quad A \times (B \times C) = (A\,.\,C)\, B - (A.\,B)\, C.$

Or, to put in the determinant form, we have

$$A \times (B \times C) = \begin{vmatrix} B & C \\ A.B & A.C \end{vmatrix}$$

And, proceeding in the same manner, we have

$$B \times (C \times A) = (B\,.\,A)\, C - (B\,.\,C)\, A$$

and $\quad C \times (A \times B) = (C\,.\,B)\, A - (C\,.\,A)\, B$

As a mnemonic or aid to memory, we might call the three vectors, the *first*, the *second* and the *third*, taken in order. Then, the vector triple product

= first × (second × third) = (first.third second – (first.second) third.

SCALAR TRIPLE PRODUCT

Let the three vectors A, B and C form a parallelopiped. Then, if B and C be the *magnitudes* of vectors B and C, we have *area of the base of the parallelopiped* = BC sinϕ = B × C.

Thus, the *magnitude of the vector product* B × C *is the area of the base of the parallelopiped, and its direction perpendicular to this area,* as shown.

If θ be the angle that the direction of the vector (B × C) makes with vector A (forming one of the edges of the parallelopiped), we have *altitude of the parallelopiped,* h = A cos θ, where A is the magnitude of vector A.

And, therefore, A. (B × C) = A cos θ (B × C) = *vertical height of parallelopiped* × *area of its base* = *volume of the parallelopiped,* V, the sign of V being positive if θ be an acute angle, *i.e.*, if A, B and C form a right-handed system of vectors.

Thus, we see that *the scalar triple product* A.(B × C) *represents the volume of a parallelopiped, with the three vectors forming its three edges.* Obviously, since any one of the faces of the parallelopiped may be taken to be its base, its volume V is also given by B (C × A) and C. (A × B), *with the cyclic order of* A, B *and* C *maintained.*

In case this cyclic order be altered, the sign of the product is changed since, as we know, B × C = – C × B. Further, since the order of the terms in a scalar product is quite immaterial, we have

$$V = A.(B \times C) = (B \times C).A = -A.(C \times B) = -(C \times B).A$$
$$= B.(C \times A) = (C \times A).B = -B.(A \times C) = -(A \times C).B$$
$$= C.(A \times B) = (A \times B).C = -C.(B \times A) = -(B \times A).C.$$

It will thus be seen that *the value of a scalar triple product is depends on the cyclic order of the vectors and is quite independent of the positions of the dots and the crosses, which may be interchanged as desired.* It is, therefore, usual to denote a scalar triple product of vectors A, B and C by [A, B, C], or [A, B, C] putting the three vectors in their cyclic order but without any dots or crosses.

Further Points to be Noted About a Scalar Triple Product

(i) *In case the three vectors be coplanar, their scalar triple product is zero,* because the vector (B × C) being then perpendicular to vector A, their scalar product is zero.

Thus, the *condition for coplanarity of three vectors* is that *their scalar triple product* [ABC] *should vanish. This condition is satisfied when two of the vectors are parallel.*

(ii) *If two of the vectors be equal, the scalar triple product is zero.* For, if the three vectors be A, A and B, we have [AAB] = (A × A)B. And, since A × A = 0 [the vector product of a vector with itself being zero. We have [AAB] = 0.

(iii) *If two of the vectors be parallel, the scalar triple product is zero.* For, of the three vectors A, B and C, if A and B be parallel,, B = kA, where k is some scalar, We therefore, have

$$[ABC] = (A \times B).C = (A \times kA).C$$
$$= k\,(A \times A).C = k\,[AAC] = 0.$$

(iv) In terms of the Cartesian components of the three vectors, we have

$$B \times C = (B_y C_z - B_z C_y)i + (B_z C_x - B_x C_z)j + (B_x C_y - B_y C_x)k.$$

$$\therefore A.(B \times C) = [ABC] = (B_y C_z - B_z C_y)i + (B_z C_x - B_x C_z)j + (B_x C_y - B_y C_x)k].\,(A_x i + A_y j + A_z k)$$

$$= A_x (B_y C_z - B_z C_y) + A_y (B_z C_x - B_x C_z) + A_z(B_x C_y - B_y C_x).$$

Or,
$$[ABC] = \begin{vmatrix} A_x & A_y & A_z \\ B_x & B_y & B_z \\ C_x & C_y & C_z \end{vmatrix}$$

which is the familiar expression for the volume of a parallelopiped with one of its corners at the origin.

(v) *The scalar triple product of the orthogonal vector triad is unity.* For, [i j k] = (i × j).k = k.k = 1.

(vi) *Since the distributive law holds good for both scalar and vector products, it also holds good for scalar triple products.* Thus, for instance,

[A, B + D, C + E] = [ABC] + [ABE] +[ADC] + [ADE],

the cyclic order of the vectors being maintained in each term.

SCALAR PRODUCT

The scalar product of two vectors A and B is denoted by A.B and is read as A dot B. It is also, therefore, known as the *dot Product* of the two vectors (also sometimes referred to as the *direct product*).

It is defined as the product of the magnitudes of the two vectors A *and*

B *and the cosine of their included angle* θ, *irrespective of the coordinate system used.* Thus,

A.B = AB cos θ or AB cos (A, B).

Since cos (A, B) is equal to cos (B, A), *the scalars product is clearly commutative and* A.B = B.A

The order of the factors may thus be reversed without in any way affecting the value of the product.

Further, since B cos θ is the projection (or the resolute) of B in the direction of A and A cos θ, the projection (or the resolute) of A in the direction of B.

A.B = A.B cos θ = B.A cos θ, *i.e., the scalar product of two vectors* (A *and* B) is the *product of the magnitude of either vector and the projection (or the resolute) of the other in its direction.*

It also follows from the above that the angle between the two vectors, *i.e.*, θ, is given by the relation cos θ = A.B/AB.

IMPORTANT POINTS ABOUT SCALAR PRODUCT

The following points in regard to scalar product of two vectors may usefully be remembered :

(i) *if the two vectors* (A *and* B) *be perpendicular to each other, i.e.*, θ = π/2 *and, therefore,* cos θ = cos (A, B) = 0, *i.e., the scalar product* A.B = AB cos θ = AB cos (A, B) = 0. *This is, therefore, the condition for two vectors to be perpendicular (or orthogonal) to each other.*

(ii) *If the two vectors have the same direction, clearly* θ = 0 and cos θ = cos (A, B) = 1. *So that, the scalar product* A.B = AB cos θ = AB cos (A, B) = AB cos 0° = AB, *i.e., the scalar product is equal to the product of the magnitudes of the two vectors.*

And, if A = B, *i.e., if the two vectors be equal,* we have

$$A.A = A.A = A^2. \text{ Or, } A^2 = A^2,$$

i.e., the square of a vector is equal to the square of its magnitude (or modulus).

(iii) *If the two vectors have opposite direction,* θ = π *and* ∴ cos θ = cos π = – 1. *So that,* A. B = – AB, *i.e., the scalar product is equal to the negative product of their magnitudes.*

(iv) *It follows from (ii) above that square of any unit vector, like* i, j or k *is unity. Thus,* $i^2 = j^2 = k^2$ = (1) (1) cos 0° = 1. or,

$$i. i = j . j = k. k = 1.$$

(v) *Since* i, j *and* k *are orthogonal (or perpendicular to each other), we have, from (i) above,*

$$i. j = j. k = k. i = (1)(1) \cos 90^\circ = 0.$$

(vi) *If either vector* (A or B) *is multiplied by a scalar or a number m, the scalar product too is multiplied by that number. Thus,*

$$(mA) .B = A. (mB) = m (A.B) = mAB \cos \theta = mAB \cos (A. B).$$

(vii) *If* A. B = 0, *it means that either* A or B = 0 or A *and* B *are perpendicular to each other.*

(viii) *The scalar product obeys the distributive law.* This may be seen from the following:

The scalar product of vector A and the resultant (B + C) of B and C is given by A. (B + C) = A. (*projection or resolute of* (B + C) *in the direction of* A), *i.e.*,

$$A.(B + C) = A \times ON = A(OM + MN)$$
$$= A.OM + A.MN.$$

Since OM is the projection (or resolute) of B in the direction of A and MN, the projection (or resolute) of C in the direction of A, we have

$$A.OM = A. B$$

and $$A.MN = A.C.$$

Substituting these values in relation (i) above we have

$$A.(B + C) = A. B + A. C,$$

Clearly showing that the distributive law holds good.

It thus follows that (A + B). (C + D) = A.(C + D) + B.(C + D)

$$= A.C + A.D = B.C + B.D$$

(ix) Repeated application of the result obtained in (viii) above shows that the *scalar product of two sums of vectors may be expanded as in ordinary algebra.* Thus.

$$(A + B)^2 = A^2 + 2AB + B^2 \text{ and } (A + B) (A - B) = A^2 - B^2 \text{ etc..}$$

(x) *The distributive law also enables us to obtain an expression for the scalar product of two vectors in term of their rectangular components.* For, if $A = A_x i + A_y j + A_z k$ and $B = B_x i + B_y j + B_z k$, we have, remembering that i, j and k are mutually perpendicular.

$$A. B = A_x B_x + A_y B_y + A_z B_z,$$

i.e., the scalar product of two vectors is equal to the sum of the products of their corresponding rectangular components.

And, since A. B = AB cos θ, we have angle θ between the two vectors given by the relation cos

$$\theta = \frac{A_xB_x + A_yB_y + A_zB_z}{\sqrt{A_x^2 + A_y^2 + A_z^2}\sqrt{B_x^2 + B_y^2 + B_z^2}}$$

Further if A = B, we have

$$A.A = A^2 = A^2 = A_xA_x + A_yA_y + A_zA_z = A_x^2 + A_y^2 + A_z^2,$$

i.e., the square of a vector is equal to the sum of the squares of its rectangular components.

(xi) The magnitudes of the components of vector A along the three coordinate axes may be written as A_x = A.i, A_y = A.j, A_z = A.k, for, as we know,

$$A.i = (A_xi + A_yj + A_zk)i$$
$$= A_xi.i + A_yi.j + A_zi.k.$$
$$= A_x + 0 + 0 = A_x,$$

Similarly, A_y = A.j and A_z = A.k.

And, if U*l* be the unit vector along a line L, the component of vector A along this line can be shown to be A.U_l.

(xii) *A scalar product, being a number, can occur as a numerical coefficient of a vector.* Thus, for example.

(i) (A. B) C is a vector of magnitude (A. B) C in the direction of C.

(ii) (A. B) (C. D) is a number obtained by the multiplication of two numbers (A.B and (C. D).

RADIAL AND TRANSVERSE COMPONENTS OF VELOCITY

As we have just seen above, v = dr/dt,

where r = r $\hat{r}$. with r standing for the magnitude of r and r for the *unit vector along its direction.* We, therefore, have

$$v = \frac{dr}{dt} = \frac{dr}{dt}\hat{r} + r\frac{d\hat{r}}{dt}.$$

The vector $\frac{dr}{dt}\hat{r}$ is called the *radial component of velocity* v of the particle, its magnitude being obviously dr/dt.

And the vector $r\frac{d\hat{r}}{dt}$ is called the *transverse component of velocity* v *of the particle, because it is perpendicular to* r, as may be seen from the following:

Since $\hat{r}$ is a *unit vector*, its scalar product with itself is equal to 1, *i.e.,* $\hat{r}\,\hat{r} = 1$. So that, differentiating with respect to t, we have

$$\frac{d\hat{r}}{dt}r + \hat{r}\,\frac{d\hat{r}}{dt}\;2\hat{r}\frac{d\hat{r}}{dt} = 0. \text{ Or, } \hat{r}\frac{d\hat{r}}{dt} = 0,$$

indicating that $\hat{r}$ and $d\hat{r}/dt$ are perpendicular to each other.

And, since $\hat{r}$ lies along r, it follows that $d\hat{r}/dt$ and hence $r\,d\hat{r}/dt$ is perpendicular to r, its magnitude being obviously $r\left|\frac{dr}{dt}\right|$.

RADIAL AND TRANSVERSE VELOCITY AND ACCELERATION OF A PARTICLE MOVING IN A PLANE—CASE OF MOVING AXES

This is really a case of the axes themselves moving in the plane of motion of the particle, so that we obtain components of velocity and acceleration *along these moving axes.*

Let OX and OY be the reference system of rectangular axes in the plane of motion of the particle and OX' and OY' another system of rectangular coordinate axes moving relative to the former, with its position at any instant given by that of the moving particle at the time.

Components of Velocity

Let the position vector $r = \overrightarrow{OP}$ of the particle along OX' make an angle θ with OX. Then, if i and j be the unit vectors along the axes OX and OY and i' and j' along the axes OX' and OY' respectively, we have r = xi + yj. Or, r = r cos θi + r sin θ'j.

Since r = ri', where r is the *magnitude* or vector r, we have

$$ri' = r\cos\theta i + r\sin\theta j,$$

$$\text{whence, } i' = \cos\theta i + r\sin\theta j. \qquad \text{...(i)}$$

Similarly, since vector $n = \overrightarrow{OQ} = nn'$, where n is the magnitude of n, we have $j' = -\cos(\pi/2 - \theta)i + \sin(\pi/2 - \theta)j$

$$= -\sin\theta\, i + \cos\theta\, j \qquad \text{...(ii)}$$

As we know, *velocity* $v = \frac{dr}{dt} = \frac{d}{dt}(r\cos\theta i + r\sin\theta j)$

$$= \left(\frac{dr}{dt}\cos\theta - r\sin\theta\frac{d\theta}{dt}\right)i + \left(\frac{dr}{dt}\sin\theta + r\cos\theta\frac{d\theta}{dt}\right)j$$

$$= \frac{dr}{dt}(\cos\theta i + \sin\theta j) + \frac{rd\theta}{dt}(-\sin\theta i + \cos\theta j).$$

Or, $$v = \frac{dr}{dt} i + r\frac{d\theta}{di} j. \qquad \text{...(iii)}$$

Thus, *radial component of* v is $\frac{dr}{dt}$ i', having a magnitude $\frac{dr}{dt}$ and transverse *component of* v is $r\frac{d\theta}{dt}$j', having a magnitude $r\frac{d\theta}{dt}$.

Components of acceleration. As we know,

acceleration $$a = \frac{dv}{dt} = \frac{d}{dt}\left(\frac{dr}{dt}i' + r\frac{d\theta}{dt}j'\right)$$

$$= \frac{d^2r}{dt^2}i' + \frac{dr}{dt}\frac{di}{dt} + \frac{dr}{dt}\frac{d\theta}{dt}j' + r\frac{d^2\theta}{dt^2}j' + r\frac{d\theta}{dt}\frac{dj}{dt} \qquad \text{...(iv)}$$

Since from relations (i) and (ii) above, $\frac{di}{dt} = (\sin\theta i + \cos\theta j)\frac{d\theta}{dt}$

$= j\frac{d\theta}{dt}$ and $\frac{dj'}{dt} = (-\cos\ \theta i - \sin\ \theta j)\frac{d\theta}{dt} = -i'\frac{d\theta}{dt}$,

we have from relation (iv),

$$a = \frac{d^2r}{dt^2}i' + 2\frac{dr}{dt}\frac{d\theta}{dt^2}j' + r\frac{d^2\theta}{dt^2}j' - r\left(\frac{d\theta}{dt}\right)^2 i'$$

Or $$a = \left[\frac{d^2r}{dt^2} - r\left(\frac{d\theta}{dt}\right)^2\right]i' + 2\left(\frac{dr}{dt}\frac{d\theta}{dt} + r\frac{d^2\theta}{dt^2}\right)j' \qquad \text{...(v)}$$

Thus, the first term on the right hand side gives the *radial acceleration* of the particle, with its magnitude equal to

$$\frac{d^2r}{dt^2} - r\left(\frac{d\theta}{dt}\right)^2.$$

And, the second term gives the *transverse acceleration* of the particle, with its magnitude equal to

$$2\frac{dr}{dt}\frac{d\theta}{dt} + r\frac{d^2\theta}{dt^2} = \frac{1}{r}\frac{d}{dt}\left(r^2\frac{d\theta}{dt}\right).$$

DIFFERENTIAL COEFFICIENTS OF SUMS OF VECTORS

The differential coefficient of the sum A + B of two differentiable vectors, both of which are functions of t, is equal to the sum of their individual differential coefficients, as show below:

If, for an increase δt in the value of t, the corresponding increases in the values of A and B be δA and δB respectively, we have

$$\delta(A + B) = (A + \delta A + B + \delta B) - (A + B) = \delta A + \delta B$$

and hence the quotient $\dfrac{\delta(A+B)}{\delta t} = \dfrac{\delta A}{\delta t} + \dfrac{\delta B}{\delta t}$.

Or, taking the limiting values on either side as $\delta t \to 0$, we have

$$\frac{d}{dt}(A+B) = \frac{dA}{dt} + \frac{dB}{dt}.$$

What applies to the sum of two vectors applies equally well to the sum of any number of them.

DIFFERENTIAL COEFFICIENTS OF PRODUCTS OF VECTORS

As in the case of an algebraic product, so also here, the differential coefficient of a product of vectors in the sum of the quantities obtained by differentiating one single vector at a time, leaving the other unchanged.

(i) *Product of a scalar and a vector*: Let us consider the product ur of a vector r with a scalar u, both being differentiable functions of the variable t.

If, for the increase in the value of t from t to t + dt, the corresponding increases in the values of u and r be du and dr, we have

$$\delta(ur) = (u + \delta u)(r + \delta r) - ur = \delta u r + u\delta r + \delta u \delta r.$$

Dividing throughout by dt, we have

$$-\frac{\delta(ur)}{\delta t} = \frac{\delta u}{\delta t} r + u\frac{\delta r}{\delta t} + \frac{\delta u}{dt}\delta r.$$

Or, in the limit, $\dfrac{d}{dt}(ur) = \dfrac{du}{dt} r + u\dfrac{dr}{dt}$. ...(i)

From relation (i) coupled with the results obtained earlier, it follows that *the components of the differential coefficient (or derivative) of a vector are the derivatives of its components for fixed directions.*

Thus, if vector r be expressed in terms of its rectangular component vectors, we have r = xi + yj + zk (where i, j and k are, as we know, constant vectors and x, y, z functions of time), and the differential coefficient or r is given by

$$v = \frac{dr}{dt} = \frac{dx}{dt} i + \frac{dy}{dt} j + \frac{dz}{dt} k,$$

with dx/dt, dy/dt and dz/dt respectively representing the magnitudes of the components of vector v, *here representing velocity.*

And, in the same manner, we have

acceleration $$a = \frac{dv}{dt} = \frac{d^2r}{dt^2} = \frac{d^2x}{dt^2}i + \frac{d^2y}{dt^2}j + \frac{d^2z}{dt^2}k,$$

where d^2x/dt^2, d^2y/dt^2 and d^2z/dt^2 represent respectively the magnitudes of the components of acceleration a.

(ii) *Scalar product of two vectors:* Let us now try to obtain the differential coefficient of a scalar product A.B. Proceeding as in case (i), if δA and δB be the increments in A and B respectively corresponding to an increase δt in t, we have increase in the scalar product given by

$$\delta(A.B) = (A + \delta A).(B + \delta B) - A.B = \delta A.B + A.\delta B + \delta A.\delta B$$

$$= \delta A.B + \delta B.A,$$

neglecting δA.δB compared with other terms.

So that, dividing throughout by dt and proceeding to the limit $\delta t \rightarrow 0$, we have

$$\frac{d}{dt}(A.B) = \frac{dA}{dt}.B + A.\frac{dB}{dt},$$

where the order of the factors in any of the terms is quite immaterial.

(iii) Vector product of two vectors: Let us now consider the vector product A × B. Proceeding exactly as above, we have

$$\delta(A \times B) = (A + \delta A) \times (B + \delta B) - A \times B$$

$$= \delta A \times B + A \times \delta B + \delta A \times \delta B.$$

$$= \delta A \times B + A \times \delta B,$$

whence, as before, $\frac{d}{dt}(A \times B) = \frac{dA}{dt} \times B + A \times \frac{dB}{dt}$, *where the order of the factors in each term must* not *be changed unless accompanied also by change of sign.*

(iv) *Triple Products:* In the case of triple products too, we proceed as before to obtain the differential coefficients. Thus,

(a) in the case of a *scalar triple product* [ABC], we have

$$\frac{d}{dt}[ABC] = \left[\frac{dA}{dt}BC\right] + \left[A\frac{dB}{dt}C\right] + \left[AB\frac{dC}{dt}\right],$$

where the cyclic order in each term must be maintained.

And (b) in the case of a ***vector triple product*** A × (B × C), we have

$$\frac{d}{dt}[A \times (B \times C)] = \frac{dA}{dt} \times (B \times C) + A\left(\frac{dB}{dt} \times C\right) + A \times \left(B \times \frac{dC}{dt}\right),$$

where, again, the order of the factors in each term must be maintained.

CIRCULAR MOTION

Let us try to obtain expressions for the velocity and acceleration of a particle P moving at a constant speed along a angular path in the x.y plane, say, of constant radius r.

The position vector r, a function of the scalar variable t, moving with constant angular velocity (or angular frequency) of magnitude ω may, at any given instant t, be expressed in terms of its components along the axes of x and y. Thus, if i and j be the *unit vectors* along the two axes respectively, we have

$$r = xi + yj,$$

i.e., $r = r \cos \omega t i + r \sin \omega t j$

$$= r (\cos \omega t\ i + \sin \omega t\ j)$$

The velocity v of partiale P is thus given by

$$v = \frac{dr}{dt} = r\frac{dr}{dt} + \hat{r}\frac{dr}{dt}$$

where $\hat{r}$ is the unit vector along the direction or r.

Now, r being constant for the circular path, dr/dt = 0. So that, we have

$$v = r\frac{dr}{dt} = r\left(\frac{d}{dt}\cos \omega t\, i + \frac{d}{dt}\sin \omega t\, j\right)$$

Or, $v = r(-\omega \sin \omega t\ i + \omega \cos wt\ j)$

$$= \omega r (-\sin \omega t\ i + \cos \omega t\ j).$$

Therefore, with v as the magnitude of the velocity of P, we have

$$v^3 = v.v = \omega r(-\sin \omega t\ i + \cos \omega t\ j).\ \omega r(-\sin \omega t\ i + \cos \omega t\ j),$$

which, from our knowledge of scalar products and the fact that i.j = 0, gives

$$v^2 = \omega^2 r^3 (\sin^2 \omega t + \cos^2 \omega t) = \omega^2 r^3, \text{ whence, } v = \omega r.$$

Now, since acceleration a is the rate of change of velocity, we have

$$a = dv/dt = \omega^2 r(-\cos \omega t\ i - \sin \omega t\ j) = -\omega^2 r(\cos \omega t\ i + \sin \omega t\ j)$$

And since $r(\cos \omega t\ i + \sin \omega t\ j) = r$, we have

$$a = -\ \omega^2 r.$$

Or, the *magnitude of the acceleration of particle* P, *i.e.*, $a = \omega^2 r$, its direction being—r or *towards the centre of the, circle*—hence the name centripetal (*i.e.*, centre-seeking) acceleration given to it.

Or, since $\omega r = v$, we may also put $a = r\,(v/r)^2 = v^2/r$.

Hence the force, called *centripetal force*, acting on the particle is given by *mass* × *acceleration* = mv^2/r and is also directed towards the centre of its circular path.

Further, since the angle swept out by the radius vector in one complete revolution = 2π radians, we have *time period* of the particle given by $T = 2\pi/\omega$ and its frequency by $n = 1/T = \omega/2\pi$, whence, $\omega = 2\pi n$.

ANGULAR VELOCITY VECTOR

The angular velocity of a particle, having magnitude as well as direction, is obviously a *vector.* To obtain its value, we note that the linear velocity v of the particle is, at any instant, perpendicular to both the angular velocity ω and the radius vector r. So that, the vector equation corresponding to the relation $v = \omega r$ comes out to be $v = \omega \times r$, showing that *the linear velocity of the particle is the cross product of the angular velocity vector and the position vector with respect to a fixed point on the axis of rotation.*

The cross product of this with r gives

$$r \times v = r \times (\omega \times r) = \omega\,(r.r) - r\,(r.\omega) = r^2\omega,$$

(because $r.\omega = 0$, r and ω being perpendicular to each other),

whence, $\omega = r \times v/r^2$ and can be easily evaluated, its direction being parallel to the axis of rotation in the positive sense relative to the direction of rotation (*i.e.*, in accordance with the right-handed screw rule).

SCALAR AND VECTOR FIELDS

A physical quantity, expressed as a continuous function of the position of a point in a region of space is referred to as *point function* (or *function of position*). The region of space concerned is known as a *scalar field* or a *vector field* according as the physical quantity in question is a scalar or a vector one, expressed as a continuous single-valued function, $\phi\,(x, y, z)$ or $F\,(x, y, z)$ respectively, of position in that region. In actual practice, *the two functions of positions themselves are spoken of as the scalar and vector fields respectively.*

As examples of scalar fields may be mentioned the distribution of temperature or magnetic and electrostatic potentials in a given region of space. The field here consists of a series of surfaces, (isothermal or equipotential, as the case may be), each having a fixed and definite value of temperature or magnetic or electrostatic potential, *i.e.*, of the function ϕ (x, y, z). Such surfaces are referred to as *level surfaces*, each level surface having its own constant value of ϕ (x, y, z) all over.

And, as examples of vector fields, we have distribution of magnetic and electric intensity or the distribution of velocity in a moving (continuous) fluid. At any given point, in this case, the single-valued vector function F (x, y, z) is specified by a vector having a definite magnitude and direction, both of which continuously change from point to point throughout the field.

Starting from any desired point in the field and proceeding through infinitesimal distances from point to point in the direction of the field, we obtain, in general, a curved line, called the *line of flow*, the *flux line* or the *vector line,* the tangent to which at any point gives the direction of the vector at that point. The measure or magnitude of the vector is given by the number of flux lines passing per unit area of a surface normal to their direction.

PARTIAL DERIVATIVES—GRADIENT

Partial derivatives. The differentiation of a function such as $\phi(x_1, x_2, x_3 ...)$ of two or more independent variables, x_1, x_2, x_3 etc., with respect to one of them, keeping the others constant, is called *partial differentiation.* The derivatives thus obtained are referred to as *partial derivatives* and are denoted by the symbol ∂ instead of d.

Thus, if we have a continuously differentiable scalar point function ϕ (x, y, z), *i.e.*, a function of the coordinates x, y, z its partial derivatives along the three coordinate axes are

$$\frac{\partial\phi}{\partial x}, \frac{\partial\phi}{\partial y} \text{ and } \frac{\partial\phi}{\partial z}$$

Gradient. The vector function $i\frac{\partial\phi}{\partial x} + j\frac{\partial\phi}{\partial y} + k\frac{\partial\phi}{\partial z}$ is called the *gradient* of the scalar point function ϕ and is denoted by grad ϕ.

Thus,

$$\text{grad}\,\phi = i\frac{\partial\phi}{\partial x} + j\frac{\partial\phi}{\partial y} + k\frac{\partial\phi}{\partial z}.$$

$$\therefore \quad \text{differential } d\phi = \frac{\partial\phi}{\partial x}dx + \frac{\partial\phi}{\partial y}dy + \frac{\partial\phi}{\partial z}dz$$

$$= (idx + jdy + kdz)\left(i\frac{\partial\phi}{\partial x} + j\frac{\partial\phi}{\partial y} + k\frac{\partial\phi}{\partial z}\right)$$

Or, $d\phi = dr \text{ grad } \phi$.

It can easily be show that grad ϕ at any point is quite independent of the choice of the coordinate axes. It follows, therefore, that the *gradient of a scalar point function is a vector point function.*

If ϕ be a constant, obviously the partial derivatives $\partial\phi/\partial x$, $\partial\phi/\partial y$ and $\partial\phi/\partial z$ will all be zero and hence grad $\phi = 0$. And, converse, if grad $\phi = 0$, the partial derivatives are all zero and hence function ϕ is a constant. Thus, *grad $\phi = 0$ only if ϕ be constant.*

THE OPERATOR ∇

We have in vector algebra a *differential operator,* denoted by ∇ (*i.e.*, inverted Δ). It is called 'del' or 'nabla'. *It operates distributively and is formally assum?l to have the character of a vector.* And since it is assumed to have the character of a vector, *its product $\nabla\phi$ with a scalar ϕ is a vector.*

If we use Cartesian orthogonal coordinates, we may write

$$\nabla = i\frac{\partial}{\partial x} + j\frac{\partial}{\partial y} + k\frac{\partial}{\partial z}.$$

So that, grad $\phi = i\dfrac{\partial\phi}{\partial x} + j\dfrac{\partial\phi}{\partial y} + k\dfrac{\partial\phi}{\partial z}$

$$= \left(i\frac{\partial}{\partial x} + j\frac{\partial}{\partial y} + k\frac{\partial}{\partial z}\right)\phi = \nabla\phi, \text{a vector.}$$

MAGNITUDE AND DIRECTION OF $\nabla\phi$

Let us consider two level surface L_1 and L_2 through two close points A and C, distance dr apart, with the values of the scalar function ϕ and $(\phi + d\phi)$ respectively. And let AB = dn be the normal to the surface L_1 at A. Since L_2 is a level surface, the value of the scalar function is the same [*viz.*, $(\phi + d\phi)$] at B as at C. Obviously, therefore, the *rate of change of ϕ* along the normal AB is the highest (this being the shortest distance between the two surfaces) and is equal to $\partial\phi/\partial n$.

Now, we may put $dn = dr\cos\theta = \hat{n}\, dr$, where $\hat{n}$ is the unit vector normal to the surface L_1 at A. So that,

$$d\phi = \frac{\partial\phi}{\partial n}dn = \frac{\partial\phi}{\partial n}\hat{n}.dr.$$

As we know, $\nabla\phi.dr = \left(i\dfrac{\partial\phi}{\partial x} + j\dfrac{\partial\phi}{\partial y} + k\dfrac{\partial\phi}{\partial z}\right)(idx + jdy + kdz)$

$$= \frac{\partial\phi}{\partial x}dx + \frac{\partial\phi}{\partial y}dy + \frac{\partial\phi}{\partial z}dz = d\phi.$$

Or, $$d\phi = \nabla\phi.dr = \frac{\partial\phi}{\partial n}\hat{n}dr,$$

whence, $$\nabla\phi = \left(\frac{\partial\phi}{\partial n}\right)\hat{n},$$

i.e., vector $\nabla\phi$ or grad ϕ has a magnitude equal to the maximum rate of change of ϕ and is directed along this maximum rate of change, *i.e.*, along the normal to the level surface L_1 (with ϕ constant).

RECIPROCAL SYSTEM OF VECTORS

If vectors a', b' and c' be defined by the equations

$$a' = \frac{b \times c}{[abc]},\ b' = \frac{c \times a}{[abc]} \text{ and } c' = \frac{a \times b}{[abc]},$$

they are said to be *reciprocal vectors* to a, b and c which are assumed to be *non-coplanar* so that the product [abc] = 0. The planes of a', b' and c' are respectively perpendicular to those of b, c; c, a and a, b.

Since, obviously, the dot products a.a' = b.b' = c.c' = 1, the name *reciprocal vectors* for a', b' and c' is only natural.

The dot product of any other pair, one from each system, is zero. Thus, a.b' = 0 because a.b' = aca/abc and the numerator, being a triple product containing two equal or identical vectors (a and a), is equal to zero. Similarly,

$$\begin{aligned} c\ b' &= b.a' = b.c' \\ &= c.a' = a.c' = 0. \end{aligned}$$

It can be shown that

$$a = \frac{b' \times c'}{[a'b'c']},\ b = \frac{c' \times a'}{[a'b'c']} \text{ and } c = \frac{a' \times b'}{[a'b'c']},$$

So that, a, b, c, and a', b', c' are really reciprocal system of vectors to each other. They, therefore, possess the same sign. *i.e.*, they are either both right-handed or both left-handed. The unit vectors i, j and k along the three coordinate axes are, however, their own reciprocals.

SCALAR AND VECTOR QUANTITIES

Physical quantities may, in general, be divided into two main classes:

(i) Scalar quantities or scalars, and

(ii) Vector quantifies or vectors.

(1) *Scalar quantities or scalars* are those physical quantities which possess *only magnitude and no direction* in space as, for example, *mass, volume, temperature, time, speed of light* etc.

A scalar quantity can, therefore, be completely specified simply by its magnitude or numerical value, indicating how many times its unit is contained in it. Thus, for example, when we speak of a mass of 50 g, we mean that the unit of mass, the *gram* (g), is contained 50 times in it.

Obviously, having no direction, the magnitude or measure of a scalar quantity is quite independent of any coordinate system. All such quantities are, however, subject to the ordinary algebraical laws of addition and multiplication, *viz.*, the following:

(i) *Law of commutation*, according to which the result of addition or multiplication of a number of scalars is quite independent of the order in which they may be taken. For example,

$$A + B \equiv B + A \text{ and } A \times B \equiv B \times A.$$

(ii) *Law of association*, according to which the sum or the product of a finite number of scalars is quite independent of the manner in which they may be grouped or associated. Thus,

$$A + B + C \equiv (A + B) + C \equiv A + (B + C) \equiv (A + C) + B.$$

And, similarly, $A\times (B \times C) \equiv (A \times B) \times C \equiv (A \times C) \times B$,

clearly indicating that in the case of continuous sums and products the brackets are really superfluous.

(iii) *Law of distribution*, which states that in expressions involving both addition and multiplication the result in the same as the sum of the individual term-wise products. For example,

$$A \times (B + C) \equiv A \times B + A \times C$$

and $$(A + B) \times C \equiv A \times C + B \times C.$$

(2) *Vector quantities or vectors* are those physical quantities which *possess magnitude as well as, direction and are subject, to the parallelogram law of addition.* This latter condition is important because a quantity may have magnitude as well as direction and may still not be a vector quantity. For instance the finite rotation of a rigid body about a given axis has magnitude (*viz.*, the angle of rotation), also direction (*viz.*, the direction of the axis) but it is not a vector quantity. This is so for the simple reason that two finite rotations of the body do not add up in accordance with the vector law of addition, *i.e.*, the resultant rotation is not the vector sum of two finite rotations.

However, if the rotation be small or infinitesimal, it may be regarded as a vector quantity; for, then, the arc described by the body in a, small interval of time is more or less a straight line and is thus representable as a vector, and so also the transverse and angular velocities of the body. Thus, whereas small rotations, angular velocity and angular momentum are vector quantities, large or finite ones are not.

It follows, therefore, that to qualify as a vector, a physical quantity must not only possess magnitude and direction but must also satisfy the .parallelogram law of vector addition.

Vectors associated with a linear or directional effect are called *polar vectors* or, usually, simply *as vectors*, and those associated with rotation about an axis are referred to as *axial vectors*. Thus, force, linear velocity and linear momentum are *polar vectors* and a couple, angular velocity and angular momentum are axial vectors.

VECTOR NOTATION

In ordinary writing, whether on a black board or a paper, a *vector quantity* A, say, is represented by putting a wavy line underneath the letter A or a small arrowhead above it (indicating its direction), as $\vec{A}$. In print, however, it is represented by the bold-faced type letter A.

The *magnitude* or *modulus* of a vector, being a scalar quantity, is denoted by a light-faced italic letter. Thus, the magnitude of a vector A is denoted by A. It may also be denoted by | A | or *mod* A.

UNIT AND ZERO VECTORS

A vector of unit magnitude is called a *unit vector* and the notation for it in the direction of A, is $\hat{A}$ read as 'A, hat or A, caret'. It is also often denoted by bold-faced lower case type letter a. Thus, we have

$$A = \hat{A}A$$

or $\quad aA,$

A unit vector, as will be readily seen, merely indicates direction.

A *vector of zero magnitude is called* a zero or a null vector, *denoted by 0.*

All zero or null vectors are taken to be equal and their directions are quite arbitrary and, indeed, quite immaterial.

Vectors other than null vectors are referred to as *proper vectors*.

GRAPHICAL REPRESENTATION OF A VECTOR

Graphically a vector is represented by an arrow drawn to a chosen scale, parallel to the direction of the vector. The length and the direction of the arrow thus represent; the magnitude and the direction of the vector respectively. Thus, the in Fig. 5(a) represents a vector A, parallel to the x-y plane and making an angle θ with the axis of x or a line parallel to it. If the magnitude of vector A be 7 units *i.e.*, if A = 7, the arrow representing it is drawn 7 times (the length of the arrow representing unit vector A, as shown in Fig. 5(b).

The *negative of vector* A, is the vector –A, having the *same magnitude* as A bat being oppositely directed to it. It is therefore, represented by an arrow equal in length to that representing A but, antiparallel to it, as shown in Fig. 5(c).

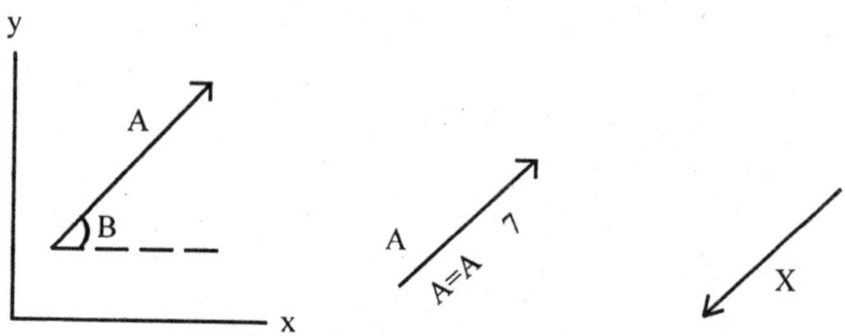

Fig. 5

MULTIPLICATION AND DIVISION OF VECTORS BY SCALARS

From the law of addition of vectors it follows that the sum A+ A + A + ... to *m terms* gives a vector of magnitude m times that of A in the same direction with it, *i.e.*, equal to mA. So that, the product of a vector A and a scalar m is a vector mA whose magnitude is m times the magnitude of A and which is similarly or oppositely directed to A according as the scalar m is positive or negative. Thus,

$$| m A | = mA.$$

Further, if m and n be two scalars and A and B, two vectors, we have

$$(m + n)A = mA + nA \text{ and } m(nA) = n(mA) = (mn) A,$$

showing clearly that *the multiplication of a vector by a scalar is distributive.*

The division of vector A by a non-zero scalar as is defined as the multiplication of vector A by 1/w. Thus, if A = $\hat{A}$A, we have $\hat{A}$ = A/A.

RECTANGULAR COMPONENT OF VECTOR

Two or more vectors which, when compounded in accordance with the parallelogram law or the polygon law of vectors-give a vector A, are said to be the *components* of vector A. The most important components with which we are concerned are mutually perpendicular or rectangular ones along the three coordinate axes OX, OY and OZ respectively. Let us deduce their values.

With O as the origin, let vector OD be R. Then on OD as diagonal, let us construct a *rectangular parallelopiped* with its three edges along the three coordinate axes, as shown.

Let the vector intercepts of the three components of R (also spoken of as the *resolutes*, or the *resolved parts,* of R) along the three axes respectively be R_x, R_y and R_z.

Then, we have $$R = R_x + R_y + R_z \qquad \text{...(i)}$$

Let i, j and k be *unit vectors along the three coordinate axes,* chosen as the *base vectors* or forming what is called the *orthogonal triad of vectors.* Then, we have

$$R_x = R_x i,\ R_y = R_y j \text{ and } R_z = R_z K.$$

So that, substituting in relation (i), we have

$$R = R_x i + R_y j + R_z k. \qquad \text{...(ii)}$$

Here, $R_x i$, $R_y j$ and $R_z k$ are the orthogonal projections of R on the directions of i, j and k respectively.

We have

$$OD^2 = OH^2 + OF^2 + OB^2$$

i.e. $$R^2 = R_x^2 + R_y^2 + R_z^2, \qquad \text{...(iii)}$$

where R is the *magnitude* (or *modulus*) of vector R.

Thus, *the square of the magnitude* (or *modulus*) *of a vector is equal to the sum of the squares of its rectangular components*

$$\therefore \qquad R = \sqrt{R_x^2 + R_y^2 + R_z^2}. \qquad \text{...(iv)}$$

And, clearly, $R_x = R\ cos(R, x)$ $R_y = R\ cos(R, y)$ and $R_z = R \cos (R, z)$, where the three cosines, *viz., cos*(R, x) = R_x/R, *cos* (R, y)=R_y/R and cos(R, z) = R_z/R are referred to as the *direction cosines* of vector R, since they help determine its direction. They are quite often represented by the letters *l*, m and n respectively.

Dividing relation (iii) by R^2, we have

$$1 = R_x^2/R^2 + R_y^2/R^2 + R_x^2/R_y^2$$

$$= \cos^2 (R, x) + \cos^2 (R, y) + \cos^2 (R, z)$$

$$= l^2 + m^2 + n^2, \qquad \text{...(v)}$$

i.e., the sum of the squares of the three direction cosines of a vector is equal to unity.

Clearly, the unit vector along R is given by

$$R = \frac{R}{R} = \frac{R_x}{R}i + \frac{R_y}{R}j + \frac{R_z}{R}k,$$

where $R_x/R = \cos(R, x)$, $R_y/R = \cos(R, y)$ and $R_z/R = \cos(R, z)$ are the *direction cosines* of R. We therefore have

$$\textit{unit vector } R = \cos(R, x)\, i + \cos(R, y) j + \cos(R, z) k, \quad \text{...(vi)}$$

Further, if we have a number of vectors R_1, R_2, R_3 etc., their sum may be expressed in terms of their rectangular components in the form

$$\sum R = \left(\sum R_x\right)i + \left(\sum R_y\right)j + \left(\sum R_z\right)k.$$

So that, $\sum R_x$, is the resolute of $\sum R$ in the direction of i and $\sum R_y$ and $\sum R_z$, its resolutes in the directions of j and k respectively.

Since the directions of i, j and k may be chosen arbitrarily, we have the result that the *resolute of a sum of vectors in any direction is equal to the sum of the resolutes of the individual vectors in that direction*

POSITION VECTOR

The position of a point P from any assigned point, such as the origin O of the Cartesian coordinate system, for example, is uniquely specified by the vector $\overrightarrow{OP} = r$ called the *position vector of point* P *relative to* O.

The coordinates of point P being (x, y, z), we have

$$r = xi + yj + zk, \quad \text{...(vii)}$$

$$\text{where } r = \sqrt{x^2 + y^2 + z^2}.$$

The direction cosines of r are obviously x/r, y/r and z/r.

Such a point, with a position vector r, is often spoken of as the point r.

It follows from the above that if there be two points P_1 and P_2, with (x_1, y_1, z_1) and (x_2, y_2, z_2) (as their respective coordinates

and $\overrightarrow{OP_1} = r_1$ and $\overrightarrow{OP_2} = r_2$

as their respective position vectors, we have

$$r_1 = x_1 i + y_1 j + z_1 k$$

and $$r_2 = x_2 i + y_2 j + z_2 k$$

and therefore, $\overrightarrow{P_1P_2} = \overrightarrow{P_1O} + \overrightarrow{OP_2} = -r_1 + r_2 = r_2 - r_1$

$$= (x_2 - x_1)i + (y_2 - y_1)j + (z_2 - z_1)k.$$

Now, keeping the direction of A unchanged, *i.e.*, keeping it fixed in its position (with respect to any of the so called fixed objects) imagine the reference frame to be rigidly rotated about the origin, say, in the anticlockwise direction, with the coordinate axes now taking up the positions OX', OY' and OZ'. Then, if the new orthogonal triad of unit vectors be formed by i', j' and k' and the components of A along the new coordinate axes be A_x', A_y' and A_z',

$$A = A_x'i' + A_y'j' + A_y'k'. \qquad ...(ii)$$

Since the length of vector A must obviously be quite independent of the orientation of the coordinate system, or its frame of reference, we have

$$A^2 = Ax^2 + Ay^2 + Az^2 = Ax'^2 + Ay'^2 + Az'^2.$$

Or, $$A = \sqrt{A_x^2 + A_y^2 + A_z^2} = \sqrt{A_x'^2 + A_y'^2 + A_z'^2}$$

In other words rigid rotation of the Cartesian system (or the frame of reference) of a vector brings about no change in its form of magnitude.

PRODUCT OF TWO VECTORS

From the manner in which two vectors enter into combination in physics we come across two distinct kinds of products: (i) *a number or a scalar*, called the *scalar product,* and (ii) a vector, called the *vector product.* The process by which we obtain a product is called *multiplication* and each of the two vectors is called a *factor* of the product.

The scalar and the vector products are also called in *inner* and the *outer* products, after Grassmann.

VECTOR PRODUCT

The vector product (or the *outer product*) of two vectors A and B is denoted by A × B, read as 'A cross B'. It is, therefore, also called the *cross product* of the two vectors.

It may also be denoted simply as [A, B]. *It is defined as a vector* R *whose magnitude (or modulus) is equal to the product of the magnitudes of the two vectors* A *and* B *and the sine of their included angle* θ. This vector R is normal to the plane of A and B and points in the direction in which a right-handed screw would advance when rotated about an axis perpendicular to the plane of the two vectors in the direction from A to B through the smaller angle θ between them.

Or, alternatively, we might state the rule thus:

If the fingers of the right hand be curled in the direction in which vector A *must be turned through the smaller included angle θ to coincide with the direction of vectors* B, *the thumb points in the direction of* R.

Either of these rules is referred to as the *right handed screw rule.*

Thus, if $\hat{n}$ be the unit vector which gives the direction of vector R, we have

$$R = A \times B = AB \sin\theta\ \hat{n}, \qquad (0 \leq \theta \leq 180^{\circ} \text{ or } \pi,)$$

IMPORTANT POINTS ABOUT VECTOR PRODUCT

(i) Since $0 \geq \theta \geq \pi$, $\sin\theta$ *and, therefore,* $|R|$ or R *cannot be negative.*

And, clearly, $$\sin\theta = \frac{A \times B}{|A|\,|B|}$$

(ii) *A change in the order of factors* A *and* B *in the cross product obviously reverses its sign on account of the reversed sense of rotation of the screw.* So that, the *cross product* $B \times A = -A \times B = -R = -AB \sin\theta\ \hat{n}$, indicating that *the vector product is not commutative.*

(iii) *If either vector* (A *or* B) *be multiplied by a number or a scalar m, their vector product too is multiplied by the same number. Thus*

$$(mA) \times B = A \times (mB) = (mAB) \sin\theta\ \hat{n}.$$

Again, $(mA) \times (nB) = (mA\ nB \sin\theta)\hat{n} = nA \times mB$

$= mnA \times B = A \times (mnB)$, where n is another scalar.

Thus, we see that the *vector product is associative.*

(iv) $\hat{A}$ and $\hat{B}$ be unit vectors, we have $A = B = 1$ and, therefore,

$$\hat{A} \times \hat{B} = AB \sin\theta\hat{n} = \sin\theta,$$

i.e., the magnitude of $\hat{A} \times \hat{B}$ *is the sine of the angle of inclination of the two.*

(v) *If the two vectors* (A *and* B) *be orthogonal or perpendicular to each other, we have* $\theta = 90^{\circ}$ or $\pi/2$ *and therefore,* $\sin\theta = 1$. *So that.* $A \times B = AB\hat{n}$.

The vectors A, B *and* $(A \times B)$ *thus form a right-handed system of mutually perpendicular vectors.*

It follows at once from the above that in the *case of the orthogonal triad of unit vectors,* i, j and k (*each perpendicular to the other*), *we have*

$$i \times j = -j \times i = k;$$

$$j \times k = -k \times j = i$$

and $$k \times i = -i \times k = j.$$

(vi) *If the two vectors* (A *and* B) *be collinear or parallel,* $\theta = 0$ or π *and, therefore, sin* $\theta = 0$.

So that, $$A \times B = AB \sin \theta \hat{n} = 0.$$

It follows at once, therefore, that $A \times A = 0$, indicating that *the vector product of two parallel or equal vectors is zero* and , clearly therefore, *the vector product of a vector with its ownself is zero.*

Obviously, therefore, i*n the case of orthogonal triad of unit vectors, we have*

$$i \times i = j \times j = k \times k = 0$$

Conversely, *if the vector product* $A \times B = 0$, we have $AB \sin \theta \hat{n} = 0$. So that, either $A = 0$ or $B = 0$ or $\sin \theta$ and , therefore, $\theta = 0$ or π, *i.e., either one of the vectors is a null (or zero) vector or the two vectors are parallel to each other.*

(vii) *The distributive law holds good for the cross product of vectors.* Thus,

$$A \times (B + C) = A \times B + A \times C.$$

This may be seen from the following: Putting $U = A \times (B + C) - A \times B - A \times C$ and taking the scalar product with an arbitrary vector V, we have

$V. U = V.[A \times (B + C) - A \times B - A \times C] = V. A \times (B + C) - V. (A \times B) - V. (A \times C)$.

As we shall see under § 1.22, $A. (B \times C) = (A \times B).C$. Therefore, we have

$$V. U = (V \times A). (B + C) - (V \times A).B - (V \times A).C.$$

Or, because scalar products are distributive, we have

$V. U = (V \times A).B + (V \times A).C - (V \times A).B - (V \times A).C = 0$

Obviously, therefore, either $V = 0$ or perpendicular to U or $U = 0$.

Since V is an arbitrary vector, it can be chosen to be *non-zero* and *non-perpendicular* to U. Hence, $U = 0$ and we, therefore, have

$$A \times (B + C) = A \times B + A \times C$$

in accordance with the distributive law.

(viii) *A vector product can be expressed in terms of rectangular components of the two vectors and put in the determinant form*, as may be seen from the following:

Writing A and B in terms of their rectangular components, we have

$$A = A_x i + A_y j + A_z k \text{ and } B = B_x i + B_y j + B_z k$$

And ∴ $A \times B = (A_x i + A_y j + A_z k) \times (B_x i + B_y j\ B_z k)$

$$= (A_x B_x) i \times i + (A_x B_y) i \times j + (A_x\ B_z) i \times k$$
$$+ (A_y B_x)\ j \times i + (A_y B_y)\ j \times j + (A_y B_z) j \times k$$
$$+ (A_z B_x)\ k \times i + (A_z B_y)\ k \times j + (A_z B_z) k \times k.$$

Since $i \times i = j \times j = k \times k = 0$ and $i \times j = k = -j \times i$ etc.,[see (v) above], we have

$A \times B = (A_x\ B_y)k - (A_z\ B_z)j - (A_y\ B_x)k + (A_y\ B_z)i + (A_z\ B_x)j - (A_z\ B_y)$.

This may be grouped together as

$A \times B = (A_y\ B_z - A_z\ B_y)i + (A_z\ B_x - A_x\ B_z)j + (A_x\ B_y - A_y\ B_x)k.$

Or, putting it in the easily remembered *determinant form*, we have

$$A \times B = \begin{vmatrix} i & j & k \\ A_x & A_y & A_z \\ B_x & B_y & B_z \end{vmatrix}$$

It may be noted that the *scalar components of the first vector A occupy the middle row of the determinant.*

(ix) if $B = C + nA$, where n is a number or a scalar, we have

$$A \times B = A \times (C + nA) = A \times C.$$

Conversely, however, if $A \times B = A \times C$, it *does not necessarily mean that* B = C. *All it implies is that* B *differs from* C *by some vector parallel to* A *which may or may not be a zero vector.*

SOME ILLUSTRATIVE APPLICATIONS OF VECTOR PRODUCT

(i) *Torque of moment of a force:* Let a force F be acting on a body free to rotate about O and let r be the *position vector* of any point P on the line of action of the force. Then, since torque = *force × perpendicular distance of its line of action from 0,* we have *torque* (*or moment*) *of force F about O i.e.,*

$$T = Fr \sin \theta.$$

Now, as we know, $Fr \sin \theta$ is the magnitude of the cross or vector product $r \times F$. So that, in vector notation, we have

$$\text{torque } (\tau) = r \times F$$

Its direction being perpendicular to the plane containing F *and* r.

If we draw a set of three coordinate axes through O, as shown, we shall have $r = xi + yj + zk$,

Where xi, yj and zk are the rectangular components of r along the three axes respectively.

Similarly, $F = F_x i + F_y j + F_z k$

and $T = T_x i + T_y j + T_z k$ ..(i)

the symbols having their usual meanings.

$$\text{Now,} \quad T = r \times F = \begin{vmatrix} i & j & k \\ x & y & z \\ F_x & F_y & F_z \end{vmatrix}$$

which, on expansion, gives

$$T = i\,(F_z y - F_y z) + j\,(F_x z - F_z x) + k\,(F_y x - F_x y). \quad \text{...(ii)}$$

Comparing expressions (i) and (ii), we have

$$T_x = F_z y - F_y z,$$

$$T_y = F_x z - F_z x$$

and $$T_z = F_y x - F_z y,$$

where T_x, T_y and T_z are the respective scalar components of T along the three coordinate axes through O.

It will be easily seen that the scalar components T_x, T_y and T_z of torque Tare given by the dot or the scalar products of T and the respective unit vectors along the three coordinate axes. Thus,

$$T_x = T.i, \; T_y = T.j$$

and $$T_z = T.k.$$

For, $$T.i = (T_x i + T_y j + T_z k).i$$

$$= T_x i.i + T_y j.i + T_z k.i$$

And since $i.\,i = 1$, $j.\,i = 0$ and $k.\,i = 0$ we have $T_x = T.i$

And, Similarly, $T_y = T.\,j$ and $T_z = T.\,k$.

(ii) *Co* R = A × B ɔle, as we know, is a combination of two equal, opposite and parallel forces. Let F and –F be two such forces acting at points P and Q and let the position vectors of P and Q with respect to O be r_1 and r_2 respectively.

Then, since *moment of the couple* C with respect to O is equal to the sum of the moments, (with respect to O) of the two forces constituting the couple, we have

$$C = r_1 \times F + r_2 \times (-F) = (r_1 - r_2) \times F.$$

Since $r_1 - r_2 = a$, where a lies in the same plane with F, we have

$$C = a \times F$$

Thus, *the moment of the couple,* C, *is a vector lying in a plane perpendicular to that containing the two forces.*

The *magnitude of couple* C $= |a \times F|$ or a F sin θ.

And since a sin θ = d, *the perpendicular distance between the two forces, we have magnitude of the moment of the couple, i.e., C = Fd = (one of the forces) (perpendicular distance between the forces).*

(iii) *Area of a parallelogram*: Let vectors A and B form the adjacent sides of a parallelogram OPQR, inclined to each other at an angle θ. Then, if OD be the perpendicular dropped from O on to PQ, we have *area of the parallelogram*

$$= P\,Q\,(O\,D) = B\,(A \sin\theta).$$
$$= AB \sin\theta = AB \sin(A, B),$$

which is obviously twice the area of the triangle OPQ with the same adjacent sides A and B.

Clearly, AB sin θ or AB sin (A, B) is *the magnitude* of the vector product A × B = C, say, *whose direction is perpendicular to the plane containing* A *and* B, *i.e., to the plane of the parallelogram.*

Now, an area, by itself, has no sign but may be regarded positive or negative in relation to the direction in which its boundary is described. A × B = C, *therefore, represents a vector area which gives both the magnitude and the orientation of the area of the parallelogram.*

The direction of vector C, drawn normal to the plane (*i.e.*, the parallelogram here) bears the same relation to the direction in which the boundary of the figure is described as the direction of advance of a right-handed screw does to its direction of rotation. Thus, with the parallelogram described as shown (A being taken first and B next), the normal vector C = A × B is drawn pointing upwards, the area of the parallelogram being regarded as positive in relation to this direction of C.

With this convention, therefore, any vector area (*i.e.*, the area of any plane figure) may be represented by a vector drawn normal to the plane of the figure in a direction relative to which the area of the figure is regarded as positive.

(iv) *The law of sines in a triangle*: Consider a triangle vector A, B and C such that A + B = C.

Taking vector product of both sides of the relation with A, we have

$$A \times A + A \times B = A \times C.$$

Or, since A× A = 0. We have A × B = A × C. So that, taking magnitudes of the two sides, we have

AB sin (A, B) = AC sin (A, C).

Or, B sin (A, B) = C sin (A, C)

whence, $$\frac{B}{\sin(A,C)} = \frac{C}{\sin(A,B)}$$

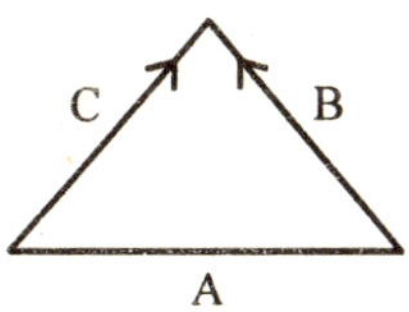

Fig. 6

which is the familiar *law of sines in a triangle.*

(v) *Force on a moving charge in a magnetic field*: Imagine a charge q to be moving with velocity v at an angle θ with a magnetic field B at any given instant. Then, the force acting on it in a direction perpendicular to B as well as v is F = qv B sin θ, where F, v and B are the *magnitudes of the force, velocity and the magnetic field* respectively. In vector form, therefore, we may put it as

F = qv × B *emu* or SI *units*.

This is called *Lorentz force law*, with the force itself referred to as the *Lorentz force*.

In most cases, q is taken in *esu* and B in *emu* (*i.e., gauss*). Since 1 *esu* charge = (1/c) *emn*, where c is the velocity of light in vacuo, we have q *esu* = q/c *emu* and, therefore,

$$F = \frac{q}{c} v \times B.$$

In case the charge also simultaneously passes through an electric field E, an additional force qE acts upon it and the Lorentz force law then takes the form $F = qE + \frac{q}{c} v \times B$.

IMPORTANT FORMULAS

If ϕ is a scalar point function and $\vec{F}, \vec{u}, \vec{v}, \vec{w}$ are vector point functions then.

1. $\frac{d}{dt}\left(\vec{u} + \vec{v}\right) = \frac{d\vec{u}}{dt} \pm \frac{d\vec{v}}{dt}$.

2. $\frac{d}{dt}\left(\vec{u} . \vec{v}\right) = \vec{u} . \frac{d\vec{v}}{dt} + \vec{v} . \frac{d\vec{u}}{dt}$.

3. $\frac{d}{dt}\left(\vec{u} \times \vec{v}\right) = \vec{u} \times \frac{d\vec{v}}{dt} + \vec{v} \times \frac{d\vec{v}}{dt}$.

4. $\frac{d}{dt}\left(\phi \vec{u}\right) = \phi \frac{d\vec{u}}{dt} + \vec{u} \frac{d\phi}{dt}$.

5. $\frac{d}{dt}\left[\vec{u}, \vec{v}, \vec{w}\right] = \left[\frac{d\vec{u}}{dt}, \vec{v}, \vec{w}\right] + \left[\vec{u}, \frac{d\vec{u}}{dt}, \vec{w}\right] + \left[\vec{u}, \vec{v}, \frac{d\vec{w}}{dt}\right]$.

6. $\frac{d}{dt}\left[\vec{u} \times \vec{v} \times \vec{w}\right] = \left(\frac{d\vec{u}}{dt} \times \vec{v}\right) \times \vec{c} + \left(\vec{u} \times \frac{d\vec{v}}{dt}\right) \times \vec{c} + \left(\vec{u} \times \vec{v}\right) \times \frac{d\vec{c}}{dt}$.

7. $\nabla = \hat{i}\frac{\partial}{\partial x} + \hat{j}\frac{\partial}{\partial y} + \hat{u}\frac{\partial}{\partial z}$.

SOLVED EXAMPLES

1. *If a = (sin θ) i + (cos θ) j + θ k, b = (cos θ) i = (sin θ) j = 3 k, c = 2i + 3 ;j = k,*

find $\frac{d}{d\theta}$ *{a × (a × c)} at θ = 0.*

Solution. We know a × (b × c) = (a . c) – (a . b) e

Here a . b = [(sinθ) i + (cosθ) j + θ k] . [cosθ) i = (sinθ) j – 3 k]

= sinθ cosθ = cosθ sinθ = 3 i^2 = 1, [∵ i j = 0 etc.]

= – 3θ

and a . c. = [(sinθ) i + (cosθ) j + θ k] . (2i + 3j = k)

= 2 sinθ + cosθ – θ as aove.

$\therefore$ From (i) we have

$$a \times (b \times c) = (2 \sin\theta + 3 \cos\theta - \theta) [(\cos\theta)\, i - (\sin \theta)\, j - 3k] - (-3\theta)(2i + 3j - k)$$

$$= (2 \sin\theta \cos\theta + 3 \cos^2\theta - \theta \cos\theta + 6\theta)\, i$$

$$= (2 \sin\theta + 3 \sin\theta \cos\theta - \theta \sin\theta - 9\theta)\, k - (6 \sin\theta + 9 \cos\theta)\, k$$

Differentiating both sides with respect to θ we get

$$\frac{d}{d\theta} = \{a \times (b \times c) = 2 \cos 2\theta - 6 \cos\theta \sin\theta - \cos\theta + \theta \sin\theta - 6)i$$

$$- (4 \sin\theta \cos\theta + 3/2 \,.\, 2 \cos 2\theta - \sin \theta)$$

$$= (2 - 1 - 6)\, i - 3 - 9)\, j$$

$$= (6)\, k \text{ when } \theta = 0$$

$$= = 5i + 5j - 6k$$ **Ans.**

2. *Evaluate (i)* $\frac{d}{dt}[(a \times b) \times c]$;

(ii) $\frac{d}{dt}\left[\left(a \times \frac{da}{dt}\right) \times \frac{d^2a}{dt^2}\right]$;

(iii) $\frac{d^2}{dt^2}\left[\left(a \times \frac{da}{dt}\right) \times \frac{d^2a}{dt^2}\right]$

Solution. (i) $\frac{d}{dt}[(a \times b) \times c]$;

$$= (a \times b) \times \frac{dc}{dt} + \left[\frac{d}{dt}(a \times b)\right] \times c,$$

$$= (a \times b) \times \frac{dc}{dt} + \left(a \times \frac{db}{dt} + \frac{da}{dt}\right) \times c,$$

$$= (a \times b) \times \frac{dc}{dt} + \left(a \times \frac{db}{dt}\right) \times c + \left(\frac{da}{dt} \times b\right) \times c \quad \ldots (a)$$

(ii) Substituting $\frac{da}{dt}$ and $\frac{d^2a}{dt^2}$ for b and c respectively in this result we get

$$\frac{d}{dt}\left[\left(a \times \frac{d^2a}{dt^2}\right) \times \frac{d^2a}{dt^2}\right] = \left(a \times \frac{da}{dt}\right) \times \frac{d}{dt}\left(\frac{d^2a}{dt^2}\right)$$

$$+\left[a \times \frac{d}{dt}\left(\frac{da}{dt}\right)\right] \times \frac{d^2a}{dt^2} + \left(\frac{da}{dt} \times \frac{da}{dt}\right) \times \frac{d^2a}{dt^2}$$

$$= \left(a \times \frac{da}{dt}\right) \times \frac{d^3a}{dt^2} + \left(a \times \frac{d^2a}{dt^2}\right) \times \frac{d^2a}{dt^2}; \qquad \left[\because \frac{da}{dt} \times \frac{da}{dt} = 0\right]$$

(iii) $$\frac{d^2}{dt^2}\left[\left(a \times \frac{da}{dt}\right) \times \frac{d^2a}{dt}\right] = \frac{d}{dt}\left[\left(a \times \frac{da}{dt}\right) \times \frac{d^3a}{dt^3}\right] + \frac{d}{dt}\left[\left(a \times \frac{d^2a}{dt^2}\right) \times \frac{d^2a}{dt^2}\right].$$

differentiating w.r. to t the result ofpart (ii)

$$= \left\{\left(a \times \frac{da}{dt}\right) \times \frac{d^4a}{dt^3} + \left(a \times \frac{d^2a}{dt^2}\right) \times \frac{d^3a}{dt^3} + \left(\frac{da}{dt} \times \frac{da}{dt}\right) \times \frac{d^3a}{dt^3}\right]$$

$$= \left\{\left(a \times \frac{d^2a}{dt^2}\right) \times \frac{d^3a}{dt^3} + \left(a \times \frac{d^2a}{dt^2}\right) \times \frac{d^2a}{dt^2} + \left(\frac{da}{dt} \times \frac{d^2a}{dt^2}\right) \times \frac{d^2a}{dt^2}\right]$$

$$+ \left\{\left(a \times \frac{d^2a}{dt^2}\right) \times \frac{d^3a}{dt^3} + \left(a \times \frac{d^2a}{dt^2}\right) \times \frac{d^2a}{dt^2} + \left(\frac{da}{dt} \times \frac{d^2a}{dt^2}\right) \times \frac{d^2a}{dt^2}\right] \text{ using (a)}$$

$$= \left(a \times \frac{da}{dt}\right) \times \frac{d^2a}{dt^2} + 2\left(a \times \frac{d^2a}{dt^2}\right) \times \frac{d^3a}{dt^3} + \left(a \times \frac{d^3a}{dt^3}\right) \times \frac{d^2a}{dt^2}$$

$$+ \left(\frac{da}{dt} \times \frac{d^2a}{dt^2}\right) \times \frac{d^2a}{dt^2} \quad \textbf{Ans.}$$

3. *Show that* $\hat{r} \times d\hat{r} = (r \times dr)\, r^2$. where $r = r\,\hat{r}$

Solution. Since $\hat{r} = \frac{1}{r} r$

$$\therefore \qquad d\hat{r} = d\left(\frac{1}{r} r\right) = \frac{1}{r} dr + r\left(-\frac{1}{r^2} dr\right)$$

or $$\hat{r} \times d\hat{r} = \hat{r} \times \left(\frac{1}{r} dr - \frac{r}{r^2} dr\right)$$

or $$\hat{r} \times d\hat{r} = \frac{r}{r} \times \left(\frac{1}{r} dr - \frac{r}{r^2} dr\right) \qquad \left[\because \hat{r} = \frac{r}{r}\right]$$

or $$\hat{r} \times d\hat{r} = \frac{r}{r^2} r \times dr - \frac{r \times r}{r^2} dr = \frac{r \times dr}{r^2} \qquad [\because r \times r = 0]$$

4. *Find the directional directive of the function in the direction of the vector $2i + j - k$ at the point (2, 3, 1).*

Solution : If $\hat{a}$ be the unit vector in the given direction, then

$$\hat{a} = \frac{2i + j - k}{\sqrt{[2^2 + 1^2 + (-1)^2]}} = \frac{2i + j - k}{\sqrt{6}}$$

and grad $\phi = i \frac{\partial \phi}{\partial x} + j \frac{\partial \phi}{\partial y} + k \frac{\partial \phi}{\partial z}$

$$= i\,(yz^2) + j\,(xz^2) + k\,(2x\,y\,z)$$

$\therefore$ grad f at (2, 3, 1)

$= i\,[3\,.\,1^2] + j\,[2\,.\,1^2] + k\,[2\,.\,2.\,3.\,1]$

$= 3i + 2j + 12k$

$\therefore$ The required directional derivative at (2, 3, 1)

$= \frac{"d\phi"}{ds} = \hat{a}\,.\,\text{grad}\,\phi$ at (2, 3, 1)

$= \frac{2i + j - k}{\sqrt{6}}\,.\,(3i + 2j + 12k)$

$= (1/\sqrt{6})\,(2i + j - k)\,.\,(3i + 2j + 12k)$

$= (1/\sqrt{6})\,(6 + 2 - 12)$

$= -4/\sqrt{6} = -\,(2/\sqrt{6})/3$ **Ans.**

5. *Find the directional derivative of $\phi = x^2yz + 4xz^2$ at (1, –2, –1) in the direction 2i –j – 2k.*

Solution : Here $\hat{a}$ = the unit vector in the direction of 2i = j = 2k

$$= \frac{2i - j - 2k}{\sqrt{[2^2 + (-1)^2 + (-2)^2]}} = \frac{2i - j - 2k}{3}$$

And grad ϕ at (1, –2, –1)

$$= i \frac{\partial \phi}{\partial x} + j \frac{\partial \phi}{\partial y} + k \frac{\partial \phi}{\partial z} \text{ at } (1, -2, -1)$$

$= i\,(2xyz + 4z^2) + j\,(x^2z) + k\,(x^2y + 8xz)$

at (1, –2, –1), $[\because\ = x^2\,yz + 4xz^2]$

$= i\,[8] + j\,[-1] + k\,[-10] = 8i = j - 10k$

$\therefore$ The required directional derivative at (1, –2, –1)

$$= \frac{"d\phi"}{ds} = \hat{a}\ .\ \text{grad f at } (1, -2, -1)$$

$$= \frac{1}{3}\ (2i - j - 2k)\ .\ (8i - j - 10k)$$

$$= \frac{1}{3}\ (16 + 1 + 20)$$

$$= \frac{1}{3}\ (37) \qquad \textbf{Ans.}$$

6. *That* $\nabla^2 (r^n r) = n\,(n + 3)\, r^{n-2}\, r.$

Solution : $\nabla^2 (r^n r) = \left(\frac{\partial^2}{\partial x^2} + \frac{\partial^2}{\partial y^2} + \frac{\partial^2}{\partial z^2}\right)(r^n r) = \Sigma \frac{\partial^2}{\partial x^2}(r^n r)$

$$= \Sigma \frac{\partial}{\partial x}\left[\frac{\partial}{\partial x}(r^n r)\right] = \Sigma \frac{\partial}{\partial x}\left[n\, r^{n-1} \frac{\partial r}{\partial x} r + r^n \frac{\partial^2 r}{\partial x}\right]$$

$$= \Sigma \frac{\partial}{\partial x}\left[n r^{n-1}\left(\frac{x}{r}\right) r + r^n\ (i)\right], \quad \left[\because \frac{\partial r}{\partial x} = \frac{x}{r}, \frac{\partial r}{\partial x} = i, \text{ as above}\right]$$

$$= \Sigma \frac{\partial}{\partial x}\left[n r^{n-2}\ r + r^n\ i\right]$$

$$= \Sigma\left[n\left\{r^{n-2}\ r + x\,(n-2) r^{n-3}\frac{\partial r}{\partial x} r + x\, r^{n-2}\frac{\partial r}{\partial x}\right\} + n r^{n-1}\frac{\partial r}{\partial x}\, i\right]$$

$$= n\ \Sigma\ [r^{n-2}\ r + (n - 2)\ x^2\ r^{n-4}\ r + xi\ r^{n-2} + r^{n-2}\ xi],$$

$$\left[\because \frac{\partial r}{\partial x} = \frac{x}{r}, \frac{\partial r}{\partial x} = i\right]$$

$$= n\ [3\ r^{n-2}\ r + (n - 2)\ r^{n-4}\ r\ (x^2 + y^2 + z^2) + r^{n-2}\ (xi + yj + zk)$$
$$+ r^{m-2}\ (xi + yj + zk)]$$

$$= n\ [3\ r^{n-2}\ r + (n - 2)\ r^{n-2}\ r + r^{n-2}\ r + r^{n-2}\ r^{n-2}], \ [\because r = xi + yj + zk]$$

$$= n\ [3 + (n - 2) + 1 + 1]\ r^{n-2}\ r$$

$$= n\ (n + 3)\ r^{n-2}\ r. \qquad \text{Hence Proved}$$

7. *A motor boat, with its engine on running river and blown over by a horizontal wind is observed to travel at 20 Km/hour in a direction 53° east of north. The velocity of the boat with its engine on in still water and blown over by the horizontal wind is 4 Km/hour eastward and the velocity of the boat with its engine on over the running river, in the absence of wind is 8 Km/hour due south. Calculate*

   (a) *the velocity of the boat in magnitude and direction, over still water in the absence of wind,*

(b) the velocity of the wind in magnitude and direction.

Solution:

Suppose $\vec{v}_B$ = velocity of boat over still water in the absence of wind.

$\vec{v}_W$ = velocity of wind with respect to ground

$\vec{v}_R$ = velocity of river with respect to ground

Now, $\vec{v}_B + \vec{v}_W + \vec{v}_R$ = 20 Km/hour,

East of north = $\hat{i}\,(20\sin 53^o) + \hat{j}\,(20\cos 53^o)$ ...(1)

$\vec{v}_B + \vec{v}_W$ = 4 Km/hour, due east

$= 4\hat{i} + 0\hat{j}$ ...(2)

and $\vec{v}_B + \vec{v}_R$ = 8 Km/hour, due south

$= \hat{i}\times 0 + 8(-\hat{j})$ ...(3)

(i) $(\vec{v}_B + \vec{v}_W) + (\vec{v}_B + \vec{v}_R) - (\vec{v}_B + \vec{v}_W + \vec{v}_R) = \vec{v}_B$

$= -12\,\hat{i} - 20\hat{j}$

i.e., $|\vec{v}_B| = \sqrt{[(12)^2 + (20)^2]} = 23.32$ Km/hour

Its direction is given by

$$\tan\theta = \frac{-20}{-12} = \frac{5}{3}$$

or $\theta = \tan^{-1}\left(\frac{5}{3}\right)$ = 59° south of west

(ii) $\vec{v}_W = \vec{v}_B + \vec{v}_W - \vec{v}_B = \hat{i}\times 16 + \hat{j}\times 20$

$\therefore\ |\vec{v}_W| = \sqrt{[(16)^2 + (20)^2]} = 25.61$ Km/hour

Its direction is given by

$$\tan\beta = \tan^{-1}\left(\frac{20}{16}\right)$$

$$= \tan^{-1}\left(\frac{5}{4}\right) = 51.34^o \text{ North of East.}$$

8. *If $\vec{A}\times\vec{B} = 0$ and $\vec{A}.\vec{B} = 0$ does it imply that one of the vectors $\vec{A}$ or $\vec{B}$ must be a null vector?*

Solution: $\vec{A}\times\vec{B} = (AB\sin\theta)\,\hat{n}$, where θ is the angle between vectors

$\vec{A}$ and $\vec{B}$ and $\hat{n}$ is a unit vector perpendicular to both $\vec{A}$ and $\vec{B}$

Hence $\vec{A} \times \vec{B} = 0$ means that

$(AB \sin \theta)\ \hat{n} = 0$

i.e. either A = 0 or B = 0 or $\sin \theta = 0$ ...(1)

Now, $\vec{A} . \vec{B} = AB \cos \theta$

Hence, $\vec{A} . \vec{B} = 0$ implies that

$$AB \cos \theta = 0 \qquad ...(2)$$

If $\vec{A}$ and $\vec{B}$ are not null vectors then from the conditions (1) and (2), it follows that $\sin \theta$ and $\cos \theta$ both should be zero simultaneously. But it can not be possible. Therefore, *it is essential that either* $\vec{\mathbf{A}} = 0$ *or* $\vec{\mathbf{B}} = 0$ *i.e. one of the vectors* $\vec{\mathbf{A}}$ **or** $\vec{\mathbf{B}}$ *must be a null vector.*

9. *A particle moves in the x – y plane under the action of a force $\vec{F}$ such that the value of its linear momentum $\vec{p}$ at any time t is $P_x = 2 \cos t$ and $P_y = 2 \sin t$. What is the angle θ between $\vec{F}$ and $\vec{p}$ at a given time t ?*

Solution: As $\vec{p} = \hat{i} p_x + \hat{j} p_y = \hat{i}\,(2 \cos t) + \hat{j}\,(2 \sin t)$

$\therefore\ |\vec{p}| = \sqrt{[2 \cos t)^2 + (2 \sin t)^2]} = 2\sqrt{2}$.

Now, according to Newton's second law of motion

$\vec{F} = d\vec{p}/dt$

Hence $\vec{F} = d/dt\ [\hat{i}\,(2 \cos t) + \hat{j}\,(2 \sin t) = \hat{i}\ (-2 \sin t) + \hat{j}\,(2 \cos t)$

with $|\vec{F}| = \sqrt{[(-2 \sin t)^2 + (2 \cos t)^2]} = 2\sqrt{2}$

Now as $\vec{F} . \vec{p} = F_x p_x + F_y p_y$

$= (-2 \sin t)\ (2 \cos t) + (2 \cos t)\ (2 \sin t) = 0$

$$\therefore \quad \theta = \cos^{-1}\left[\frac{\vec{F} . \vec{p}}{F p}\right]$$

$$= \cos^{-1}\left[\frac{0}{2\sqrt{2} \times 2\sqrt{2}}\right] = \cos^{-1}(0) \quad i.e.\ \theta = 90^\circ$$

10. *The x and y components of vector $\vec{A}$ are 4 and 6 m respectively. The x and y components of vector $\vec{A} + \vec{B}$ are 10 and 9 m respectively.*

Calculate for the vector $\vec{B}$ the following :

(a) its x and y components

(b) its length, and

(c) the angle its makes with x-axis.

Solution: In terms of components

$$\vec{A}+\vec{B} = (\hat{i}A_x + \hat{j}A_y) + (\hat{i}B_x + \hat{j}B_y)$$

i.e. $\vec{A}+\vec{B} = \hat{i}(A_x + B_x) + \hat{j}(A_y + B_y)$

Given that : $A_x + B_x = 10$ and

$A_x + B_x = 9$ and

$A_x = 4$ and $A_y = 6$

Hence,

(a) $B_x = 10 - 4 = 6$ m and $B_y = 9 - 6 = 3$ m

(b) $B = \sqrt{(B_x^2 + B_y^2)} = \sqrt{(6^2 + 3^2)} = \sqrt{(45)}$ m $= 3\sqrt{(5)}$m

(c) $\theta = \tan^{-1}(B_y/B_x) = \tan^{-1}(3/6) = \tan^{-1}(1/2) = 26.6°$.

11. *If* $|\vec{A}+\vec{B}| = |\vec{A}-\vec{B}|$, *then find the angle between* $\vec{A}$ *and* $\vec{B}$.

Solution : Here, $|\vec{A}+\vec{B}| = |\vec{A}-\vec{B}|$

Suppose θ be the angle between $\vec{A}$ and $\vec{B}$. Dot product of a vector with itself is equal to square of its magnitude.

$|A + B|^2 = (\vec{A}+\vec{B}).(\vec{A}+\vec{B})$

$= \vec{A}.\vec{A}+\vec{A}.\vec{B}+\vec{B}.\vec{A}+\vec{B}.\vec{B}$

$= A^2 + 2AB\cos\theta + B^2$

Again, $|\vec{A}-\vec{B}|^2 = (\vec{A}-\vec{B}).(\vec{A}-\vec{B})$

$\vec{A}.\vec{A} - \vec{A}.\vec{B} - \vec{B}.\vec{A} + \vec{B}.\vec{B}$

$= A^2 - 2AB\cos\theta + B^2$

As $|\vec{A}+\vec{B}| = |\vec{A}-\vec{B}|$

hence, $|\vec{A}+\vec{B}|^2 = |\vec{A}-\vec{B}|^2$

or $A^2 + B^2 + 2AB\cos q = A^2 + B^2 - 2AB\cos\theta$

or $4AB\cos\theta = 0$

If we assume that $\vec{A}$ and $\vec{B}$ are nonzerc vectors, then 4 AB is nonzero and hence

$\cos\theta = 0$ *i.e.* $\theta = 90^\circ$

12. *Find the components of vector* $\vec{A} = 2\hat{i} + 3\hat{j}$ *along the directions of* $\hat{i} + \hat{j}$ *and* $\hat{i} - \hat{j}$

Solution: Here $\vec{A} = 2\hat{i} + 3\hat{i}$

In order to find the component of $\vec{A}$ along the directions of $\hat{i} + \hat{j}$, let us find out the unit vector along $\hat{i} + \hat{j}$. If $\hat{a}$ is the unit vector along $\hat{i} + \hat{j}$, then

$$\hat{a} = \frac{\hat{i}+\hat{j}}{|\hat{i}+\hat{j}|} = \frac{\hat{i}+\hat{j}}{\sqrt{2}}$$

Hence, the magnitude of the component vector of $\vec{A}$ along $\hat{i} + \hat{j}$

$$= \vec{A}\cdot\hat{a} = (2\hat{i}+3\hat{j}).\frac{\hat{i}+\hat{j}}{\sqrt{2}} = \frac{1}{\sqrt{2}}(2+3) = \frac{5}{\sqrt{2}}$$

Therefore, component vector of A along $\hat{i}+\hat{j}$

$$= (\vec{A}.\hat{a})\,\hat{a} = \frac{5}{\sqrt{2}}\left(\frac{\hat{i}+\hat{j}}{\sqrt{2}}\right) = \frac{\mathbf{5}}{\mathbf{2}}(\hat{\mathbf{i}}+\hat{\mathbf{j}})$$

Similarly, if $\hat{b}$ is the unit vector along the direction of $\hat{i}-\hat{j}$, then magnitude of the component vector of $\vec{A}$ along $\hat{i}-\hat{j}$

$$= \vec{A}.\hat{b} = (2\hat{i}+3\hat{j}).\frac{(\hat{i}-\hat{j})}{|\hat{i}-\hat{j}|}$$

$$= (2\hat{i}+3\hat{j}).\frac{(\hat{i}-\hat{j})}{\sqrt{2}} = \frac{(2-3)}{\sqrt{2}} = -\frac{1}{\sqrt{2}}$$

$\therefore$ Component vector of $\vec{A}$ along $\hat{i}-\hat{j}$

$$= (\vec{A}.\hat{b})\,\hat{b} = -\frac{1}{\sqrt{2}}\left(\frac{\hat{i}-\hat{j}}{\sqrt{2}}\right) = -\frac{\mathbf{1}}{\mathbf{2}}(\hat{\mathbf{i}}-\hat{\mathbf{j}})$$

13. *If* $|\vec{A}+\vec{B}| = |\vec{A}-\vec{B}|$, *then find the angle between* $\vec{A}$ *and* $\vec{B}$.

Solution: Here, $|\vec{A}+\vec{B}| = |\vec{A}-\vec{B}|$

Suppose θ be the angle between $\vec{A}$ and $\vec{B}$. Dot product of a vector with itself is equal to square of its magnitude.

$$A + B|^2 = (\vec{A}+\vec{B}).(\vec{A}+\vec{B})$$

$$= \vec{A}.\vec{A}+\vec{A}.\vec{B}+\vec{B}.\vec{A}+\vec{B}.\vec{B}$$

$= A^2 + 2AB\cos\theta + B^2$

Again, $|\vec{A}-\vec{B}|^2 = (\vec{A}-\vec{B}).(\vec{A}-\vec{B})$

$\vec{A}.\vec{A} - \vec{A}.\vec{B} - \vec{B}.\vec{A} + \vec{B}.\vec{B}$

$= A^2 - 2AB\cos\theta + B^2$

As $|\vec{A}+\vec{B}| = |\vec{A}-\vec{B}|$

hence, $|\vec{A}+\vec{B}|^2 = |\vec{A}-\vec{B}|^2$

or $A^2 + B^2 + 2AB\cos q = A^2 + B^2 - 2AB\cos\theta$

or $4\,AB\cos\theta = 0$

If we assume that $\vec{A}$ and $\vec{B}$ are nonzero vectors, then 4 AB is nonzero and hence

$\cos\theta = 0$ i.e. $\theta = 90^\circ$

14. *Find the components of vector $\vec{A} = 2\hat{i} + 3\hat{j}$ along the directions of $\hat{i} + \hat{j}$ and $\hat{i} - \hat{j}$*

Solution:

Here $\vec{A} = 2\hat{i} + 3\hat{i}$

In order to find the component of $\vec{A}$ along the directions of $\hat{i} + \hat{j}$, let us find out the unit vector along $\hat{i} + \hat{j}$. If $\hat{a}$ is the unit vector along $\hat{i} + \hat{j}$, then

$$\hat{a} = \frac{\hat{i}+\hat{j}}{|\hat{i}+\hat{j}|} = \frac{\hat{i}+\hat{j}}{\sqrt{2}}$$

Hence, the magnitude of the component vector of $\vec{A}$ along $\hat{i} + \hat{j}$

$$= \vec{A}\cdot\hat{a} = (2\hat{i}+3\hat{j}).\frac{\hat{i}+\hat{j}}{\sqrt{2}} = \frac{1}{\sqrt{2}}(2+3) = \frac{5}{\sqrt{2}}$$

Therefore, component vector of A along $\hat{i}+\hat{j}$

$$= (\vec{A}.\hat{a})\,\hat{a} = \frac{5}{\sqrt{2}}\left(\frac{\hat{i}+\hat{j}}{\sqrt{2}}\right) = \frac{\mathbf{5}}{\mathbf{2}}(\hat{\mathbf{i}}+\hat{\mathbf{j}})$$

Similarly, if $\hat{b}$ is the unit vector along the direction of $\hat{i}-\hat{j}$. then magnitude of the component vector of $\vec{A}$ along $\hat{i}-\hat{j}$

$$= \vec{A}.\hat{b} = (2\hat{i}+3\hat{j}).\frac{(\hat{i}-\hat{j})}{|\hat{i}-\hat{j}|}$$

$$= (2\hat{i}+3\hat{j}).\frac{(\hat{i}-\hat{j})}{\sqrt{2}} = \frac{(2-3)}{\sqrt{2}} = -\frac{1}{\sqrt{2}}$$

$\therefore$ Component vector of $\vec{A}$ along $\hat{i}-\hat{j}$

$$= (\vec{A}.\hat{b})\,\hat{b} = -\frac{1}{\sqrt{2}}\left(\frac{\hat{i}-\hat{j}}{\sqrt{2}}\right) = -\frac{1}{2}(\hat{i}-\hat{j}).$$

15. *If $r(t) = (5t^2)\ i + t\ j = t\ k$ prove that*

$$\int_1^2 r \times \frac{d^2r}{dt^2}\, dt = -14i + 75j - 15k$$

Solution. Given $r(t) = (5t^2)\ i + (t)\ j - (t^3)\ k$

$$\therefore \quad \frac{dr}{dt} = (10\ t)\ i + (1)\ j = (3t^2)\ k \text{ and } \frac{d^2r}{dt^2}$$

$$= (10)\ i + (0)\ j = (6t)\ k$$

$$\therefore \quad r \times \frac{d^2r}{dt^2} = (5t^2\ i - t\ j = t^3\ k) \times (10\ i = 6\ t\ k)$$

$$= -30\ t^3\ i \times k + 10\ t\ j \times i - 6\ t^2\ j \times k - 10\ t^3\ k \times i,$$

$[\because\ i \times j = k\ -j \times i \text{ etc.}]$

$$= -30\ t^3\ (-j) + 10\ t\ (-k) - 6\ t^2\ (i) - 10\ t^3\ (j)\ ,$$

$[\because\ i \times j = k\ -j \times i \text{ etc.}]$

$$= -6t^2\ i + 20\ t^3\ j - 10\ r\ k$$

$$\therefore \quad \int_1^2 r \times \frac{d^2r}{dt^2}\, dt = -6i \int_1^2 t^2\, dt + 20\, j \int_1^2 t^3\, dt - 10\, k \int_1^2 t\, dt$$

$$= -6i\left(\frac{1}{3}t^3\right)_1^2 + 20\, j\left(\frac{1}{4}t^1\right)_1^2 - 10\, k\left(\frac{1}{2}t^2\right)_1^2$$

$$= -2i\ (8-1) + 5j\ (16-1) - 5k\ (4-1)$$

$$= -14\ i + 75\ j - 15\ k.$$

16. *Solve* $\frac{d^2r}{dt^2} = at + b$, *where 'a' and 'b' are constant vector, given that*

$r = 0$ and $\frac{dr}{dt} = 0$; when $t = 0$.

Solution. Given $\frac{d^2r}{dt} = \int(at+b)\,dt = \int at\,dt + \int b\,dt$

Intergrating both sides with respect to t, we have

$$\frac{dr}{dt} = \int(a t + b)\,dt = \int a t\,dt + \int b\,dt$$

$$= a\left(\frac{1}{2}r^2\right) + b\,t + c, \qquad \ldots \text{(i)}$$

where c is an arbitrary constant vector

Given $\frac{dr}{dt} = 0$ when t = 0, so from (i) we have

$$0 = a\,(0) + b\,(0) + c \text{ or } c = 0$$

So from (i) we have $\frac{dr}{dt} = a\left(\frac{1}{2}r^2\right) + b\,(t)$

Intergrating again with respect to t, we get

$$r = \int\left(\frac{1}{2}a\,r^2 + bt\right)dt$$

$$= \frac{1}{2}a\int t^2\,dt + b\int t\,dt$$

or $$r = \frac{1}{2}a\left(\frac{1}{3}t^3\right) + b\left(\frac{1}{2}t^2\right) + d$$

where d is an arbitrary constant vector.

Given r = 0 when t = 0, so from (ii) we have

$$0 = \frac{1}{2}a(0) + b\,(0) + d \text{ or } d = 0$$

$\therefore$ From (ii) we get $r = \frac{1}{6}a\,t^3 + \frac{1}{2}b\,t^2$ **Ans.**

17. *Find the unit vector perpendicular to each of the vector a = 3i + j + 2k and b = 2i – 2j + 2k. Also determine the sine of the angle between a and b. What will be the vector perpendicular to a and b and having a magnitude* $4\sqrt{3}$?

Solution: By its very definition, the vector product a × b is perpendicular to the plane of a and b. So that, we have *unit vector perpendicular to a and* b, say, $\hat{n} = \frac{a \times b}{|a \times b|}$.

Now, $a \times b = (3i + j + 2k) \times (2i - 2j + 4K) = 8i - 8j - 8k$

and ∴ $|a \times b| = \sqrt{64+64+64} = 8\sqrt{3}.$

Hence unit vector $\hat{n}$, perpendicular to a and b

$$= \frac{8i-8j-8k}{8\sqrt{3}} = \frac{1}{\sqrt{3}}(i-j-k)$$

Finally, since the unit vector perpendicular to a and b is $\hat{n}$ equal to $\frac{1}{\sqrt{3}}$ $(i - j - k)$, we have *vector of magnitude* $4\sqrt{3}$ and perpendicular to a and b = $4\sqrt{3}\,\hat{n}$

$$= 4\sqrt{3}\left[\frac{1}{\sqrt{3}}(i-j-k)\right] = 4i-4j-4k = 4(i-j-k).$$

18. *(a) Find the area of the parallelogram determined by the vectors $a = 3i + 2j$ and $b = 2j - 4k$.*

(b) Find the area of the triangle whose vertices are (1, – 1, – 3), (4, – 3, 1) and (3, – 1, 2).

Solution:

(a) Here, *vector area of the parallelogram*

$$= a \times b = (3i + 2j) \times (2j - 4k)$$
$$= 8i - 12j + 6k.$$

∴ *magnitude of the area of the parallelogram*

$$= |8i+12j+6k|$$
$$= \sqrt{64+144+36} = 2\sqrt{61}.$$

(b) Clearly, the position vectors of the vertices A, B and C of the triangle ABC are $a = (i - j - 3k)$ $b = (4i - 3j + k)$, and $c = (3i - j - 2k)$.

∴ $\overrightarrow{BA} = a - b = -3i + 2j - 4k$ and $\overrightarrow{BC} = c - b = -i + 2j + k$.

Hence *vector area of the triangle* $= \frac{1}{2}\overrightarrow{BA} \times \overrightarrow{BC}$

$$= 1/2\,(-3i + 2j - 4k) \times (-i + 2j + k)$$
$$= 1/2\,(10i + 7j - 4k)$$

and *magnitude of the area of the triangle* $= \frac{1}{2}|10i+7j-4k|$

$$= \frac{1}{2}\sqrt{100+49+16} = \frac{1}{2}\sqrt{165}.$$

19. *Prove that the vector area of a triangle whose vertices are a, b, c is 1/2 (b × c + c × a + a × b). What is the condition for the collinearity of the vertices ?*

Solution:

As in example 18 above, let the triangle be ABC. Then, we have

$$\overrightarrow{BC} = (c - b) \text{ and } \overrightarrow{BA} = (a - b)$$

∴ *vector area of the triangle* $= 1/2\,\overrightarrow{BC} \times \overrightarrow{BA} = 1/2\,[(c - b) \times (a - b)]$

$$= 1/2\,(c \times a - c \times b - b \times a + b \times b]$$

Now, $b \times b = 0,\ - c \times b = b \times c$ and $- b \times a = a \times b$

So that, the *vector area of the triangle*

$$= 1/2\,(b \times c + c \times a + a \times b).$$

If the vertices of the triangle be collinear, we shall have area of the triangle equal to zero and, therefore vector area of the triangle too equal to zero. Thus, the *condition for collinearity of the vertices* is that 1/2 (b × c + c × a + a × b) = 0.

20. *Show that if the vector area of each face of a tetrahedron has the direction of the outward normal, the sum of their vector areas is zero.*

Solution: Let DABC be the tetrahedron, such that taking D as the origin, the position vectors of A, B and C are a, b and c respectively. Then, we clearly have *vector area of face* DBC = 1/2 b × c, *of face* DAB = 1/2 a × b, *of face* DCA = 1/2 c × a and of face BAC = 1/2 (b – a) (b – c) = 1/2 (b × b – b × c – a × b + a × c) = – 1/2 (b × c + a × b + c × a), all directed along their respective outward normals.

∴ *sum of the vector areas of all the faces of the tetrahedron*

$$= 1/2\,(b \times c + a \times b + c \times a) - 1/2\,(b \times c + a \times b + c \times a) = 0.$$

21. *A rigid body is rotating with angular velocity 3 radians/second about an axis passing through the point 2i – j – k and parallel to i – 2j + 2k. Find the magnitude of the velocity of the point (of the rigid body) whose position vector is 2i + 3j – 4k.*

Solution: Obviously, the unit vector, $\hat{n}$, say, in the direction of

$$i - 2j + 2k = \frac{i-2j+2k}{|i-2j+2k|}$$

$$= \frac{i-2j+2k}{\sqrt{1+4+4}} = \frac{1}{3}(i-2j+2k).$$

angular velocity of the rigid body, $\omega = \omega\hat{n} = 3\ [1/2\ (j - 2j + 2k)]$

$= i - 2j + 2k.$

Let the point whose velocity is desired to be determined be P. Then, its positive with respect to the point 2i – j – k on the axis given by

$$r = (2i + 3j - 4k) - (2i - j - k) = 4j - 3k.$$

Hence, if v be the velocity of point P, we have $v = \omega \times r$

$$= (i-2j+2k)\times(4j-3k) = \begin{vmatrix} i & j & k \\ 1 & -2 & 2 \\ 0 & 4 & -3 \end{vmatrix} = -2i+3j+4k.$$

So that, *magnitude of the velocity* $= \sqrt{(-2)^2 + 3^2 + 4^2} = \sqrt{29}$ units.

22. *A force F = – 2i + 3j +4k is acting at a point 5i + 4j + 3k. Obtain the moment of the force about the origin.*

Solution: Here, r = 5i + 4j + 3k. Therefore, *moment of the force about the origin* $r \times F = (5i + 4j + 3k) \times (-2i + 3j + 4k)$

$= i\ (16 - 9) + j\ (-6 - 20) + k\ (15 + 8) = 7i - 26j + 23k.$

23. *A = 4i – 5j + 3k, B = 2i – 10j – 7k and C = 5i + 7j – 4k calculate the following: (i) A × B . C, (ii) A × (B × C).*

Solution: We have

$$\text{(i) } A\times B.C \text{ or } [ABC] = \begin{vmatrix} 4 & -5 & 3 \\ 2 & -10 & -7 \\ 5 & 7 & -4 \end{vmatrix}$$

$$= 4(40+49) - 5(-35+8) + 3(14+50) = 356+135+192 = 683.$$

(ii) $A \times (B \times C) = B\ (A.C) - C\ (A.B)$

$= (2i - 10j - 7k)\ 20 - 35 - 12) - (5i + 7j - 4k)\ (8 + 50 - 21)$

$= -54i + 270j + 189k - 185i - 259j + 148k = -239i + 11j + 337k.$

24. *(a) Prove that a × (b × c) + b × (c × a) + c × (a × b) = 0.*

(b) Prove that the four points (4i + 5j + k), – (j + k), (3i + 9j + 4k) and (4 – i – j – k) are coplanar.

Solution:

(a) As we know $a \times (b \times c) = (a.c)b - (a.b)c$ ...(i)

$-b \times (c \times a) = (a.b)c - (b.c)a$ ...(ii)

and $c \times (a \times b) = (b.c)a - (c.a)\ b.$ ...(iii)

So that, adding up the three, we have

$a \times (b \times c) + b \times (c \times a) + c \times (a \times b) = 0.$

(b) If $4i + 5j + k$, $-(j + k)$, $3i + 9j + 4k$ and $4(-i + j + k)$ be the position vectors of four points A, B, C and D with reference to an origin O, clearly the points will be coplanar if the vectors $\vec{BA}$, $\vec{BC}$ and $\vec{CD}$ be coplanar.

Now, $\vec{BA} = \vec{OA} - \vec{OB} = (4i + 5j + k) - [-(j + k)] = 4i + 6j + 2k,$

$\vec{BC} = \vec{OC} - \vec{OB} = (3i + 9j + 4k) - [-(j + k)] = 3i + 10j + 5k$

and $\vec{CD} = \vec{OD} - \vec{OC} = (-4i + 4j + 4k) - (3i + 9j + 4k) - 7i - 5j.$

And, these vectors ($\vec{BA}, \vec{BC}$ and $\vec{CD}$) will be coplanar if their scalar product is zero, Let us see if its is so.

The triple scalar product of the vectors is clearly

$$[\vec{BA}\ \vec{BC}\ \vec{CD}] = \begin{vmatrix} 4 & 6 & 2 \\ 3 & 10 & 5 \\ -7 & -5 & 0 \end{vmatrix} = 4(25) + 6(-35) + 2(55) = 0$$

The three vectors $\vec{BA}$ $\vec{BC}$ and $\vec{CD}$, and hence the four points A, B, C and D, are thus coplanar.

25. *The edge of a parallelopiped are given by the vectors $i + 2j + 3k$, $5j$ and $4j + mk$. What should be the value of m in order that the volume of the parallelopiped be 20 units?*

Solution: The volume of a parallelopiped, as we know is given by the scalar triple product of the vectors representing its three edges. So that, we have:

$$\text{volume of the parallelopiped} = \begin{vmatrix} 1 & 2 & 3 \\ 0 & 5 & 0 \\ 0 & 4 & m \end{vmatrix} = 1(5m) + 2(0) + 3(0) = 5m.$$

Since the volume is given to be 20 units, we have $5m = 20$, whence, $m = 4$.

26. *Show that (i) $(a \times b) \times (c \times d) = [abd]c - [abc]d$.*
(ii) $(a \times b) \times (c \times d) = [acd]b - [bcd]a$.

Solution: (i) Let $a \times b = m$.

Then, $(a \times b) \times (c \times d) = m \times (c \times d)$

$$= (m \,.\, d)\, c - (c \,.\, m)d$$

$$= [(a \times b).d]\, c - [c.\,(a \times b)]d.$$

Since in the triple products $(a \times b).d$ and $c.(a \times b)$, the positions of the dots and crosses are quite immaterial and may be interchanged or omitted, as desired, we have $(a \times b) \times (c \times d) = [abd]c - [abc]d$.

(ii) Again, putting $(c \times d) = n$, we have

$$(a \times b) \times (c \times d) = (a \times b) \times n = (a.n)b - (b.n)a$$

$$= [a.(c \times d)]\, b - [b.(c \times d)]a$$

$$= [acd]b - [bcd]a.$$

27. *Obtain the value of the product $(A \times B).(C \times D)$ and show that $A \times [B \times (C \times D)] = (A \times C)\,(B.D) - (A \times D)\,(B.C)$*

Solution: Since in the case of scalar products, the positions of dots and crosses are freely interchangeable, we have

$$(A \times B).\,(C \times D) = A.[B \times (C \times D)] = A.[(B.D)C - (B.C)D]$$

$$= (A.C)\,(B.D) - (A.D)\,(B.C)$$

Again, $\quad A \times [B \times (C \times D)] = A \times [(B.D)C - (B.C)D]$

$$= (A \times C)(B.D) - (A \times D)(B.C).$$

28. *Establish the kinematic relation $v = u + at$, $S = ut + 1/2at^2$ and $v^2 - u^2 = 2aS$.*

Solution: We know that *acceleration* $a = dv/dt$. So that, $dv = a.dt$, which, on integration gives $v = at + C_1$ where C_1 is a constant of integration.

Since at $t = 0$, $v = C_1$, clearly C_1 stands for the initial velocity u.

We, therefore, have $\quad v = at + u$

or $\quad v = u + at.$

Now, $\quad v = dS/dt = u + at.$

Or $\quad dS = (u + at)dt.$

Integrating, we have $S = ut + 1/2at^2 + C_2$, where C_2 is another constant of integration.

Since at $t = 0$, $S = 0$ and $\therefore$ $C_2 = 0$, we have $S = ut + 1/2\, at^2$

Finally, taking the scalar product of a with v, we have

$$v.a = v.\frac{dv}{dt} \quad \text{Or,.} \quad \frac{dS}{dt}.a = v.\frac{dv}{dt}.\left[\because a = \frac{dv}{dt} \text{ and } v = \frac{dS}{dt}.\right.$$

Integration gives $S.a = 1/2\, v.v + C_2$ (a being a constant).

where C_3 is yet another constant of integration.

Since v.v = v^2 = v^2. we have S.a = $1/2v^2 + C_3$.

Again, when S = 0, v = u and we have 0 = $1/2v^2 + C_3$,

whence, $C_3 = -u^2/2$.

So that, S.a = $1/2v^2 - 1/2u^2 = 1/2v^2 - 1/2u^2$, whence, $v^2 - u^2 = 2aS$

[$\because u^2 = v^2$ and $u^2 = u^2$.

29. *If the acceleration of moving particle at any instant be given by $\mu r + v/r^3$ show that (i) it has a constant speed and that (ii) its angular momentum consists of two components, one constant in magnitude and direction and the other constant in magnitude in the direction of, r.*

Solution: (i) Clearly, the equation of motion of the particle is

$$\frac{d^2r}{dt^2} = \frac{\mu r \times v}{r^3} = \frac{\mu r}{r^3} \times \frac{dr}{dt}$$

Taking the scalar product of relation (i) with dr/dt, we have

$$\frac{dr}{dt} \cdot \frac{d^2r}{dt^2} = \frac{\mu r}{r^3} \times \frac{dr}{dt} \cdot \frac{dr}{dt} = 0.$$

This, on integration, given $(dr/dt)^3$ = constant, indicating that *the speed* dr/dt *of the particle is constant.*

(ii) Obtaining the vector product of (i) with r, we have

$$r \times \frac{d^2r}{dt^2} = \frac{\mu}{r^3} r \times \left(r \times \frac{dr}{dt}\right) = \frac{\mu}{r^3}\left[\left(r.\frac{dr}{dt}\right)r - (r.r)\frac{dr}{dt}\right]$$

$$\frac{\mu}{r^3}\left[r\frac{dr}{dt}r - r^2\frac{dr}{dt}\right] = \mu\frac{d}{dt}\left(\frac{r}{r}\right),$$

which, on integration, gives $r \times \frac{dr}{dt} = -\mu\frac{r}{r} + a$, where a is a *constant vector.*

Clearly, $r \times \frac{dr}{dt}$ is the angular velocity of the particle and hence its angular momentum is given by $m\left(r \times \frac{dr}{dt}\right) = -\frac{m\mu r}{r} + ma.$

It clearly consists of two components, vis (i) a component—$m\mu r/r$ whose magnitude is the constant $|-m\mu|$ and whose direction is that of r and (ii) the component ma, which is constant both in magnitude and direction.

30. *A rigid body is spinning with an angular velocity of 4 radians/about an axis parallel to 3j – k passing through the point i + 3j – k. Find* *the velocity of the particle at the point 4i – 2j +k.*

Solution: Let $\hat{n}$ be the unit vector in the direction of 3j – k. Then,

$$\hat{n} = \frac{3j-k}{\sqrt{0+9+1}} = \frac{1}{\sqrt{10}}(3j-k),$$

∴ angular velocity of the particle P, say, $\omega = \omega\hat{n} = \frac{4}{\sqrt{10}}(3j-k)$.

The position vector of P with reference to point i + 3j – k, *i.e.,*

$$r = (4i - j + k) - (i + 3j - k) = 3i - 5j + 2k.$$

Hence *velocity (linear) of particle* $P = r \times \omega = (3i-5j+2k) \times \frac{4}{\sqrt{10}}(3i-k)$

This may be seen from the following: Consider, a particle P in a rigid body, revolving around an axis with an angular velocity ω. Let r be the position vector of the particle with respect to any point O on the axis. Then, since the particle describes a circle of radius r sinθ, lying in the plane perpendicular to the axis or rotation, we have

magnitude of its linear velocity (or of the velocity vector) v = ωr sinθ.

But by the definition of cross product, ω × r has also the same magnitude ωr sin θ and is perpendicular to the plane containing r and ω.

We, therefore have v = r × ω.

$\frac{4}{\sqrt{10}}(i+3j+9k)$ and its magnitude $= \frac{4}{\sqrt{10}}\sqrt{1+9+81} = 4\sqrt{\frac{91}{10}}$

= 12 *units* very nearly.

31. *A particle describes a circle in the i, j plane with a uniform speed in 12 seconds in the direction i to j. If the initial position of the radius vector relative to the centre of the circle be i, determine its position at the end of the 3rd second from the start. What will be its velocity vector then?*

Solution: Let OX and OY be the rectangular coordinates in the plane of the circle through its centre O and let i and j be the unit vectors along the two axes respectively. Then, taking the circle to be of unit radius, we have OA = OB =1, where A and B are the points in which the circle meets the two axes respectively.

Let A be the initial position of the particle and P, its position after t *second* from the start. So that, angular velocity of the particle being 2π/12 or π/6 radian/s, angle XOP = (π/6)t.

If PM be the perpendicular dropped from P on to OA, we have

$$\overrightarrow{OP} = \overrightarrow{OM} + \overrightarrow{MP}. \text{ Or, } \overrightarrow{OP} = \left(OP \cos\frac{\pi}{6}t\right)i + \left(OP \sin\frac{\pi}{6}t\right)j.$$

Or, since OP = 1 and t = 3, we have $\overrightarrow{OP} = \left(\cos\frac{3\pi}{6}\right)i + \left(\cos\frac{3\pi}{6}\right)j$

$$= (\cos \pi/2)\, i + (\sin \pi/2) j = j$$

Thus, the position of the radius vector will, after 3 seconds, be along OB. And, the velocity vector of the particle will, therefore, be $-\pi i/6$, the velocity v being tangential to the circle at B, as shown.

32. *Find grad r^m where r is the distance of any point from the origin, Also show that differential $d\phi$ = dr grad ϕ.*

Solution: we have $\phi(x, y, z) = r^m = (x^2 + y^2 + z^2)^{m/2}$

$$\left[\begin{array}{l} \because r^2 = x^2+y^2+z^2 \\ \text{and } \therefore r = (x^2+y^2+z^2)^{1/2} \end{array}\right]$$

$$\therefore \qquad \frac{\partial\phi}{\partial x} = mx\left(x^2+y^2+z^2\right)^{\frac{m}{2}-1} = mxr^{m-2}$$

Similarly, $\dfrac{\partial\phi}{\partial y} = myr^{m-2}$ and $\dfrac{\partial\phi}{\partial z} = mzr^{m-2}$

Hence grad

$$r^m \left(\text{or } \nabla r^m\right) = \sum i\frac{\partial\phi}{\partial x} = mr^{m-2}\sum ix = mr^{m-2}r, \qquad \text{...(i)}$$

where r is the *position vector* of the point

Or, since $r = r\hat{r}$, where r is the magnitude of r and r, the unit vector along r, we have $\hat{r} = r/r$.

So that, grad r^m also $= mr^{m-1}(r/r) = mr^{m-1}\hat{r}$ ...(ii)

Now, differential $d\phi = \dfrac{\partial\phi}{\partial x}dx + \dfrac{\partial\phi}{\partial y}dy + \dfrac{\partial\phi}{\partial z}dz$

$$= (idx + jdy + kdz)\left(i\frac{\partial\phi}{\partial x} + j\frac{\partial\phi}{\partial y} + k\frac{\partial\phi}{\partial z}\right) = dr \text{ grad } \phi.$$

33. *Verify Stokes' theorem for the function F = x (ix + jy), integrated round the square, in the plane z = 0, whose sides are along the lines x = 0, y = 0, x = a, y = a,*

Solution: We clearly have

$$\oint_C F.dl + \int_{OA} F.dl + \int_{AB} F.dl + \int_{BC} F.dl + \int_{CO} F.dl,$$

where, $\displaystyle\int_{OA} F.dl = \int_0^a x(ix + jy).idx = \int_0^a x^2 dx = \frac{1}{3}a^3,$

$$\int_{AB} F.dl = \int_{o}^{a} x(ix + jy).jdy = \int_{o}^{a} aydy = \frac{1}{2}a^3,$$

$$\int_{BC} F.dl = \int_{a}^{o} x(ix + jy).idx = -\int_{o}^{a} x^2dx = -\frac{1}{3}a^3 \text{ and}$$

$$\int_{CO} F.dl = \int_{a}^{o} x(ix + jy).jdx = 0$$

So that, $\oint_c F.dl = \frac{1}{3}a^3 + \frac{1}{2}a^3 - \frac{1}{3}a^3 + 0 = \frac{1}{2}a^3.$

Now, in accordance with Stokes' theorem, $\oint c$ F.d l = $\iint$S curl FdS.

Since curl x (ix + jy) = ky, we have

$$\iint_S \text{curl FdS} = \iint_S \text{curl}\, x(ix + jy)dS = \int_o^a\int_o^a ky.kdxdy$$

$$= \int_o^a\int_o^a ydxdy = \frac{1}{2}a^3.$$

Stokes' theorem thus stands verified.

34. *If r = a cos wt + b sin wt show that*

$$r\frac{dr}{dt} = w\ a \times b \text{ and } \frac{d^2r}{dt^2} = -w^2r$$

Solution. Given r = a cos wt + b sin wt . . . (i)

$$\therefore \frac{dr}{dt} = -a\ w \sin wt + b\ w \cos wt \quad ...(2)$$

$$\therefore r \times \frac{dr}{dt} = (a \cos wt + b \sin wt) \times (-aw \sin wt + bw\ bw \cos wt)$$

= a × bw $\cos^2$ wt – b × aw $\sin^2$ wt, [∵ a × a = 0 etc.]

= (w $\cos^2$ wt + w $\sin^2$ wt) a × b, [∵ –b × a = a × b]

= w a × b.

Also from (ii) we get

$$\frac{d^2r}{dt^2} = -aw^2 \cos wt - bw^2 \sin wt$$

= – w^2 (a cos wt + b sin wt) = – w^2r, from (i).

35. *If r = (cos nt) i + (sin nt) j, where n is constant and t varies, then show that*

$$r \times \frac{dr}{dt} = nk \text{ and } r\frac{dr}{dt} = 0$$

Solution. Given r = (cos nt) i + (sin nt) j

$\frac{dr}{dt} = (-n \sin nt)\, i + (n \cos nt)\, j$ . . . (ii)

$\therefore r \times \frac{dr}{dt} = [(\cos nt)\, i\, (\sin nt)\, j\,] \times [\,-n\, \{(\sin nt)\, i = \cos nt)\, j\}]$ from (i), (ii)

$= -n\, [\,-\cos^2 nt)\, i \times j + (\sin^2 nt)\, j \times i\,],$ $[\because\ i \times i = 0]$

$= n\, [\cos^2 nt + \sin^2 nt)\, i \times j$ $[\because\ j \times i = -\, i \times j]$

$= n\, k,$ $[\because\ i \times j = k]$ Hence proved.

Also $r.\frac{dr}{dt} = [(\cos nt)\, i + (\sin nt)\, j]\ .\ [-n\, \{(\sin nt)\, i - (\cos nt)\, j\}],$

from (i), (ii)

$= -\, n \cos nt \sin nt + n \sin nt \cos nt,$ $[\because\ i.i = 1,\ i.j = 0$ etc.$]$

$= 0.$ Hence proved.

36. *If $r = (a \cos t)\, i + (a \sin t)\, j + tk$, find* $\left|\frac{d^2r}{dt^2}\right|$

Solution. Here $\frac{dr}{dt} = (-a \sin t)\, i + (a \cos t)\, j + (1)\, k$

and $\frac{d^2r}{dt^2} = (-a \cos t)\, i + (= a \sin t)\, j + (0)\, k$

$\left|\frac{d^2r}{dt^2}\right|$ $\sqrt{[(-a \cos^2 t)^2 + (-a \sin t)^2]}$

$= \sqrt{[(a^2 \cos^2 t + \sin^2 t)}\ \sqrt{a^2} = a$ **Ans.**

37. *If $u = (t)^2\, i - tj + (2t + 1)\, k$ and $v = (2t = 3)\, i + j - tk$. find $\frac{d}{dt}$ $(u\, .\, v)$ and $\frac{d}{dt}$ $(u \times v)$ when $t = 1$*

Solution. $\frac{d}{dt}(u) = \frac{d}{dt}[(t^2)\, i\, j + (2t + 1)\, l] = (2t)\, i - j + 2k$

and $\frac{d}{dt}(v) = \frac{d}{dt}[(2t - 3)\, i + j - t\, k] = 2\, i + 0\, j - k = 2i - k$

$\therefore$ $\frac{d}{dt}(u\ v) = u\,\frac{dv}{dt} + v.\frac{du}{dt}$

$= [(t^2) - i\, j + (2t + 1)\, k]\ .\ (2\, i - k) + [(2t - 3)$

$$i + j - i\,k] \, . \, [(2t)\, i - j + 2\, k]$$
$$= [2t^2 - (2t + 1)] + [(2t - 3)\,(2t) - 1 - 2t],$$
$$[\because i^2 = 1 \, . \, i \, . \, j \, . = 0 \text{ etc.}]$$
$$= 2t^2 - 2t - 1 + 4t^2 - 6t - 1 - 2t$$
$$= 6t^2 - 10t - 2 = -6, \text{ when } t = 1 \textbf{ Ans.}$$

Again $\frac{d}{dt}(u \times v)\; u \times \frac{dv}{dt} + \frac{du}{dt} \times v$

$$= [t^2\, i - i\, j + (2t + 2)\, k] \times (2i - k)$$
$$+ (2t\, i - j + 2k) \times [(2t - 3)\, i + j - t\, k]$$
$$= [-t^2\, i \times k - t\, j \times i + t\, j \times k + 2\,(2t + 1)\, k \times i]$$
$$+ (2t\, i \times j - 2t^2\, i \times k - (2t - 3)\, j \times i + t\, j \times k]$$
$$+ 2\,(2t - 3)\, k \times i + 2k \times j]$$
$$[\because i \times j = 0 \text{ etc.}]$$
$$= [t^2\, j + 2\, t\, k + t\, i + 2\,(2t + 1)\, j] + [2\, t\, k + 2\, t^2\, j$$
$$+ (2t - 3)\, k + t\, i + 2\,(2t - 3)\, j - 2i]$$
$$[\because i \times j = k = -j \times i \text{ etc.}]$$
$$= 2\,(t - 1)\, i + (3t^2 + 8t - 4)\, j + 3\,(t - 1)\, k$$
$$= 7\, j, \text{ when } t = 1 \textbf{ Ans.}$$

38. *(a) Obtain the magnitudes and direction cosines of vectors (A+B) and (A–B), if $A = 3i + 2j + k$ and $B = i - 2j + 3k$.*

(b) A force $F = 2.5i + 4.5j - 5k$ newton acts through the origin. What is the magnitude of this force and what angles does it make with the three coordinate axes?

Solution:

(a) Obviously $(A + B) = 4i + 4k$ and $(A - B) = 2i + 4j - 2k$. So that, their *magnitudes* are

$$|A + B| = \sqrt{4^2 + 4^2} = 4\sqrt{2} \text{ and } |A - B| = \sqrt{2^2 + 4^2 + 2^2} = \sqrt{24}.$$

Hence the *direction cosines* of $(A + B) =$ are $\frac{4}{4\sqrt{2}} = \frac{1}{\sqrt{2,0}}$ and $\frac{4}{4\sqrt{2}} = \frac{1}{\sqrt{2}}$

and the *direction cosines* of (A – B) are

$$\frac{2}{\sqrt{24}}, \frac{4}{\sqrt{24}} \text{ and } \frac{2}{\sqrt{24}}.$$

(b) Clearly, the *magnitude of the force is* given by

$|F| = \sqrt{(2.5)^2 + (4.5)^2 + (-5^2)} = 7.176$ *newton and its direction cosines* are

(i) $\cos\theta_x = 2.5/7.176 = 0.3433$ and $\therefore \theta_x = 69.6°$,

(ii) $\cos\theta_j = 4.5/7.176 = 0.6270$ and $\therefore \theta_y = 51.2°$ and

(iii) $\cos\theta_z = 5/7.176 = -0.6068$ and $\therefore \theta_z = 124.2°$.

39. *Deduce the necessary condition for two vectors $A = a_1i + b_1j + c_1k$ and $B = a_2i + b_2j + c_2k$ to be collinear.*

Solution: Obviously, to be collinear (or parallel), the direction cosines of vector A must be equal to the respective direction cosines of vector B, *i.e.*, we should have $a_1/A = a_2/B$, $b_1/A = b_2/B$ and $c_1/A = c_2/B$, which gives $a_1/a_2 = b_1/b_2 = c_1/c_2 = A/B$.

Thus, the condition for collinearity of the two vectors works out to be that *the ratio between the scalars* a_1 and a_2, b_1 and b_2 and c_1 and c_2 *must be the same, i.e.*, $a_1/a_2 = b_1/b_2 = c_1/c_2$.

Alternatively, we could obtain the same condition by taking the cross product of the two vectors and equating it to zero. Thus

$A \times B = (b_1c_2 - c_1b_2)i + (c_1a_2 - a_1c_2)j + (a_1b_2 - b_1a_2)k = 0,$

which gives $b_1c_2 - c_1b_2 = 0$, $c_1a_2 - a_1c_2 = 0$ and $a_1b_2 - b_1a_2 = 0$,

whence, $a_1/a_2 = b_1/b_2 = c_1/c_2$, as before.

40. *In a tetrahedron, if two pairs of opposite edges are perpendicular, show that the third pairs are also perpendicular to each other, and the sum of the squares on two opposite edges is the same for each pair.*

Solution: Let ABCD be a tetrahedron such that taking D as the origin of reference, the position vectors of points A, B and C are a,b and c respectively.

Then clearly, $\overrightarrow{AB} = (b - a)$. $\overrightarrow{AC} = (c - a)$ and $\overrightarrow{CB} = (b - c)$.

If the edge BD is perpendicular to CA, we have b. (c – a) = 0. *i.e.*,

$$b.c - b.a = 0, \text{ r, } b.c = a.b \quad \text{...(i)}$$

Similarly, if edge DA is perpendicular to BC, we have

$$a.b\,(b - c) = 0, \; i.e.,$$

$$a.b - a.c = 0.$$

or, $$a.b = c.a \quad \text{...(ii)}$$

From relation (i) and (ii) therefore,

a. b = b. c = c. a

i.e., b. c = c. a

or, b. c – c. a = 0.

or, c(b – a) = 0, *showing that DC is perpendicular to BA.*

Now, the sum of the squares on the opposite pair of edges BD and CA

$$= b^2 + (c - a)^2 = b^2 + c^2 + a^2 - 2ac$$

and since a. b = b. c = c. a, the sum of the square on the other two pairs of opposite edges (viz; DA, BC and DC. AB) will also be the same.

41. *The position vectors of two particles ejected simultaneously from the same source are $r_1 = 3i + 4j + 5k$ and $r_2 = 2i + 6j + 8k$. Obtain (i) the displacement r of particle two with respect to particle one, (ii) the magnitudes or r_1. r_2 and r, (iii) angles between r_1 and r_2 and r_1 and r and r_2 and r and (iv) projection of r on r_1.*

Solution:

(i) Here, obviously, *displacement of particle 2 with respect to particle* 1,

$$i.e.,\ r = r_2 - r_1 = -i + 2j + 3k.$$

(ii) Magnitude of $r_1 = |r_1| = \sqrt{9+16+25} = 5\sqrt{2} = 7.07$,

Magnitude of $r_2 = |r_2| = \sqrt{4+36+64} = \sqrt{104} = 10.20$

and Magnitude of $r = |r| = \sqrt{1+4+9} = \sqrt{14} = 3.74$.

(iii) We have $r_1.r_2 = |r_1||r_2| \cos\theta_{r_1r_2}$, whence, $\cos\theta_{r_1r_2} = \dfrac{r_1r_2}{|r_1||r_2|}$

$$= \frac{6+24+40}{(7.07)(10.20)} = 0.9706 \text{ and } \therefore \theta_{r_1r_2} = 14.0^\circ.$$

Similarly,

$$\cos\theta_{r_1r} = \frac{r_1.r}{|r_1||r|} = \frac{-3+8+15}{(7.07)(3.74)} = 0.7583 \text{ and } \therefore \theta_{r_1r} = 49.9^\circ.$$

and

$$\cos\theta^\circ_{r_2r} = \frac{r_2.r}{|r_2||r|} = \frac{-2+12+24}{(10.20)(3.74)} = 0.8913 \text{ and } \therefore \theta_{r_2r} = 27.8^\circ.$$

(iv) And, projection of r on r_1 = r.r_1 (where r_1 is the unit vector in the direction of r_1) $\frac{r.r_1}{|r_1|} = \frac{20}{7.07} = 2.83.$

42. *(a) Obtain the cross product of P = 3i – 2j + 8k, and Q = – i – 2j – 3k.*

(b) Show that if a, b and c be three non-zero vectors, such that

$$a \times b = b \times c = 0, \text{ then, } a \times c = 0.$$

Solution:

(a) We have *cross product* $P \times Q = \begin{vmatrix} i & j & k \\ 3 & -2 & 8 \\ -1 & -2 & -3 \end{vmatrix}$

$= i[(-2)(-3) - (-2)8] + j[(-1)8 - (-3)3] + k[3(-2) - (-2)(-1)]$

$= 22i + j - 8k.$

(b) Since a × b = 0, it is clear that a is parallel to b and, similarly, since b × c = 0, b is parallel to c. Obviously, therefore, a is parallel to c and hence

$$a \times c = 0$$

43. *(a) Express the magnitude of a × b in terms of scalar products.*
(b) If two vectors are 6i + 0 3j – 5k, i – 4.2j + 2.5k, find their vector product and by calculation prove that the new vector is perpendicular to the two.

Solution:

(a) Clearly, $(a \times b)^2 = (a \times b).(a \times b) = |a \times b|^2$

$$= |a|^2 |b|^2 \sin^2\theta = |a|^2 |b|^2 (1 - \cos^2\theta)$$

$$= a^2b^2 - |a|^2 |b|^2 \cos^2\theta = a^2b^2 - (a.b)^2.$$

(b) Let the two given vectors be called a and b respectively. Them,

their *vector product c, say,* $= \begin{vmatrix} i & j & k \\ 6 & 0.3 & -5 \\ 1 & -4.2 & 2.5 \end{vmatrix}$

$= i[(0.3)(2.5) - (-5)(-4.2)] + [j(-5)1 - 6(2.5)] + k[6(-4.2)$

$- (0.3)\ 1]$

$= -20.25i - 20j - 25.5k.$

Now, taking the scalar product of the new vector c with vector a, we have

$$c.a = (-20.25i - 20j - 25.5k).\ (6i - 0.3j + 5k)$$

$$= -121.5 - 6 + 127.5 = 0$$

showing that c is perpendicular to a

similarly, $c.b = (-20.25i - 20j - 25.5k).\ (i - 4.2j + 2.5k)$

$$= -20.25 + 84 - 63.75 = 0,$$

showing that c is also perpendicular to b.

44. *(a) The position vectors of four points A, B, C and D are $a = 2i + 3j + 4k$, $b = 3i + 5j + 7k$, $c = i + 2j + 3k$ and $d = 3i + 6j + 9k$ respectively. Examine whether vectors $\overrightarrow{AB}$ and $\overrightarrow{CD}$ are collinear*

(b) Define the condition that vectors $5i + 7j - 3k$ and $2i - bj + ck$ may be parallel.

Solution: (a) Clearly, vector $\overrightarrow{AB} = (b - a) = i + 2j + 3k$ and vector $\overrightarrow{CD} = (d - c) = 2i + 4j + 6k$. For collinearity of vectors, as we have seen in example 2 above, the condition is that $a_1/a_2 = b_1/b_2 = c_1/c_2$. Here, $a_1/a_2 = 1/2$, $b_1/b_2 = 2/4 = 1/2$ and $c_1/c_2 = 3/6 = 1/2$, *i.e.*, the condition for collinearity is satisfied. The two vectors are thus collinear.
(b) For the two given vectors to be parallel (or collinear), we must have $5/2 = -7/b = -3/c$, whence, $b = -14/5$ and $c = -6/5$, which is thus the required condition.

45. *If a position vector $r = 2i + 2j + k$, obtain (i) its direction cosines and (ii) its components in the xz and yz planes.*

Solution: Clearly, *magnitude of* r is given by

$$r = \sqrt{2^2 + 2^2 + 1^2} = \sqrt{9} = 3 \text{ units, and}$$

(i) it *direction cosines* are cos (r, x) = 2/3, cos (r, y) = 2/3 and cos (r, z) = 1/3 and

(ii) its components in the xz and yz planes are $r_{xz} = 2i + k$ and $r_{yz} = 2j + k$, with their magnitudes equal to

$$\sqrt{2^2 + 1^2} = \sqrt{5} \text{ and } \sqrt{2^2 + 1^2} = \sqrt{5} \text{ respectively.}$$

46. *Show that the vectors a – 2b + 3c, – 2a + 3b – 4c and – b + 2c are coplanar, where a,b and c are unit (or any) vectors.*

Solution: We have the condition for coplanarity of three vectors, namely, that the

determinant $\begin{vmatrix} A_X & B_X & C_X \\ A_Y & B_Y & C_Y \\ A_Z & B_Z & C_Z \end{vmatrix} = 0.$

Here, the determinant is $\begin{vmatrix} 1 & -2 & 3 \\ -2 & 3 & -4 \\ 0 & -1 & 2 \end{vmatrix}$

So that, its value is 1 (6 – 4) + (– 2) (+ 4) + 3(2) = 2 – 8 + 6 = 0, indicating that the three vectors are coplanar.

Alternatively, we could express one of the vectors in terms of the other two, equate the coefficients of a, b and c, solve any two of the three equations thus obtained and see if the same solution satisfies the third equation too. If it does, the three vectors are coplanar.

47. *The projection velocity of a rocket is expressed as* $V = 5\hat{V}_x + 7\hat{V}_y + 9\hat{V}_z$, *where* $\hat{V}_x$, $\hat{V}_y$ *and* $\hat{V}_z$ *are unit velocity vectors along east, north and vertical directions respectively. Calculate the magnitude of the horizontal and vertical components of the velocity. Also deduce the change in the angle of projection if the vertical component be doubled.*

Solution: Obviously, the magnitude of vector V, *i.e.*,

$V = \sqrt{5^2 + 7^2 + 9^2} = \sqrt{155}$, and *horizontal component of the vector* is $5\hat{V}_x + 7\hat{V}_y$, with *its magnitude* $\sqrt{5^2 + 7^2} = \sqrt{74}$ and *vertical component of the vector* is $9\hat{V}_x$, with *its magnitude* = 9.

If this latter (*i.e.*, the vertical) component be *doubled*, the new vector will be, say, $V' = 5\hat{V}_x + 7\hat{V}_y + 18\hat{V}_x$, with magnitude

$V = \sqrt{5^2 + 7^2 + 18^2} = \sqrt{398}$. Then, clearly, *angle between vectors* V and V' *will directly give the required change in the direction of velocity.* Now, V.V' = VV' cos (V, V'). whence, cos (V, V') = V.V'/VV'.

Since V.V' = 5(5) + 7(7) + 9(18) = 236 and $VV' = \left(\sqrt{155}\right)\left(\sqrt{398}\right)$, we have

$$\cos(V, V') = \frac{236}{(\sqrt{155})(\sqrt{398})} = 0.95.$$

Hence, angle (V, V') or *the change in the direction of velocity* $= \cos^{-1}(0.95) = 18.2°$.

48. *A particle is displaced from the point whose position vector is 5i – 5j – 7k to the point 6i + 2j – 2k under the action of forces 10i – j + 11k, 4i + 5j + 6k, – 2i + j – 9k. Find the total work done.*

Solution: Clearly, *total work done* = F.d, where F is the *resultant force on the particle* and d, its *displacement*

Here. F = (10i – j + 11k) + (4i + 5j + 6k) + (– 2i + j – 9k) = 12i + 5j + 8k, and *displacement* d = *position vector of second point—position vector of first point.*

$$= (6i + 2j - 2k) - (5i - 5j - 7k) = i + 7j + 5k.$$

$\therefore$ *total work done* = F.d = (12i + 5j + 8k) . (i + 7j + 5k)

$$= 12(1) + 5(7) + 8(5)$$

$$= 12 + 35 + 40 = 87 \textit{ units.}$$

49. *Evaluate* (i) $\frac{d}{dt}|a\ b\ c|$; (ii) $\frac{d}{dt}\left[a\ \frac{da}{dt}\ \frac{d^2a}{dt^2}\right]$; (iii) $\frac{d^2}{dt^2}\left[a\ \frac{da}{dt}\ \frac{d^2a}{dt^2}\right]$.

Solution. (i) $\frac{d}{dt}[a\ b\ c] = \frac{d}{dt}(a.b\times c)$

$$= a.\frac{d}{dt}(b\times c) + (b\times c).\frac{da}{dt},$$

$$= a.b\times\frac{dc}{dt} + a.\frac{db}{dt}\times c + (b\times c).\frac{da}{dt},$$

$$= \left[a.b\frac{dc}{dt}\right] + \left[a\frac{db}{dt}c\right] + \left[bc\frac{da}{dt}\right]$$

$$= \left[\frac{da}{dt}bc\right] + \left[a\frac{db}{dt}c\right] + \left[ab\frac{dc}{dt}\right]$$

50. *Evaluate* $\left|\frac{dt}{dt}\times\frac{d^2r}{dt^2}\right|$ and $\left|\frac{dt}{dt}\ \frac{d^2r}{dt^2}\ \frac{d^3r}{dt^3}\right|$, *where*

r = (a sin t) i + (a cos t) j + (at tan θ) k

Solution. Given r = (a cos t0 j + (at tan θ) k

$\therefore \quad \frac{dr}{dt} = (a \cos t)\ i = (-\ a \sin t)\ j + (a \tan \theta)\ k.$

$\frac{d^2r}{dt^2} = (-a \sin t)\ i + (-a \cos t)\ j + (0)\ k$

$\frac{d^3r}{dt^3} = (-a \cos t)\ i + (a \sin t)\ j$

$$\therefore \quad \frac{dr}{dt} \times \frac{d^2r}{dt^2} = \begin{vmatrix} i & j & k \\ a \cos t & -a \sin t & a \tan \theta \\ -a \sin t & --a \cos t & 0 \end{vmatrix}$$

$= (a^2 \cos t \tan \theta)\ i$

$= (a^2 \sin t \text{ than } \theta)\ j - a^2\ k$

$$\therefore \quad \left|\frac{dr}{dt} \times \frac{d^2r}{dt^2}\right| = \sqrt{[\ a^4\cos^2 t \tan^2\theta + a^4 \sin^2 t \tan^2 \theta + a]}$$

$= a^2 \sqrt{[\tan^2 \theta + 1]} = a^2 \sec \theta$

Also we know $[\ a\ b\ c] = \begin{vmatrix} a_1 & a_2 & a_3 \\ b_1 & b_2 & b_3 \\ c_1 & c_2 & c_3 \end{vmatrix}$ where $a = a i + a_2\ j + a_3\ k$ etc.

$$\therefore \quad \left[\frac{dr\ d^2r\ d^3r}{dt\ dt^2\ dt^3}\right] = \begin{vmatrix} a \cos t & -a \cos t & a \tan \theta \\ -a \sin t & -a \cos t & 0 \\ -a \cos t & a \sin t & 0 \end{vmatrix}$$

$= a^3 \tan \theta$ **Ans.**

51. $\frac{da}{dt} = c \times a$ and $\frac{db}{dt} = c \times b$ *then prove that*

$\frac{d}{dt} = (a \times b) = c \times (a \times b)$

Solution. $\frac{d}{dt}(a \times b) = \frac{da}{dt} \times b + a \times \frac{db}{dt}$

$= (c \times a) \times b + a \times (c \times b)$, given

$= -b \times (c \times a) + a\ (c \times b)$, $[\because u \times v = -\ v \times u]$

$= -[(b\ .\ a)\ c - (b\ .\ c)\ a] + [(a\ .\ b)\ c - (a\ .\ c)\ b]$

$= (b\ .\ c)\ a - (a\ .\ c)\ b = (c\ .\ d)\ a - (c\ .\ a)\ b.$

52. *Evaluate* $\int_s F\ .\ n\ dS$, *where* $F = a\ x\ i + b\ y\ j + c\ z\ k$ *and S is the* *surface of sphere* $x^2 + y^2 + z^2 = 1$.

Solution : Here div F = ∇ . F

$$= \left(i \frac{\partial}{\partial x} + j \frac{\partial}{\partial y} + \frac{\partial}{\partial z} \right) . (axi + byj + czk)$$

$$= \frac{\partial}{\partial x}(ax) + \frac{\partial}{\partial y}(by) + \frac{\partial}{\partial z}(cz) = a + b + c$$

$\therefore$ By Gauss' theorem we have

$$\int_S F . n\, dS = \int_V \text{div } F\, dV,$$

where V is the volume enclosed by the surface S

$= \int_V (a + b + c)\, dV = (a + b + c) \int_V dV$

$= (a + b + c)$ V, where V = volume enclosed by S = $4/3\ \pi\ (1)^3$

$= 4/3\ \pi\ (a + b + c)$. Hence Proved

53. *If* $f = (2x^2 y - x^4)\, i + (e^n - y \sin x)\, j + (x^2 \cos y)\, k$ *then verify that*

$$\frac{\partial^2 f}{\partial y\, \partial x} = \frac{\partial^2 f}{\partial x\, \partial y}.$$

Solution. Given

$f = (2x^2y - x^2)\, i + (e^n - y \sin x)\, j + (x^2 \cos y)\, k$ . . . (i)

$$\therefore \quad \frac{\partial f}{x} = \frac{\partial}{\partial x}(2x^2y - x^2)\, i + \frac{\partial}{\partial x}(e^n - y \sin x)\, j + \frac{\partial}{\partial x}(x^2 \cos y)\, k$$

$$= (4xy - 4x^3)\, i + (ye^n - y \cos x)\, j + \frac{\partial}{\partial x}(x^2 \cos y)\, k$$

$$\therefore \quad \frac{\partial^2 f}{\partial x\, \partial y} = \frac{\partial}{\partial y}(4xy - 4x^3)\, i + \frac{\partial}{\partial y}(ye^n - y \cos x)\, j + \frac{\partial}{\partial y}(2x \cos y) k$$

$= (4x)\, i + (yxe^n + e^n - \cos x)\, j + (-2\, x \sin y)\, k$ . . .(ii)

Again from (i) we get

$$\frac{\partial f}{\partial y} = \frac{\partial}{\partial y}(2x^2y - x^4) + \frac{\partial}{\partial y}(e^n - y \sin x)\, j + \frac{\partial}{\partial y}(x^2 \cos y) k$$

$= (2x^2)\, i + (xe^n - \sin x)\, j + (-x^2 \sin y)\, k$

$$\frac{\partial^2 f}{\partial x\, \partial y} = \frac{\partial}{\partial x}(2x^2)\, i + \frac{\partial}{\partial x}(xe^n - \sin x)\, j + \frac{\partial}{\partial x}(-x^2 \sin y) k$$

$= (4x)\, i + (xye^n + e^n - \cos x)\, j + (-\, 2x \sin x)\, k$

$\therefore$ From (ii) and (iii) we have $\dfrac{\partial^2 f}{\partial y\, \partial x} = \dfrac{\partial^2 f}{\partial x\, \partial y}$

54. *If $A = x^2 yz\, i = 2xz^3\, j + xz^2\, k$, $B = 2z\, i + y\, j - x^2$ find the value of $\dfrac{\partial^2}{\partial x\, \partial y}$ $(A \times B)$ at $(1, 0 -2)$.*

Solution. $A \times B = (x^2yzi - 2xz^3\, j + xz^2\, k) \times (2zi + yj - x^2k)$

$= x^2y^2zi \times j\ x^4yzi \times k - 4xz^4\, j \times i\ 2x^3z^3j \times k$

$+ 2xz^3k\ 5\ i + xyz^2\, k \times j,$

$[\because i \times i = 0 \text{ etc.}]$

$= x^2y^2zk \times x^4yzj + 4xz^4k + 2x^3z^3\, i + 2xz^3j - xyz^2z\, i,$

$[\because i \times j = k = -j \times i]$

or $A \times B = (2x^3z^3 - xyz^2)\, i + (x^4yz + 2xz^3)\, j + (x^2y^2z + 4xz^4)k$

$\therefore \dfrac{\partial}{\partial y}(A \times B) = (2x^3z^3 - xz^2)\, i + (x^4yz + 2xz^3)\, j + (x^2y^2z + 4xz^4)k$

$\dfrac{\partial^2}{\partial x\, \partial y}(A \times B) = (-z^2)\, i + (4x^3z)\, j + (4\, xyz)\, k$

$\therefore$ At $(1, 0, -2)$ we have

$\dfrac{\partial^2}{\partial x\, \partial y}(A \times B) = [-(-2)^2]\, i + [4\,(1)^3\,(-2)]\, j + [4\,(1)\,(0)\,(-2)]\, k$

$= -4i - 8j = -4\,(i + 2j)$ **Ans.**

55. *Evaluate* $\int_1^2 [a \cdot (b \times c) + a \times (b \times c)]\, dt.$

where $a = t\, i = 3\, j + 2\, t\, k$, $b = i - 2\, j + 2\, k$. $c = 3\, i + t\, j - k$

Solution. $b \times c = (i - 2j + 2k) \times (3i + t\, j - k)$

$= t\, i \times j - i \times k - 6j \times i + 2j \times k + 6\, k \times i + 2\, t\, k \times j$

$[\because i \times i = 0 \text{ etc.}]$

$= t\, k - (-j) - 6\,(-k) + 2i + 6j + 2\, t\,(-i),\ [\because i \times j = k = -j \times i \text{ etc.}]$

$= 2\,(1 - t)\, i + 7\, j + (t + 6)\, k$

$\therefore\quad a \cdot b \times c = (t\, i - 3\, j + 2\, t\, k) \cdot [2\,(1 - t)\, i + 7\, j + (t + 6)\, k]$

$= 2t\,(1 - t) - 21 + 2\, t\,(t + 6),\ [\because i \times i = 1,\ i \cdot j = 0 \text{ etc.}]$

$= 2t - 2t^2 - 21 + 2t^2 = 12t = 7\,(2t - 3)$

$$\int_1^2 (a.b \times c)\, dt = \int_1^2 7\,(2t - 3)\, dt = 7\left[r^2 - 3t\right]_1^2$$

$$= 7\,[\{2^2 - 3(2)\} - \{1^2 - 3(1)\}] = 7[0] = 0 \qquad \ldots (i)$$

Again $\quad b \times c = \begin{vmatrix} i & j & k \\ 1 & -2 & 2 \\ 3 & t & 1 \end{vmatrix} = -(2 + 2t)\, i + 5j + (t + 6)\, k$

$$\therefore \quad a \times (b \times c) = \begin{vmatrix} i & j & k \\ t & -3 & 2t \\ -(2+2t) & 5 & t+6 \end{vmatrix}$$

$$\int_1^2 [a \times (b \times c)]\, dt = \int_1^2 [(-13t - 18)\, i - 5\,(t^2 + 2t)\, j\,(t+6)k]\, dt$$

$$= \left[-\left(\frac{13}{2}t^2 + 18t\right) i - 5\left(\frac{1}{3}t^3 + t^2\right) j - \left(\frac{1}{2}t^2 + 6t\right) k\right]_1^2$$

$$= \left[\left\{-(26+36)i - 5\left(\frac{8}{3}+4\right)j - (2+12k\right\}\right.$$

$$\left. - \left\{-\left(\frac{13}{2} + 18\right) i - 5\left(\frac{1}{3} + 1\right) j - \left(\frac{1}{2} + 6\right) k\right\}\right]$$

$$= -\,(55/2)\, i = (80/3)\, j - (15/2)\, k \qquad \ldots (ii)$$

$$\therefore \quad \int_1^2 [a.\,(b \times c) + a \times (b \times c)]\, dt$$

$$= \int_1^2 a.\,(b \times c)\, dt + \int_1^2 a \times (b \times c)\, dt$$

$= 0 + [-(55/2)\, i - (80/3)\, j - (15/2)\, k]$, from (i) and (ii)

$= -\,(55/2)\, i - (80/3)\, j - (15/2)\, k$ **Ans.**

56. *Evaluate* $\int_0^1 (e^t i + e^{-2t}\, j + t k)\, dt$

Solution. $\int_0^1 (e^t\, i + e^{-2t}\, j + t\, k)\, dt$

$$= i \int_n^1 (e^t\, dt + j \int_0^1 e^{-2t}\, dt + k \int_0^1 t\, dt$$

$$= i\left[e^t\right]_n^1 + j\left[-\frac{1}{2}e^{-2t}\right]_n^1 + k\left[\frac{1}{2}t^2\right]_n^1$$

$= (e - 1)\, i - 1/2\, (e^{-2} - 1)\, j = 1/2\, k$ **Ans.**

57. *If $r \times dr = 0$, prove that $\hat{r}$ = constant vector.*

Solution. Let r = x i + y j + z k, then dr = i (dx) + j (dy) + k (dz)

$$\therefore \quad r \times dr = \begin{vmatrix} i & j & k \\ x & y & z \\ dx & dy & dz \end{vmatrix}$$

$$= (y\,dz - z\,dy)\,i + (z\,dx - x\,dz)\,j + (x\,dy - y\,dx)\,k$$

$\therefore$ $r \times dr = 0$ gives

$(y\,dz - z\,dy)\,i + (z\,dx - x\,dz)\,j + (x\,dy = y\,dx) = 0$, whence we have

$ydz - zdy = 0,\ zdx - xdz = 0,\ xdy = ydx = 0$

or $\dfrac{dx}{x} = \dfrac{dy}{y} = \dfrac{dz}{z}$

If $\dfrac{dx}{x} = \dfrac{dy}{y}$, then $\log x = \log y + \log c_1$

or $x = c_1 y$

If $\dfrac{dy}{dy} = \dfrac{dz}{z}$, then $\log y + \log c_2 = \log z$

or $z = c_2\,y$

Now $r = xi + yi + zk = (c_1 y)\,i + (y)\,j + (c_2 y)\,k$

$= y\,(c_1\,i + j + c_2\,k)$

$$\therefore \quad \hat{r} = \frac{y\,(c_1\,i + j + c_2\,k)}{\sqrt{|(c_1\,y)^2 + (y)^2 + (c_2\,y)^2|}}$$

$$= \frac{c_1\,i + j + c_2\,k}{\sqrt{(c_1^2 + 1 + c_2^2)}} = \text{constant vector}$$

as it in independent of x, y, z.

58. *A particle moves so that its position vector is given by r = (cos wt) i + (sin wt) j, where w is constant. Show that*

(i) *the velocity v of the particle is perpecdicular to r,*

(ii) the acceleration a is directed towards the origin and has magnitude proportional to the distance from the origin,

(iii) $r \times v$ = a constant vector.

Solution. Given that r = (cos wt) i + (sin wt) i . . .(i)

$$\therefore \qquad v = \frac{d}{dt}(r) = (-w \sin wt)\, i + (\sin wt)\, j \qquad \text{. . . (ii)}$$

$$\text{and} \qquad a = \frac{d}{dt}(v) = -(w^2 \cos wt)\, i + (-w^2 \sin wt)\, j$$

$$= -w^2 [(\cos wt)\, i + (\sin wt)\, j]$$

$$\text{or} \qquad a = -w^2 r \qquad \text{. . . (iii)}$$

(i) $v \,.\, r = [(-w \sin wt)\, i + (w \cos wt)\, j] \,.\, [(\cos wt)\, i\, (\sin wt\, j)]$, from (i) and (ii)

$= (-w \sin wt \cos wt)\, i \,.\, i + (w \cos wt \sin wt)\, j \,.\, j$,

since $i \,.\, j = 0 = j \,.\, i$

$= -w \sin wt \cos wt + w \cos wt \sin wt$, $[\because i \,.\, i = 1 = j \,.\, j]$

or $v \,.\, r = 0$ i.e. v is perpendicular to r.

(ii) From (iii) we hve $a = -w^2 r$,

which shows that the vector a being negative is directed towards the origin.

Also $| a | \alpha | r |$ i.e. magnitude of acceleration a is proportional to $| ar |$ i.e. the distance from one origin.

(iii) From (i) and (ii) we get

$r \times v = [(\cos wt)\, i + (\sin wt)]\, j] \times [(-w \sin wt0\, i + (w \cos wt)\, j]$

$= (w \cos^2 wt)\, i \times j - (w \sin^2 wt)\, j \times i$, $[\because i \times i = 0 = j \times j]$

$= (w \cos^2 wt)\, i \times j + (w \sin^2 wt)\, i \times j$, $[\because j \times i = i \times j]$

$= w (\cos^2 wt + \sin^2 wt)\, i \times j$,

$\therefore \quad = | r \times v | = | w k | = w | k | = w$, $[\because | k | = 1]$

Hence $r \times v$ is a constant vector of magnitude w.

EXERCISES

1. Two particles travel along the axes of z and y with velocities $v_1 = 3i$ and $v_2 = 4j$ respectively. If at the start their position be $x_1 = -4$ cm, $y_1 = 0$ and $x_2 = 0$, $y_2 = -5$ cm, obtain the position vector r giving the relative position of the second particle with respect to the first.

   When will the two particles be the closest together ?

   $\frac{d}{dt}|r|^3 = \frac{d}{dt}\left[(3t-4)^2 + (4t-5)^2\right] = 0$, whence : can be easily calculated.]

2. What is meant by the scalar product of two vectors ? show that it *is commutative as well as distributive.*

   Use the concept of scalar product to deduce the *cosine law.*

3. Given the three vectors A = 5i – 7j + 3k, B = – 4i + 7j – 8k, C = 2i – 3j, deduce the following:

   (i) A. B × C and A × B. C, (ii) A^2 C.

4. A body of mass 500 g, free to move, is subjected to a force of i + 2j + k newton. What velocity will it acquire after 4 seconds and what distance will it have covered in that interval of time?

5. Show that in the case of a rigid body spinning about a fixed axis, the linear velocity (v) of a particle of the body is the cross product of the angular velocity ω and the position vector r of the particle relative to a fixed point on the axis.

6. A rigid body is spinning with an angular velocity of 4 radians per second about an axis parallel to (3j – k), passing through the point (1 + 3j –k). Find the velocity of the particle at the point 4i – 2j + k.

7. Obtain the values of (i) grad (1/r), (ii) [grad f (r)] × r.

8. If a and b are the vector $\overrightarrow{AB}, \overrightarrow{BC}$, determined by the adjacent sides of a regular hexagon, what are the vectors determined by the other sides, taken in order ?

9. (a) The points A, B and C are respectively i + 2j – 3k, 3i – 4j + 5k and 5i – 10j + 13k. Show that vectors $\overrightarrow{AB}$ and $\overrightarrow{BC}$ are collinear.

10. Deduce the condition for the coplanarity of three vectors A, B and C.

11. Show that the points A(4, 5, 1), B(0, – 1, – 1), C(3, 9, 4), and D(–4, 4, 4) are coplanar.

12. The x component of the resultant of several vectors
    (a) is equal to the sum of the x-components of the vectors
    (b) may be smaller than the sum of the magnitudes of the vectors
    (c) may be greater than the sum of the magnitudes of the vectors
    (d) may be equal to the sum of the magnitudes of vectors

13. A vector $\vec{\mathbf{A}}$ points vertically upward and $\vec{\mathbf{B}}$ points towards north. The vector product $\vec{\mathbf{A}} \times \vec{\mathbf{B}}$ is

    (a) along west (b) along last

(c) zero (d) vertically downwards

14. What are *direction cosines* ? Show that is a unit be vector be resolved in terms of i, j and k the coefficients of i, j and k are the direction cosine of the vector.

15. What angle does the vector i + $\sqrt{2}$ j + 3k make with the axis of z and what is the value of its component in the y — z plane?

16. If vectors A and B be respectively equal to 3i – 4j + 5k and 2i + 3j – 4k, obtain the magnitudes of vectors A + B and A – B. What will be the unit vectors parallel to A + B and A – B respectively ?

17. (a) If A = 3i + 4j +4k and B = i + 4j + 4k, obtain the direction cosines of A + B and A – B..

    (b) What should be the vector which, on addition to vectors – 2i + 3j – 4k and – 2j + 4k, will give a unit vector along the axis of y?

18. Given that the electric field is the negative of the gradient of potential, *i.e.*, E = $-\nabla$ V, obtain the value of the electric field if that of the potential be – Kxy, where K is a constant.

19. If r = xi + yj + zk, show that (i) div r = 3, (ii) curl r = 0, (iii) div (r^nr) = (n + 3)r^n and (iv) $\nabla^2 r^{-1} = 0$.

20. Show that (a) curl rϕ (r) = 0 and curl (a. r) a = 0.

21. Prove that curl grad ϕ = 0 and div curl f = 0.

22. The Laplacian Operator $\nabla^2 = \nabla.\nabla$ Why do we not have another operator $\nabla \times \nabla$?

23. What is meant by line and surface integrals ? Enunciate and prove Gauss's theorem.

24. Show that $\iint_3 F.dS = 6V$, where S is a closed surface, enclosing a volume V and F = xi + 2yj + 3zk.

25. Show that $\int_3 \left(x^2 i + y^2 j + z^2 k\right).\, ndS = 0$, where S denotes the surface of the ellipsoid $\frac{x^2}{a^2} + \frac{y^2}{b^2} + \frac{z^2}{c^2} = 1$.

26. Enunciate and prove *stokes' theorem.*

27. Show, using either Gauss's or Stokes' theorem, that $\int_3 \text{curl}\, f.ndS = 0$, where S is a closed surface.

28. Let $\vec{C} = \vec{A} + \vec{B}$

(a) $|\vec{C}|$ is always greater than $|\vec{A}|$

(b) it is possible to have $|\vec{C}| < |\vec{A}|$ and $|\vec{C}| < |\vec{B}|$

(c) C is always equal to A + B

(d) C is never equal to A + B

29. The displacement of a particle is proportional to the cube of time. Then magnitude of its acceleration

(a) increases with time

(b) decreases with time

(c) constant but not zero

(d) zero

30. Let the angle between two nonzero vectors $\vec{A}$ and $\vec{B}$ be 120° and its resultant be $\vec{C}$

(a) C must be equal to $|A - B|$

(b) C must be less than $|A - B|$

(c) C must be greater than $|A - B|$

(d) C may be equal to $|A - B|$

31. The magnitude of the vector product of two vectors $\vec{A}$ & $\vec{B}$ may be

(a) greater than AB (b) equal to AB

(c) less than AB (d) equal to zero

32. If a and b are two vectors given by $a = a_x i + a_y j + a_z k$ and $b = b_x i + b_y j + b_z k$, show that $a.b = a_x b_x + a_y b_y + a_z b_z$.

33. Obtain the scalar product of vectors (6i + 2j + 3k) and (2i – 9j + 6k) and also the cosine of the angle between them.

34. Define scalar and vector products of two vector. If i, j, k are unit vectors along x, y, and z axes respectively, show that ii = 1 and i.j = 0.

35. (a) Show that vectors A = 2i + 3j – 4k and B = 5i + 2j + 4k are perpendicular to each other.

(b) Under what condition will the vectors A = 3i – 5j + 5k and B = 5i – j + bk be perpendicular to each other ?

36. If $|A + B| = |A - B|$, Show that A and B are perpendicular to each other.

37. What conclusions can you draw from the fact that scalar product A B = 0 ?

38. Forces acting on a particle have magnitudes 5, 3, 1 lb wt and act in the directions of the vectors 6i + 2j + 3k, 3i – 2j + 6k, 2i – 3j –6k respectively. These remain constant while the particle is displaced from the point A(2, – 1, – 3) to B (5, –1, –1). Find the work done by the forces, the unit of length being 1ft.

39. Show that if the scalar product of a vector r with each of three non-coplanar vectors be zero, then r must be the zero vector.

40. Show that if a and b are equal in magnitude and not-parallel, a + b and a – b are perpendicular.

41. Find the cosine of the angle between the vectors a = 3i + j + 2k and b = 2i – 2j + 4k. Find also the unit vector perpendicular to both a and b.

42. If two vectors are 6i + 0.3j – 5k and i – 4.2j + 2.3k, find the vector product and by calculation prove that the new vector is perpendicular to the two.

43. What is the unit vector perpendicular to each of the vectors a = 2i – j + k and b = 3i + 4j – k ? Calculate the sine of the angle between these vectors.

44. Show that the vector product is (i) *associative*, (ii) *not commutative*.

45. Write down the values of (i) i × i, j × j, k × k (ii) i × j, j × k and k × i.

46 Show that if b = c + ma, where m is any scalar, a × b = a × c, but that if a × b = a × c, b is not necessarily equal to c.

47. Express the magnitude of a × b in terms of the scalar product.

48. Obtain the vector area of a triangle whose vertices are a, b and c. What is the condition for collinearity of the three points?

49. Find the area of the parallelogram determined by vectors i + 2j 3k and – 3i – 2j +k.

    What is the sine of the angle between the two vectors ?

50. Find the torque about the point i + 2j – k of a force represented by 3i + k acting through the point 2i – j + 3k.

51. Deduce the unit vector which is perpendicular to both the vectors P = 3i – 3j and Q = 6j + 5k.

52. A proton of velocity (3i + 2j) 10^8 cm/s enters a magnetic field 2000j + 3000k *gauss*. Calculate the acceleration. Also deduce the simultaneous

electric field which will keep the particle undeflected.

53. The vectors i + 2j + 3k, – 3i + 2k and – 5k represent the edges of a parallelopiped; obtain its volume.

54. Deduce the condition for coplanarity of three vectors.

55. Obtain the scalar triple product of (i) the vectors a, a, and b, (ii) the vectors a, b and c where a and b are parallel.

56. Show that if a, b, c be three vectors, we have a × (b + c) = a × b + a. × c.

57. Vectors A, B and C are respectively given by 2i + j – 3k, i + 3j + 4k and 5j + mk. What should be the value of m so that (i) the volume of the parallelopiped, with its edges represented by the three vectors, be 15 units, (ii) the three vectors may be coplanar?

68. Deduce the condition for the vectors 2i + 3j – 4k and 3i – aj + bk to be parallel (or collinear).

59. Write down the values of (i) k. k, (ii) [i j k], (iii) a × (b × c), (iv) b × (c × a) (v) c × (a × b).

60. Show that a × (b × c) + b × (c × a) + c × (a × b) = 0.

4

PARTIAL DERIVATIVES OF VECTORS

INTRODUCTION

Suppose r is a vector depending on more than one scalar variable. Let r = f(x, y, z) *i.e.*, let r be a function of three scalar variables x, y and z. The partial derivative of r with respect to r is defined as

$$\frac{\partial r}{\partial x} = \lim_{\delta x \to 0} \frac{f(x+\delta x, y, z) - f(x, y, z)}{\delta x}$$

if this limit exists. Thus $\partial r/\partial x$ is nothing but the ordinary derivative of r with respect to x provided the other variables y and z are regarded as constants. Similarly, we may define the partial derivatives

$$\frac{\partial r}{\partial y} \text{ and } \frac{\partial r}{\partial z}.$$

Higher partial derivatives can also be defined as in Scalar Calculus. Thus, for example.

$$\frac{\partial^2 r}{\partial x^2} = \frac{\partial}{\partial x}\left(\frac{\partial r}{\partial x}\right), \frac{\partial^2 r}{\partial y^2} = \frac{\partial}{\partial y}\left(\frac{\partial r}{\partial y}\right), \frac{\partial^2 r}{\partial z^2} = \frac{\partial}{\partial z}\left(\frac{\partial r}{\partial z}\right).$$

$$\frac{\partial^2 r}{\partial x\, \partial y} = \frac{\partial}{\partial x}\left(\frac{\partial r}{\partial y}\right), \frac{\partial^2 r}{\partial y\, \partial x} = \frac{\partial}{\partial y}\left(\frac{\partial r}{\partial x}\right).$$

If r has continuous partial derivatives of the second order at least, the,

$$\frac{\partial^2 r}{\partial x\, \partial y} = \frac{\partial^2 r}{\partial y\, \partial x}$$

i.e., the order of differentiation is immaterial. If r = f(x, y, z), the total differential d r of r is given by

$$dr = \frac{\partial r}{\partial x} dx + \frac{\partial r}{\partial y} dy + \frac{\partial r}{\partial z} dz.$$

THE VECTOR DIFFERENTIAL OPERATOR DEL (∇)

The vector differential operator ∇ (read as *del* or *nabla*) is defined as

$$\nabla = \frac{\partial}{\partial x}i + \frac{\partial}{\partial y}j + \frac{\partial}{\partial z}k = i\frac{\partial}{\partial x} + j\frac{\partial}{\partial y} + k\frac{\partial}{\partial z}$$

and operates distributively.

The vector operator ∇ can generally the treated to behave as an ordinary vector. It properties like ordinary vectors. The symbols

$$\frac{\partial}{\partial x}\cdot\frac{\partial}{\partial y}\cdot\frac{\partial}{\partial z}$$

can be treated as its components along i, j, k.

SOME USEFUL RESULTS

The gradient of a scalar point function ϕ is a vector point function and so is the curl of a vector point function f. We may also, therefore, obtain their divergence and curl, as given below:

(i) div gradϕ = $\nabla.\nabla\phi = \frac{\partial^2\phi}{\partial x^2} + \frac{\partial^2\phi}{\partial y^2} + \frac{\partial^2\phi}{\partial z^2} = \nabla^2.$

where ∇^2 is the *Laplacian operator* (and a *scalar*).

(ii) curl grad ϕ = $\nabla \times \nabla \ \phi = 0$, because the cross product of two equal vectors is zero, and here both the vectors or ∇'s are equal since they operate on the same function ϕ.

It follows from this that if *we have a vector field* f *such that its curl is zero, i.e.,* $\nabla \times \phi = 0$, f *is certainly the gradient of some scalar field* ϕ, *i.e.,* if $\nabla \times f = 0$, we have $\phi = \nabla\phi$.

(iii) div *curl* f = $\nabla.\nabla \times \phi = 0$, because, as we know, it is a combination similar to A.A × B, where A × B is perpendicular to A and has no component in the direction of A.

It follows, therefore that *if the divergence of a vector field* D *is zero, then,* D *is clearly the curl of some vector field f, i.e., if* ∇ D = 0, we have D = $\nabla \times \phi$.

(iv) curl f = $\nabla \times (D \times f) = \nabla\nabla.f - \nabla^2 f$ = grad div f – ∇^2f.

We may have two more combinations, *viz.*,

(v) $\nabla(\nabla.f)$ and (vi) $(\nabla`\nabla)$ f = ∇^2f. There is, however, nothing special about them. They represent just vector fields which we may possibly come across sometimes.

SOME IMPORTANT FORMULAE

1. $\nabla(\phi_1\phi_2) = \phi_1(\nabla\phi_2) + \phi_2(\nabla\phi_1)$.

2. $\nabla\left(\frac{\phi_1}{\phi_2}\right) = \frac{\phi_2\nabla\phi_1 - \phi_1\nabla\phi_2}{\phi_2^2}$, $\phi_2 \neq 0$.

3. Grad. $\phi = \nabla\phi = \left(i\frac{\partial}{\partial x} + \hat{j}\frac{\partial}{\partial y} + \hat{k}\frac{\partial}{\partial z}\right)\phi$ grad ϕ is a vector quantity.

4. If ϕ_1 and ϕ_2 are scalar point functions, then

$$\nabla(\phi_1 \pm \phi_2) = \nabla\phi_1 \pm \nabla\phi_2.$$

Divergence : If $\vec{F} = F_1\hat{i} + F_2\hat{j} + F_3\hat{k}$, then.

5. div $\vec{F} = \nabla.\vec{F} = \frac{\partial F_1}{\partial x} + \frac{\partial F_2}{\partial y} + \frac{\partial F_3}{\partial z}$ $\nabla.\vec{F}$ is a scalar quantity.

6. A vector $\vec{F}$ is called solenoided, if div $\vec{F} = 0$ *i.e.,* $\nabla.\vec{F} = 0$.

If ϕ is a vector point function and $\vec{u}, \vec{v}$ are vector point function then.

7. $\nabla.\left(\phi\vec{u}\right) = (\nabla\phi).\vec{u} + \phi\left(\nabla.\vec{v}\right)$.

8. $\nabla.\left(\phi\vec{u}\right) = (\nabla\phi).\vec{u} + \phi\left(\nabla.\vec{u}\right)$.

9. $\nabla.\vec{v} = 3$ where $\vec{r} = x\hat{i} + y\hat{j} + z\hat{k}$.

10. If $\vec{F} = F_1 i + F_2 j + F_3 k$, then

$$\text{curl } \vec{F} = \nabla \times \vec{F} = \begin{vmatrix} \hat{i} & \hat{j} & \hat{k} \\ \frac{\partial}{\partial x} & \frac{\partial}{\partial y} & \frac{\partial}{\partial z} \\ F_1 & F_2 & F_3 \end{vmatrix}$$ $\nabla \times \vec{F}$ is a vector quantity.

11. A vector $\vec{F}$ is called irrotational if curl $\vec{F} = \vec{0}$ is $\nabla \times \vec{F} = 0$.

12. Curl $\vec{r}$ *i.e.,* $\nabla \times \vec{r} = \vec{0}$ where $\vec{r} = n\hat{i} + y\hat{j} + x\hat{i}$. If $\vec{u}, \vec{v}$ are vector point function, and ϕ is a scalar point function then.

13. $\nabla \times \left(\vec{u} + \vec{v}\right) = \left(\nabla \times \vec{u}\right) + \left(\nabla \times \vec{v}\right)$.

14. $\nabla \times \left(\phi \vec{u}\right) = \phi\left(\nabla \times \vec{u}\right) + (\nabla \phi) \times \nabla \vec{u}$

Second Order Differential Operator: If ϕ is a scalar point friction and $\vec{F}$ is a vector point function then.

15. div (grad ϕ) = $\nabla (\nabla \phi) = \nabla^2\phi$.

16. curl (grad ϕ) = $\nabla \times (\nabla\phi) = \vec{0}$ ·

17. $\text{div}\left(\text{cur } \vec{F}\right) = \nabla.\left(\nabla \times \vec{F}\right) = 0$.

18. $\text{curl}\left(\text{curl } \vec{F}\right) = \text{grad}\left(\text{div } \vec{F}\right) - \nabla^2 \vec{F}$

$\Rightarrow \nabla \times \left(\nabla \times \vec{F}\right) = \nabla\left(\nabla.\vec{F}\right) + \nabla^2 \vec{F}$.

19. $\text{grad}\left(\text{div } \vec{F}\right) = \text{curl}\left(\text{curl } \vec{F}\right) + \nabla^2 \vec{F}$

$\Rightarrow \nabla\left(\nabla.\vec{F}\right) = \nabla \times \left(\nabla \times \vec{F}\right) + \nabla^2 \vec{F}$.

20. $\nabla^2 \equiv \frac{\partial^2}{\partial x^2} + \frac{\partial^2}{\partial y^2} + \frac{\partial^2}{\partial z^2}$ (∇^2 is called $\propto$ Laplacian operator).

21. $\nabla.(\nabla\phi) = \nabla^2\phi = \frac{\partial^2\phi}{\partial x^2} + \frac{\partial^2\phi}{\partial y^2} + \frac{\partial^2\phi}{\partial z^2}$.

22. $\delta\phi = \nabla\phi.d\vec{r}$ where $\vec{r} = x\hat{i} + y\hat{j} + x\hat{k}$ and ϕ is a scalar point function.

23. *Definition :* The directional derivative of a scalar field ϕ at a point P(x, y, z) in the direction of $\hat{a}$ is given by $\frac{d\phi}{ds} = \vec{a}.\text{grad } \phi = \vec{a}.(\nabla\phi)$.

24. $\frac{d\phi}{ds}$ will be maximum when $\vec{a}$ is in the direction of $\vec{n}$ where $\hat{n}$ is the unit outward drawn normal vector to the surface $\phi(x, y, z)$ = constant.

25. A necessary and sufficient condition of $\vec{F}$ to be an irrotational vector function $\pi_{dc}.\int \vec{F}.d\vec{r} = 0$ for every closed curve C is that $\nabla \times \vec{F} = \vec{0}$ identically.
26. Any one of the following three conditions characteristics an irrational field $\vec{F}$.

 (i) $\nabla \times \vec{F} = 0$

 (ii) $\vec{F} = \nabla\phi$

 (iii) $\int \vec{F}.d\vec{r} = 0$, along every closed curve C.

GRADIENT OF A SCALAR FIELD

Definition. *Let f(x, y, z) be defined and differentiable at each point (x, y, z) in a certain region of space (i.e., defines a differentiable scalar field). Then the gradient of f, written as ∇f or grad f, is defined as*

$$\nabla f = \left(\frac{\partial}{\partial x}i + \frac{\partial}{\partial y}j + \frac{\partial}{\partial z}k\right) f = \frac{\partial f}{\partial x}i + \frac{\partial f}{\partial y}j + \frac{\partial f}{\partial z}k.$$

It should be noted that ∇f is a vector whose three successive components are

$$\frac{\partial f}{\partial r}, \frac{\partial f}{\partial y} \text{ and } \frac{\partial f}{\partial z}.$$

Thus, the gradient of a scalar field defines a vector field. *If f is a scalar point function, then ∇f is a vector point function.*

FORMULAS INVOLVING GRADIENT

Theorem 1:

Gradient of the sum of two scalar point functions. If and g are two scalar point functions, then,

grad (f + g) = grad f + grad g or ∇ (f + g) = ∇f + ∇g.

Proof:

We have

$$\text{grad } (f + g) = \nabla (f + g) = \left(i\frac{\partial}{\partial x} + j\frac{\partial}{\partial y} + k\frac{\partial}{\partial z}\right)(f + g)$$

$$= i\frac{\partial}{\partial x}(f+g) + j\frac{\partial}{\partial y}(f+g) + k\frac{\partial}{\partial z}(f+g)$$

$$= i\frac{\partial f}{\partial x} + i\frac{\partial g}{\partial x} + j\frac{\partial f}{\partial y} + j\frac{\partial g}{\partial y} + k\frac{\partial f}{\partial z} + k\frac{\partial g}{\partial z}$$

$$= \left(i\frac{\partial f}{\partial x} + j\frac{\partial f}{\partial y} + k\frac{\partial f}{\partial z}\right) + \left(i\frac{\partial g}{\partial x} + j\frac{\partial g}{\partial y} + k\frac{\partial g}{\partial z}\right)$$

$$= \left(i\frac{\partial}{\partial x} + j\frac{\partial}{\partial y} + k\frac{\partial}{\partial z}\right)f + \left(i\frac{\partial}{\partial x} + j\frac{\partial}{\partial y} + k\frac{\partial}{\partial z}\right)g$$

$= \nabla f + \nabla g$ = grad f + grad g.

Similarly, we can prove that $\nabla(f-g) = \nabla f - \nabla g$.

Theorem 2:

Gradient of a constant. The necessary and sufficient condition for a scalar point function to be constant is that $\nabla f = 0$.

Proof:

If f(x, y, z) is constant, then $\frac{\partial f}{\partial x} = 0$,

$$\frac{\partial f}{\partial y} = 0,\ \frac{\partial f}{\partial z} = 0.$$

Therefore grad f $= i\frac{\partial f}{\partial x} + j\frac{\partial f}{\partial y} + k\frac{\partial f}{\partial z}$

$= 0i + 0j + 0k = 0.$

Hence the condition is necessary.

Conversely, let grad f = 0.

Then $i\frac{\partial f}{\partial x} + j\frac{\partial f}{\partial y} + k\frac{\partial f}{\partial z} = 0.$

Therefore $\frac{\partial f}{\partial x} = 0,\ \frac{\partial f}{\partial y} = 0,\ \frac{\partial f}{\partial z} = 0.$

$\therefore$ f must be independent of x, y and z.

$\therefore$ f must be a constant. Hence the condition is sufficient.

Theorem 3:

Gradient of the product of two scalar point functions. If f and g are scalar point functions, then grad (fg) = f grad g + g grad f or $\nabla (fg) = f\nabla g + g\nabla f$.

Proof:

We have $\nabla (fg) = \left(i\frac{\partial}{\partial x} + j\frac{\partial}{\partial y} + k\frac{\partial}{\partial z} \right)(fg)$

$$= i\frac{\partial}{\partial x}(fg) + j\frac{\partial}{\partial y}(fg) + k\frac{\partial}{\partial z}(fg)$$

$$= \left(f\frac{\partial g}{\partial x} + g\frac{\partial f}{\partial x} \right) + j\left(f\frac{\partial g}{\partial y} + g\frac{\partial f}{\partial y} \right) + k\left(f\frac{\partial g}{\partial z} + g\frac{\partial f}{\partial z} \right)$$

$$= f\left(i\frac{\partial g}{\partial x} + j\frac{\partial g}{\partial y} + k\frac{\partial g}{\partial z} \right) + g\left(i\frac{\partial f}{\partial x} + \frac{\partial f}{\partial y} + k\frac{\partial f}{\partial z} \right)$$

$= f\nabla g + g\nabla f =$ f grad g + g grad f.

In particular, if c is a constant, then

$$\nabla (cf) = c\nabla f + f\nabla c = c\nabla f + 0\ c\nabla f.$$

Theorem 4:

Gradient of the quotient of two scalar functions. If f and g are two scalar point functions, then

$$\nabla \left(\frac{f}{g} \right) = \frac{g\nabla f - f\nabla g}{g^2}.$$

Proof:

We have $\nabla \left(\frac{f}{g} \right) = \left(i\frac{\partial}{\partial x} + j\frac{\partial}{\partial y} + k\frac{\partial}{\partial z} \right)\left(\frac{f}{g} \right)$

$$= i\frac{\partial}{\partial x}\left(\frac{f}{g} \right) + j\frac{\partial}{\partial y}\left(\frac{f}{g} \right) + k\frac{\partial}{\partial z}\left(\frac{f}{g} \right).$$

But $\frac{\partial}{\partial x}\left(\frac{f}{g} \right) = \frac{g\frac{\partial f}{\partial x} - f\frac{\partial g}{\partial x}}{{}^2}$, $\frac{\partial}{\partial y}\left(\frac{f}{g} \right) = \frac{g\frac{\partial f}{\partial y} - f\frac{\partial g}{\partial y}}{g^2}$,

and $\frac{\partial}{\partial z}\left(\frac{f}{g} \right) = \frac{g\frac{\partial f}{\partial z} - f\frac{\partial g}{\partial z}}{g^2}$.

$$\therefore \quad \nabla \left(\frac{f}{g} \right) = \frac{1}{g^2}\left\{ i\left(g\frac{\partial f}{\partial x} - f\frac{\partial g}{\partial x} \right) + j\left(g\frac{\partial f}{\partial y} - f\frac{\partial g}{\partial y} \right) + k\left(g\frac{\partial f}{\partial z} - f\frac{\partial g}{\partial z} \right) \right\}$$

$$= \frac{1}{g^2}\left\{g\left(i\frac{\partial f}{\partial x} + j\frac{\partial f}{\partial y} + k\frac{\partial f}{\partial z}\right) - f\left(i\frac{\partial g}{\partial x} + j\frac{\partial g}{\partial y} + k\frac{\partial g}{\partial z}\right)\right\}$$

$$= \frac{1}{g^2}(g\nabla f - f\nabla g).$$

EQUIPOTENTIAL SURFACES OR LEVEL SURFACES

Let f(x, y, z) be a scalar field over a region R. The points satisfying an equation of the type

f(x, y, z) = c, (arbitrary constant)

constitute a family of surfaces in three dimensional space. The surfaces of this family are called *level surfaces*. Any surface of this family is such that the value of the function f at any point of it is the same. Therefore, these surfaces are also called *is-f-surfaces*.

Theorem 1.

Let f(x, y, z) be a scalar field over a region R. Then through any point of R there passes one and only one level surface.

Proof:

Let (x_1, y_1, z_1) be any point of the region R.

Then the level surface $f(x, y, z) = f(x_1, y_1, z_1)$ passes through this point.

Now suppose the level surfaces $f(x, y, z) = c_1$ and $f(x, y, z) = c_2$ pass through the point (x_1, y_1, z_1). Then

$$f(x_1, y_1, z_1) = c_1 \text{ and } f(x_1, y_1, z_1) = c_2.$$

Since f(x, y, z) has a unique value at (x_1, y_1, z_1), therefore we have $c_1 = c_2$.

Hence only one level surface passes through the point (x_1, y_1, z_1).

Theorem 2:

∇f is a vector normal to the surface f(x, y, z) = c where c is a constant.

Proof:

Let r xi + yj + zk be the position vector of any point P (x, y, z) on the level surface f(x, y, z) = c. Let $Q(x + \delta x, y + \delta y, z + \delta z)$ be a neighboring point on this surface.

Then the position vector of

$$Q = r + \delta r = (x + \delta x)\, i + (y + \delta y)\, j + (z + \delta z)\, k.$$

$$\therefore \quad \vec{PQ} = (r + \delta r) - r = \delta r = \delta x i + \delta y j + \delta z k.$$

As Q → P, the line PQ tends to tangent at P to the level surface.

Therefore dr = dxi + dyj + dzk lies in the tangent plane to the surface at P.

From the differential calculus, we have

$$df = \frac{\partial f}{\partial x}dx + \frac{\partial f}{\partial y}dy + \frac{\partial f}{\partial z}dz$$

$$= \left(i\frac{\partial f}{\partial x} + j\frac{\partial f}{\partial y} + k\frac{\partial f}{\partial z}\right).(dxi + dyj + dzk) = \nabla f \,.\, dr.$$

Since f(x, y, z) = constant, therefore df = 0.

∴ ∇f . dr = 0 so that ∇f is a vector perpendicular to d r and therefore to the tangent plane at P to the surface f(x, y, z) = c.

Hence Ñf is a vector normal to the surface f(x, y, z) = c.

Thus if f(x, y, z) is a scalar field defined over a region R, then ∇f at any point (x, y, z) is a vector in the direction of normal at that point to the level surface f(x, y, z) = c passing through that point.

DIRECTIONAL DERIVATIVE OF A SCALAR POINT FUNCTION

Definition. *Let f(x, y, z) define a scalar field in a region R and let P be any point in this region. Suppose Q is a point in this region in the neighbourhood of P in the direction of a given unit vector* $\hat{a}$.

Then $$\lim_{Q \to P} \frac{f(Q) - f(P)}{PQ},$$

if it exists, is called the directional derivative of f at P in the direction of $\hat{a}$.

Interpretation of directional derivative. Let P be the point (x, y, z) and let Q be the point (x + δx, y + δy, z + δz). Suppose PQ = δs. Then ds is a small element at P in the directional $\hat{a}$.

If df = f(x + δx,y + δψ, ζ + δz) – f(x, y, z) = f(Q) – f(P),

Then $\frac{\delta f}{\delta s}$ represents the average rate of change of f per unit distance in the direction of $\hat{a}$ is

$$\lim_{Q \to P} \frac{f(Q) - f(P)}{PQ} = \lim_{\delta s \to 0} \frac{\delta f}{\delta s} = \frac{\delta f}{\delta s}.$$

It represents the rate of change of f with respect to distance s at the point P in the direction of unit vector $\vec{a}$.

Theorem 1:

The directional derivative of a scalar field f at a point P (x, y, z) in the direction of a unit vector $\vec{a}$ *is given by* $\frac{df}{ds} = \nabla f \cdot \vec{a}$.

Proof:

Let f(x, y, z) define a scalar field in the region R. Let r = xi + yj + zk denote the position vector of any point P (x, y, z) in this region. If s denotes the distance of P from some fixed point A in the direction of $\vec{a}$, then δs denotes small element at P in the direction of $\vec{a}$.

Therefore, $\frac{dr}{ds}$ is a unit vector at P in this direction *i.e.*, $\frac{dr}{ds} = \hat{a}$.

But r = xi + yj + zk.

$$\therefore \quad \frac{dr}{ds} = \frac{dx}{ds}i + \frac{dy}{ds}j + \frac{dz}{ds}k = \hat{a}.$$

$$\text{Now } \nabla f \cdot \hat{a} = \left(i\frac{\partial f}{\partial x} + j\frac{\partial f}{\partial y} + k\frac{\partial f}{\partial z} \right) \cdot \left(\frac{dx}{ds}i + \frac{dy}{ds}j + \frac{dz}{ds}k \right)$$

$$= \frac{\partial f}{dx}\frac{dx}{ds} + \frac{\partial f}{dy}\frac{dy}{ds} + \frac{\partial f}{\partial z}\frac{dz}{ds}$$

$$= \frac{df}{ds}$$ = directional derivative of f at P in the direction of $\hat{a}$.

Alternative Proof:

Let Q be a point in the neighbourhood of P in the direction of the given unit vector $\hat{a}$. If l, m, n are the direction cosines of the line PQ, then li + mj + nk = the unit vector in the direction of PQ = $\hat{a}$. Further, if PQ = δs, then the co-ordinates of Q are $(x + l\delta s, y + m\delta s, z + n\delta s)$. Now the directional derivative of f at P in the direction of $\hat{a}$ is

$$= \lim_{Q \to P} \frac{f(Q) - f(P)}{PQ}$$

$$= \lim_{\delta s \to 0} \frac{f(x + l\delta s, y + m\delta s, z + n\delta s) - f(x, y, z)}{\delta s}$$

on expanding by Taylor's theorem

$$= l\frac{\partial f}{\partial x} + m\frac{\partial f}{\partial y} + n\frac{\partial f}{\partial z}$$

$$= \left(\frac{\partial f}{\partial x}i + \frac{\partial f}{\partial y}j + \frac{\partial f}{\partial z}k\right).(li + mj + nk) = \nabla f\,.\,\hat{a}\,.$$

Theorem 2:

If $\hat{n}$ *be a unit vector normal to the level surface f(x, y, z) = c at a point P (x, y, z) and n be the distance of P from some fixed point A in the direction of* $\hat{n}$ *so that* δn *represents element of normal at P in the direction of* $\hat{n}$*, then*

$$grad\, f = \frac{df}{dn}\,\hat{n}\cdot$$

Proof:

We have

$$\text{grad } f = \nabla f = \frac{\partial f}{\partial x}i + \frac{\partial f}{\partial y}j + \frac{\partial f}{\partial z}k.$$

Also grad f is a vector normal to the surface f(x, y, z) = c. Since $\hat{n}$ is a unit vector normal to the surface f(x, y, z) = c, therefore let grad f = $A\hat{n}$, where A is some scalar to be determined.

Now $\frac{df}{dn}$ = directional derivative of f in the direction of $\hat{n}$

$= \nabla f \cdot \hat{n} = A\hat{n}\,.\,\hat{n}$ $\qquad [\Theta\ \nabla f = \text{grad } f = A\hat{n}]$

$$\therefore \quad \text{grad } f = \nabla f = \frac{df}{dn}\hat{n}.$$

Theorem 3:

Grad f is a vector in the direction of which the maximum value of the directional derivative of f i.e., $\frac{df}{ds}$ *occurs.*

Proof:

The directional derivative of f in the direction of $\hat{a}$ is given by

$$\frac{df}{ds} = \nabla f \cdot \hat{a} = \left(\frac{df}{dn}\hat{n}\right)\cdot \hat{a} \qquad \left[\because \nabla f = \frac{df}{dn}\hat{n}\right]$$

$$= \frac{df}{dn}(\hat{n}.\hat{n}) = \frac{df}{dn}\cos\theta, \text{ where } \theta \text{ is the angle between } \hat{a} \text{ and } \hat{n}.$$

Now $\frac{df}{dn}$ is fixed. Therefore $\frac{df}{dn}\cos\theta$ is maximum when $\cos\theta$ is maximum *i.e.,* when $\cos\theta = 1$. But $\cos\theta$ will be 1 when the angle between $\hat{a}$ and $\hat{n}$ is 0 *i.e.,* when $\hat{a}$ is along the unit normal vector $\hat{n}$.

Therefore the directional derivative is maximum along the normal to the surface. Its maximum value is

$$= \frac{df}{dn} = |\text{grad } f|.$$

TANGENT PLANE AND NORMAL TO A LEVEL SURFACE

To find the equations of the tangent plane and normal to the surface $f(x, y, z) = c$.

Let $f(x, y, z) = c$ be the equation of a level surface. Let $r = xi + yj + zk$ be the position vector of any point P (x, y, z) on this surface.

Then $\nabla f = \frac{\partial f}{\partial x}i + \frac{\partial f}{\partial y}j + \frac{\partial f}{\partial z}k$ is a vector along the normal to the surface at P *i.e.,* ∇f is perpendicular to the tangent plane at P.

Tangent plane at P. Let $R = Xi + Yj + Zk$ be the position vector of any current point Q (x, y, z) on the tangent plane at P to the surface. The vector

$$\vec{PQ} = R - r = (X - x)\, i + (Y - y)\, j + (Z - z)k$$

lies in the tangent plane at P. Therefore it is perpendicular to the vector ∇f.

$$\therefore \quad (R - r) . \nabla f = 0$$

$$\text{or } [(X - x)\, i + (Y - y)\, j + (Z - z)k] . \left(\frac{\partial f}{\partial x}i + \frac{\partial f}{\partial y}j + \frac{\partial f}{\partial z}k\right) = 0$$

$$\text{or} \quad (X - x)\frac{\partial f}{\partial x} + (Y - y)\frac{\partial f}{\partial y} + (Z - z)\frac{\partial f}{\partial z} = 0.$$

is the equation of the tangent plane at P.

Normal at P. Let $R = Xi + Yj + Zk$

be the position vector of any current point Q (X, Y, Z)) on the normal at P to the surface. The vector $\vec{PQ} = R - r = (X - x)i + (Y - y)j + (Z - z)$

k lies along the normal at P to the surface. Therefore it is parallel to the vector ∇f.

$$\therefore \quad (R - r) \times \nabla f = 0 \qquad \text{...(2)}$$

is the vector equation of the normal at P to the given surface.

Cartesian form. The vectors

$$(X - x)\, i + (Y - y)\, j + (Z - z)\, k$$

and $$\nabla f = \frac{\partial f}{\partial x} i + \frac{\partial f}{\partial y} j + \frac{\partial f}{\partial z} k$$

will be parallel if

$$(X - x)\, i + (Y - y)\, j + (Z - z)\, k = p\left(\frac{\partial f}{\partial x} i + \frac{\partial f}{\partial y} j + \frac{\partial f}{\partial z} k\right),$$

where p is some scalar.

Equating the coefficients of i, j, k, we get

$$X - x = p\,\frac{\partial f}{\partial x} \cdot Y - y$$

$$= p\,\frac{\partial f}{\partial y} \cdot Z - z$$

$$= p\,\frac{\partial f}{\partial z} \text{ or } \frac{X - x}{\frac{\partial f}{\partial x}}$$

$$= \frac{Y - y}{\frac{\partial f}{\partial y}} = \frac{Z - z}{\frac{\partial f}{\partial z}} \qquad \text{...(3)}$$

which are the equations of the normal at P.

APPLICATIONS OF DIVERGENCE AND CURL

(i) *Rate of flow of flux (or a fluid).* Let v be a vector point function (or the value of the vector field) representing the *velocity* of a fluid at any given instant t at a point P inside a small parallelopiped, *with edges* δx, δy, and δz *parallel to the three coordinate axes.*

Let $$v = v_x i + v_y j + v_z k.$$

Then, *velocity component along the x-axis at any point of the face* ABCD *normal to this axis*

$$= v_x - \frac{1}{2}\frac{\partial vx}{\partial x}\delta x,$$

neglecting the second and the higher powers of δx.

∴ *flux (or fluid) entering per second through face* ABCD *of the parallelopiped*

= (*component of velocity normal to the face*) (*area of the fact*)

$$= \left(Vx - \frac{1}{2}\frac{\partial V_x}{\partial x}\delta x \right)\delta y \delta z.$$

Similarly, *flux (or fluid) leaving per second through the opposite face* A' B' C' D' *of the parallelopiped*

$$= \left(xx + \frac{1}{2}\frac{\partial v_x}{\partial x}\delta x \right)\delta y \delta z.$$

Thus, *volume of the flux (or fluid) passing per second through faces* ABCD *and* A' B' C' D' *of the parallelopiped, i.e., along the direction of x*

$$= \left(vx + \frac{1}{2}\frac{\partial v_x}{\partial x}\delta x \right)\delta y \, \delta z - \left(v_z - \frac{1}{2}\frac{\partial v_z}{\partial x}\delta x \right)\delta y \, \delta z = \frac{\partial v_x}{\partial x}\partial x \, \partial y \, \partial z$$

Considering the other faces of the parallelopiped also, we have

total volume of flux (or fluid) moving out of the parallelopiped per second

$$= \left(\frac{\partial v_x}{\partial x} + \frac{\partial v_y}{\partial y} + \frac{\partial v_z}{\partial z} \right) \delta x \; \delta y \; \delta z$$

$$= (\nabla \, . \, v) \; \delta x \; \delta y \; \delta z$$

or $\quad$ (div v) $\delta x \; \delta y \; \delta z$.

The volume of the parallelopiped being $\delta x \; \delta y \; \delta z$, we find that in the limit when δx, δy, and δz tend to zero, *the volume of flux (or fluid) flowing past (or diverging from) point* (x, y, z) is ∇ v or div v.

This means, in other words, that *if* v *be the velocity of the fluid at a point, its rate of divergence per unit volume from that point is given by div* v or ∇ v.

A positive value of div v many be interpreted to mean either that the fluid is undergoing expansion (with its density falling proportionately) or that the point itself is the *source* of the fluid.

Similarly, a negative value of div v may be interpreted to mean the reverse, *viz.*, that either the fluid is undergoing contraction (with its density rising proportionately) or that the point serves as a sink (or a *negative source)* of the fluid.

If, however, div v = 0, the fluid entering and leaving the element (or the parallelopiped) is the same. *i.e.*, there is no change in the density of the fluid indicating that it is *incompressible*. In this case, therefore, if ρ be the density of the fluid, we have *mass of the fluid per unit volume that flows through a point* (x, y, z) = ρ div v.

(ii) *Solenoidal vector point function, A vector point function* F *is said to be solenoidal is a region if its flux across any closed surface in that region be zero,* which will happen only when div F is equal to zero.

This means that *for a vector point function to be solenoidal, either the lines of flow of its flux should form closed curves (like, for example, the lines of force in the magnetic field of an electric current) or extend to infinity*

It can be shown that every solenoidal vector function is the curl of some function. Further, as shall see in the next article, the divergence of every curl is zero. It follows, therefore, that *the curl of every function is necessarily solenoidal.*

DIVERGENCE AND CURL

A scalar point function φ we can obtain a vector point function grad φ. Similarly, from a vector point function f we can obtain two more point function: a scalar one, called *divergence* and a vector one, called *curl* (or *rotation*).

Thus, if f represents a vector field, *i.e.*, if f be a *continuously differentiable vector point function*, the function

$$i.\frac{\partial f}{\partial x}+j.\frac{\partial f}{\partial y}+k.\frac{\partial f}{\partial z},$$

which is a *scalar,* is called *divergence of* f and is written as div f. And the function

$$i\times\frac{\partial f}{\partial x}+j\times\frac{\partial f}{\partial y}+k\times\frac{\partial f}{\partial z},$$

which is a *vector*, is called *curl of* f and is written as *curl* f.

It can easily be shown that both *div* f and *curl* f are quite independent of the orthogonal unit triad (*i.e.*, unit vectors) that may be chosen. They are thus *essentially point functions.*

Now, clearly, $\text{div } f = i.\frac{\partial f}{\partial x}+j.\frac{\partial f}{\partial y}+k.\frac{\partial f}{\partial z},$

$$=\left(i\frac{\partial}{\partial x}+j\frac{\partial}{\partial y}+k\frac{\partial}{\partial z}\right).f = \nabla f.$$

So that, *divergence of a vector point function* f *is the scalar product of the Del operator with* f. *And,*

$$\text{Curl } f = i\times\frac{\partial f}{\partial x} + j\times\frac{\partial f}{\partial y} + k\times\frac{\partial f}{\partial z}$$

$$= \left(i\frac{\partial}{\partial x} + j\frac{\partial}{\partial y} + k\frac{\partial}{\partial z}\right)\times f = \nabla\times f,$$

i.e., curl of a vector point function f *is the vector product of the Del operator with* f. The term $\nabla\times$, [read as *Del cross* (*or Nabla cross*)], is therefore also used in place of the term '*curl*'.

Both divergence and curl of f may be expressed in terms of the components of vector f, as shown below:

if f = i fx + j fy + k fz, we have

$$\text{div } f = \sum i.\frac{\partial f}{\partial x} = \sum i\left[i\frac{\partial fx}{\partial x} + j\frac{\partial fy}{\partial x} + k\frac{\partial fz}{\partial x}\right]$$

$$= \frac{\partial fx}{\partial x} + \frac{\partial fy}{\partial y} + \frac{\partial fz}{\partial z}.$$

$$\text{And curl } f = \sum i\times\frac{\partial f}{\partial x}\sum i\times\left[i\frac{\partial fx}{\partial x} + j\frac{\partial fy}{\partial y} + k\frac{\partial fz}{\partial z}\right]$$

$$= i\left(\frac{\partial fz}{\partial y} - \frac{\partial fy}{\partial z}\right) + j\left(\frac{\partial fx}{\partial z} - \frac{\partial fz}{\partial x}\right) + k\left(\frac{\partial fy}{\partial x} - \frac{\partial fx}{\partial y}\right)$$

$$= \begin{vmatrix} i & j & k \\ \frac{\partial}{\partial x} & \frac{\partial}{\partial y} & \frac{\partial}{\partial z} \\ f_x & f_y & f_z \end{vmatrix}$$

SOLVED PROBLEMS

1. *Evaluate* $\int_s$ *n . curl F dS, where F = yi + zj + xk and S is the portion of the surface* $x^2 + y^2 - 2ax + az = 0$ *above the plane z = 0 and verify Stoke's theorm.*

Solution : The equation of the given surface can be written as

$$(x - a)^2 + y^2 = -a\ (z = a)$$

which is a paraboloid with its vertex at (a, 0, 0). It meets the plane z = 0 in the circle $(x - a)^2 + y^2 = a^2$, z = 0

Now $\int_C$ F . dr = $\int_C$ (yi + zj + xk) . (idx + jdy + kdz)

$$= \int_C (ydx + zdy + xdz)$$

$= \int_C y\,dx$, since on C, z = 0, dz = 0

$= \int_{\theta=0}^{2\pi} a \sin\theta\,(-a \sin\theta)\,d\theta$, putting $x = a + a\cos\theta$,

$y = a \sin\theta$

$$= \frac{1}{2} a^2 \int_{\theta=0}^{2\pi} (\cos 2\theta - 1)\,d\theta = \frac{1}{2} a^2 \left[\frac{1}{2}\sin 2\theta - \theta\right]_0^{2\pi}$$

$$= -a^2\pi \qquad \ldots \text{(i)}$$

$\therefore$ The given integral $= -a^2\pi$

$$\text{Again curl } F = \begin{vmatrix} i & j & k \\ \dfrac{\partial}{\partial x} & \dfrac{\partial}{\partial y} & \dfrac{\partial}{\partial z} \\ y & z & x \end{vmatrix} = -i - j - k$$

Also C is the boundary for the region S as well as the plane region S_1. $\therefore n = k$

$\therefore$ $n \cdot \text{curl } F = k \cdot (-i - j - k) = -1$

$\therefore \int_S n \cdot \text{curl } F\,dS = \int_{S1} n \cdot \text{curl } F\,dS$, where $dS = r\,d\theta\,dr$

$$= \int_{\theta=0}^{2\pi} \int_{r=0}^{a} (-1)\,r\,d\theta\,dr = -\left[\frac{1}{2}r^2\right]_0^a (\theta)_0^{2\pi} = -\pi a^2 \qquad \ldots \text{(ii)}$$

Hence from (i) and (ii) Stoke's theorem is verified.

2. *Use Stoke's theorem to evaluate $\int$ curl F . n dS over the open hemi-spehricl surface $x^2 + y^2 + z^2 = a^2$, $z > 0$ where $F = yi + zxj + yk$.*

Solution : Here by Stoke's theorem if S be the given hemispherical surface then $\int_S (\text{curl } F) \cdot dS = \int_C F \cdot dr$, where C is the circle $x^2 + y^2 = a^2$, $z = 0$ (Note)

$\therefore \int_S (\text{curl } F) \cdot n\,dS$

$$= \int_C (yi + zxj + yk) \cdot (idx + jdy + kdz)$$

$$= \int_C (ydx + zxdy + ydz) \qquad \ldots \text{(i)}$$

The parametric equations of circle C are

$x = a\cos\theta$, $y = a\sin\theta$, $z = 0$, where θ varies from 0 to 2π

Also $z = 0 \Rightarrow dz = 0$

$\therefore$ From (i) we have

$$\int_C F \cdot dr = \int_C y\,dx = \int_{\theta=0}^{2\pi} (a\sin\theta)(-a\sin\theta)\,d\theta$$

$$= -a^2 \int_0^{2\pi} \sin^2\theta\,d\theta = -4a^2 \int_0^{\pi/2} \sin^2\theta\,d\theta$$

$$= -4a^2\left[\frac{1}{2}\cdot\frac{\pi}{2}\right] = -\pi a^2$$ **Ans.**

3. *Prove that curl grad $\phi = 0$ by Stoke's theorem.*

Solution : We know that

$$\int_S n \cdot \text{curl } F \, dS = \int_C F \cdot dr, \quad \ldots \text{(i)}$$

where S is any surface enclosed by a simple closed curve C.

Let F = grad ϕ

Then F . dr = (grad ϕ) . dr

or $F \cdot dr = \left(i\frac{\partial\phi}{\partial x} + j\frac{\partial\phi}{\partial y} + k\frac{\partial\phi}{\partial z}\right) \cdot (i\,dx + j\,dy\ k\,dz)$

$$= \frac{\partial\phi}{\partial x}dx + \frac{\partial\phi}{\partial y}dy + \frac{\partial\phi}{\partial z}dz = d\phi$$

$\therefore \int_C F \cdot dr = \int_C d\phi = 0,$ [$\because$ C is any closed curve]

$\therefore$ From (i) we have

$\int_S n \cdot (\text{curl grad } \phi)\, dS = 0$, for all values of S

Hence curl grad $\phi = 0$

4. *Prove that $\int_c \phi\, dr = \int_s n \times \nabla\phi\, a\, S = \int_s dS \times \nabla\phi$*

Solution : We know $\int_C F \cdot dr = \int_S n \cdot \text{curl } F\, dS$

Put F = ϕ a, where ϕ is a continuously differentiable scalar point function and a is an arbitrary constant vector.

$\therefore \quad \int_C \phi\, a \cdot dr = \int_S n \cdot \text{curl } (\phi\, a)\, dS$

or $\quad \int_C a \cdot \phi\, dr = \int_S n \cdot [\phi \text{ curl } a + \nabla\phi \times a]\, dS$, [$\because$ curl (ϕ a) = ϕ curl a + $\nabla\phi \times$ a]

$= \int_S n \cdot [\nabla\phi \times a]\, dS,$ [$\because$ curl a = 0, if a is constant vector]

$= \int_S (n \times \nabla\phi) \cdot a\, dS,$ [$\because$ a × b . c = a . b. × c]

or $a \cdot \int_C \phi\, dr = a \cdot \int_S n \times \nabla\phi\, dS,$ [$\because$ a is a constant vector]

or $a \cdot \left[\int_C \phi\, dr - \int_S n \times \nabla\phi\, dS\right] = 0$

or $\int_C \phi\, dr - \int_S n \times \nabla\phi\, dS = 0,$ [$\because$ a is arbitrary]

or $\int_C \phi\, dr = \int_S n \times \nabla\phi\, dS$ Hence proved.

Also we know that n dS = dS

$\therefore \int_C \phi\, dr = \int_S n \times \nabla\phi\, dS = \int_S dS \times \nabla\phi$ Hence proved.

5. *Prove that curl [r × (a × r)] = 3r × a, where 'a' is a constant vector.*

Solution : Curl [r × [a × r)]

$= \nabla \times [(r \,.\, r)\, a\, (r \,.\, a)\, r],\quad [\because a \times (b \times c) = (a \,.\, c)\, b - (a \,.\, b)\, c]$

$= \nabla \times [r^2\, a - (r \,.\, a)\, r],\quad [\because r \,.\, r = r^2 = r^2]$

$= \nabla \times [r^2\, a) - \nabla \times \{(r \,.\, a)\, r\},\ [\because \nabla \times (A + B) = \nabla \times A + \nabla \times B]$

$= r^2\, (\nabla \times a) + (\nabla r^2) \times a - (r \,.\, a)\, (\nabla \times r) - \{\nabla\, (r \,.\, a)\} \times r,$

$[\because \nabla \times (\phi\, V) = \phi\, (\nabla \times V) + \nabla\phi \times V]$

$= r^2\, (0) + (2r) \times a - (r \,.\, a)\, (0) - \{\nabla\, (r \,.\, a)\} \times r,$

$[\because \nabla \times a = 0$, a being constant] and $\nabla \times r = 0$

$$= 2\,(r \times a) - \left\{\Sigma i \frac{\partial}{\partial x} (r \,.\, a)\right\} \times r$$

$$= 2\,(r \times a) - \left\{\Sigma i \left(\frac{\partial}{\partial x} . \, a\right)\right\} \times r,$$

$= 2\,(r \times a) - \{\Sigma\, i\, (i \,.\, a)\} \times r,\quad [\because r = xi + yj + zk]$

$= 2\,(r \times a) - \{\Sigma\, i\, (a_4)\} \times r$, if $a = a_1 i + a_2 j + a_3 k$

$= 2\,(r \times a) - a \times r,\quad [\because a = \Sigma\, a_1 i]$

$= 2\,(r \times a) + (r \times a),\quad [\because a \times -r \times a]$

$= 3\,(r \times a).$ Hence proved.

6. *If* $f = (3r^2 - 4r^{1/2} + 6r^{-1/3})$, show that $\nabla\phi = 2\,(3 - r^{-3/2} - r^{-7/3})\, r$.

Solution: We have $r = xi + yj + zk$

and $r = |r| = \sqrt{(x^2 + y^2 + z^2)}$

so that $r^2 = x^2 + y^2 + z^2.$

Now ϕ is a function of r.

$$\therefore\quad \nabla\phi = \frac{d\phi}{dr} \nabla r \qquad [\because \nabla f(r) = f'(r)\, \nabla r]$$

$$= \left[6r - 4.\left(\frac{1}{2}\right) r^{-1/2} + 6.\left(-\frac{1}{3}\right) r^{-4/3}\right] \nabla r$$

$$= (6r - 2r^{-1/2} - 2r^{-4/3})\, \frac{1}{r} r \qquad \left[\because \nabla r = \frac{1}{r} r\right]$$

$$= (6 - 2r^{-3/2} - 2r^{-7/3})\, r$$

$$= 2\,(3 - r^{-3/2} - r^{-7/3})\, r.$$

7. *Prove that* $F\, (2xy + z^3)\, i + x^2\, j + 3xz^2\, k$ *is a conservative force field. Also find the scalar potential.*

Solution : We know that it F is a conservative field then curl F = 0.

Now here curl F = curl $[(2xy + z^2)\, i + x^2\, j + 3\, xz^2\, k]$

$$= \begin{vmatrix} i & j & k \\ \frac{\partial}{\partial x} & \frac{\partial}{\partial y} & \frac{\partial}{\partial z} \\ 2xy + z^3 & x^2 & 3xz^2 \end{vmatrix}$$

$$= \left[\frac{\partial}{\partial y}(3xz^2) = \frac{\partial}{\partial z}(x^2)\right] + j\left[\frac{\partial}{\partial z}(2xy + z^2) - \frac{\partial}{\partial x}(3xz^2)\right]$$

$$+ k\left[\frac{\partial}{\partial x}(x^2) - \frac{\partial}{\partial y}(2xy + z^2)\right]$$

$$= i\,[0] + j\,[3z^2 - 3z^2] + k\,[2x - 2x] = 0$$

Hence F is a conservative force field.

Again if ϕ be the required scalar potential, then we know that

$$F = \text{grad}\ \phi = i\frac{\partial \phi}{\partial x} + j\frac{\partial \phi}{\partial y} + k\frac{\partial \phi}{\partial z}$$

Now $F\ .\ d\,r = \left(i\frac{\partial \phi}{\partial x} + j\frac{\partial \phi}{\partial y} + k\frac{\partial \phi}{\partial z}\right).(i\,dx + j\,dy + k\,dz)$

$$= \frac{\partial \phi}{\partial x}dx + \frac{\partial \phi}{\partial y}dy + \frac{\partial \phi}{\partial z}dz = d\phi$$

$\therefore\ d\phi = F\ .\ dr = [2xy + z^3)\, i + x^2 j + 3xz^2\, k]\ .\ (i\, dx + j\, dy + k\, dz)$

$= (2xy\, dx + x^3\, dy) + (z^3\, dx + 3\, xz^2\, dz)$

$= d\,(x^2 y) + d\,(z^3 x) = d\,(x^3 y + z^3\, x)$

or $\quad \phi = x^2\, y + z^3 x$ + constant. **Ans.**

8. *Show that* $\Delta\,[r\ a\ b] = a \times b$.

Solution : As in first of all prove that

$$\nabla\,(r\ .\ a) = a \qquad \ldots \text{(i)}$$

Then $\quad \nabla\,[r\ a\ b] = \nabla\,[r\ .\ (a \times b)]$

$= a \times b$, from (i) above. Hence proved.

9. *Show that the curl of the velocity of any particle of a rigid body is equal to twice the angular velocity of the body.*

Solution: We have the relation $v = \omega \times r$, where v is the velocity of any particle of the body, with r, the *position vector* relative to a fixed point and ω, the *angular velocity* of the body.

Obviously, v is a *vector point function* and vector ω is the same for all the particles.

Now,

$$\text{curl } v = \text{curl}(\omega \times r) = \sum i \frac{\partial}{\partial x}(\omega \times r) = \sum i\left(\omega \times \frac{\partial r}{\partial x}\right) = \sum i \times (\omega \times i)$$

$$= \sum (i.i)\,\omega - \sum (i.\omega) i = 3\omega - \omega = 2\omega.$$

10. *Obtain div grad r^m. Also show that curl grad $r^m = 0$.*

Solution:

We have

grad $r^m = \nabla r^m = mr^{m-2}\, r = mr^{m-2}\,(ix + jy + kz)$

$\therefore$ div grad r^m

$$\text{or} \quad \text{div}\Delta r^m = \frac{\partial}{\partial x}\left(mr^{m-2}x\right) + \frac{\partial}{\partial y}\left(mr^{m-2}y\right) + \frac{\partial}{\partial z}\left(mr^{m-2}z\right).$$

$$\text{Now, } \frac{\partial}{\partial x}\left(mr^{m-2}x\right) = m(m-2)r^{m-3}\frac{\partial r}{\partial x} + mr^{m-2}$$

$$= m(m-2)r^{m-1}x^2 + mr^{m-2}, \text{ because } \partial r/\partial x = x/r.$$

$$\text{Similarly,} \quad \frac{\partial}{\partial y}\left(mr^{m-2}\, y\right) = m(m-2)\, r^{m-1}\, y^2 + mr.^{m-2}$$

$$\text{and} \quad \frac{\partial}{\partial y}\left(mr^{m-2}\, z\right) = m(m-2)\, r^{m-1}\, z^2 + mr.^{m-2}$$

$$\therefore \text{div grad } r^m = \text{div}\nabla\, r^m = m(m-2)\, r^{m-1}\, r^2 + 3\, mr^{m-2}$$

$$= m\,(m+1)r^{m-2}$$

$$\text{And,} \qquad \text{curl grad } r^m = \begin{vmatrix} i & j & k \\ \frac{\partial}{\partial x} & \frac{\partial}{\partial y} & \frac{\partial}{\partial z} \\ mr^{m-2}x & mr^{m-2}y & mr^{m-2}z \end{vmatrix}$$

$$\text{So that coefficient of i in curl grad } r^m = \frac{\partial}{\partial y} mr^{m-2}z - \frac{\partial}{\partial z} mr^{m-2}\, y$$

$$= m(m-2)r^{m-3}\frac{\partial r}{\partial y}\, z - m(m-2)r^{m-3}\frac{\partial r}{\partial z}\, y$$

$$= m(m-2)r^{m-3}\frac{yz}{r} - m(m-2)r^{m-3}\frac{zy}{r} = 0.$$

Hence *curl grad* $r^m = \nabla \times \nabla r^m = 0$.

11. *Show that* $\int_S F \times n dS = -\int_V \text{curl } F dV$.

Solution: Let us put f = a × F, where f is a vector function and a, an *arbitrary constant vector*. Then, applying Gauss's theorem to f, we have

$$\int_S a \times F.n dS = \int_V \text{div}(a \times F) dV. \quad ...(i)$$

Now, $$a \times F.\ n = a.\ F \times n \quad ...(ii)$$

and div (a × F) = ∇. (a × F) = – a.∇ × F. ...(iii)

So that, using relations (ii) and (iii), relation (i) may be put as

$\int_S a.F \times n dS = -\int_V a.\nabla + F dV$. Or, $a\int_S F \times n dS = -a\int_V \nabla \times dFV$.

Or, $$a\left[\int_S F \times n dS + \int_V \nabla \times F dV\right] = 0.$$

Since a is just an arbitrary vector, we have

$$\int_S F \times n dS + \int_V \nabla \times F dV = 0.$$

Or, $$\int_S F \times n dS = -\int_V \nabla \times F dV = -\int_V \text{curl } F dV.$$

12. *Show that div* $(r^n r) = (n + 3) r^n$, *where* $r = | r |$ *and* $r = x i + y i + z k$.

Solution : We know $r = x^2 + y^2 + z^2$

$$\therefore \quad 2r\frac{\partial r}{\partial x} = 2x \text{ or } \frac{\partial r}{\partial x} = \frac{x}{r} \quad ...(i)$$

Similarly $$\frac{\partial r}{\partial y} = \frac{y}{r} \text{ and } \frac{\partial r}{\partial z} = \frac{z}{r}$$

$$\text{dir } (r^n r) = \text{div } [r^n (xi + yi + zk)]$$

$$= \frac{\partial}{\partial x}(r^n x) + \frac{\partial}{\partial y} + \frac{\partial}{\partial z}(r^n z) \quad ...(ii)$$

Now $$\frac{\partial}{\partial x}(r^n x) = r^n + x n r^{n-1}\frac{\partial r}{\partial x} = r^n + n x r^{n-1}\left(\frac{x}{r}\right), \text{ from (i)}$$

$$= r^n + n x^2 r^{n-2}$$

Similarly, $$\frac{\partial}{\partial y}(r^n y) = r^n + x y^2 r^{n-2}; \frac{\partial}{\partial z}(r^n z) = r^n + n z^2 r^{n-2}$$

Substituting these values in (ii) we get

$$\text{div}\,(r^n\, r) = (r^n + n\,x^2\,r^{n-2}) + (r^n + n\,y^2\,r^{n-2}) + (r^n + n\,z^2\,r^{n-2})$$

$$= 3r^n + nr^{n-2}\,(x^2 + y^2 + z^2) = 3r^n + nr^{n-2}\,(r^2)$$

$$= (3 + n)\,r^n$$

If r^n r is solenoidal then fiv (r^n r) = 0 and hence n + 3 = 0, as $r \neq 0$ Hence proved.

13. *Evaluate div (r × a), where 'a' is a constant vector.*

Solution : $\text{div}\,(r \times a) = \left(i\frac{\partial}{\partial x} + j\frac{\partial}{\partial y} + k\frac{\partial}{\partial z}\right) . (r \times a)$

or $\text{div}\,(a \times a) = i\,.\,\frac{\partial}{\partial x}\,(r \times a) + j\,.\,\frac{\partial}{\partial y}\,(r \times a) = k\,.\,\frac{\partial}{\partial y}\,(r \times a)$

$$= \Sigma\, i\,\frac{\partial}{\partial x}\,(r \times a)$$

$$= \Sigma\, i\,.\left(\frac{\partial r}{\partial x} \times a + r \times \frac{\partial a}{\partial x}\right)$$

$$= \Sigma\, i\,.\left(\frac{\partial r}{\partial x} \times a\right), \quad \left[\because \frac{\partial a}{\partial x} = 0, \text{ a being constant}\right]$$

$$= \Sigma\, i\,.\,(i \times a); \qquad [\because r = xi + yj + zk]$$

$$= \Sigma[i\ i\ a] = 0, \text{ as } [a\ b\ c] = 0 \text{ if } a = b$$

∴ div (r × a) = 0 **Ans.**

14. *Prove that vector* $V = 3y^4z^2i + 4x^3z^2j - 3x^2y^2k$ *is solenoidal.*

Solution : $\text{div}\,V = \frac{\partial}{\partial x}\,(3y^4z^2) + \frac{\partial}{\partial y}\,(4x^3z^2) + \frac{\partial}{\partial z}\,(-3x^2y^2)$

$$= 0 + 0 + 0 = 0$$

Hence vector V is solenoidal

15. *Show that the vector* $a = (x + 3y)\,i + (y - 3z)\,j + (x - 2z)\,k$ *is solenoidal and find (a Δ) a.*

Solution : Do firs part as above

Again a . ∇ = [(x = 3y) i + (y – 3z) j + (x – 2z) k]

$$\left[i\frac{\partial}{\partial x} + j\frac{\partial}{\partial y} + k\frac{\partial}{\partial z}\right]$$

$$= (x + 3y) \frac{\partial}{\partial x} + (y - 3z) \frac{\partial}{\partial y} + (x - 2z) \frac{\partial}{\partial z}$$

$$\therefore \quad (a . \nabla) a = \left[(x + 3y) \frac{\partial}{\partial x} + (y - 3z) \frac{\partial}{\partial y} + (x - 2z) \frac{\partial}{\partial z} \right] a$$

$$= (x + 3y) \frac{\partial}{\partial x} [(x + 3y) i + (y = 3z) j + (x - 2z) k]$$

$$+ (y - 3z) \frac{\partial}{\partial y} [(x + 3y) i + (y - 3z) j + (x - 2z) k]$$

$$+ (x - 2z) \frac{\partial}{\partial z} [(x + 3y) i + (y - 3z) j + (x - 2z) k]$$

$$= (x + 3y) [i + k] + (y - 3z) [3i + j] + (x - 2z) (-3j - 2k)$$

$$= [(x + 3y) + 3 (y - 3z)] i + [(y - 3z) - 3 (x - 2z)] j + [(x + 3y) - 2 (x - 2z)] k$$

$$= (x + 6y = 9z) i + (-3x + y + 3z) j + (-x + 3y + 4z) k$$ **Ans.**

16. *Show that div* $\left[\frac{f(r)\, r}{r}\right] = \frac{1}{r^2} \frac{d}{dr}(r^2 f)$, *where r and r have their usual meanings.*

Solution : $f(r)\, r = f(r) (xi + yj + zk)$

$$\therefore \text{ div} \left[\frac{f(r)\, r}{r}\right] = \frac{\partial}{\partial x}\left\{\frac{f(r)x}{r}\right\} + \frac{\partial}{\partial y}\left\{\frac{f(r)y}{r}\right\} + \frac{\partial}{\partial z}\left\{\frac{f(r)z}{r}\right\} \quad \text{...(i)}$$

Now $$\frac{\partial}{\partial x}\left[\frac{f(r)x}{r}\right] = x \frac{\partial}{\partial x}\left\{\frac{f(r)}{r}\right\} + \frac{f(r)}{r}\frac{\partial}{\partial x}(x)$$

or $$\frac{\partial}{\partial x}\left[\frac{f(r)x}{r}\right] = x \left\{\frac{1}{r} f'(r) \frac{\partial r}{\partial x} - \frac{1}{r^2} f(r) \frac{\partial r}{\partial x}\right\} + \frac{f(r)}{r}$$

$$= \left\{\frac{x}{r} f'(r) - \frac{x}{r^2} f(r)\right\} \frac{\partial r}{\partial x} + \frac{f(r)}{r}$$

$$= \left\{\frac{x}{r} f'(r) - \frac{x}{r^2} f(r)\right\} \frac{x}{r} + \frac{f(r)}{r}, \quad \left[\because \frac{\partial r}{\partial x} = \frac{x}{r}\right]$$

$$= \frac{x^2}{r^2} f'(r) = \frac{x^2}{r^3} f(r) + \frac{f(r)}{r}$$

Similarly we can get

$$\frac{\partial}{\partial y}\left[\frac{f(r)y}{r}\right] = \frac{y^2}{r^2} f'(r) - \frac{y^2}{r^3} f(r) + \frac{f(r)}{r}$$

and
$$\frac{\partial}{\partial z}\left[\frac{f(r)z}{r}\right] = \frac{z^2}{r^2} f'(r) - \frac{z^2}{r^3} f(r) + \frac{f(r)}{r}$$

Substituting these values in (i) we get

$$\text{div}\left[\frac{f(r)z}{r}\right] = \frac{x^2+y^2+z^2}{r^2} f'(r) - \frac{x^2+y^2+z^2}{r^3} f(r) + \frac{3f(r)}{r}$$

$$= f'(r) - \frac{1}{r} f(r) + \frac{3}{r} f(r), \quad [\because x^2+y^2+z^2 = r^2]$$

$$= f'(r) + \frac{2}{r} f(r), \text{ where } f'(r) = \frac{d}{dr} f(r)$$

$$= \frac{1}{r^2}\left[r^2 f'(r) + 2rf(r)\right] = \frac{1}{r^2}\left[r^2 \frac{d}{dr} f(r) + 2rf(r)\right]$$

$$= \frac{1}{r^2}\frac{d}{dr}\left[r^2 f(r)\right]$$

17. *Prove that* $\nabla . (A \nabla B - B \nabla A) = A \nabla^2 B - B \nabla^2 A$.

Solution : As in we can prove that

$$\nabla . (\phi V) = (\nabla\phi) . V + \phi (\nabla . V)$$

Put $\phi = A$ and $V = \nabla B$

Then $\nabla . (A \nabla B) = (\nabla A) . (\nabla B) + A (\nabla . \nabla B)$

or $\nabla . (A \nabla B) = (\nabla A) . (\nabla B) + A \nabla^2 B$ . . . (i)

Interchanging A and B we get

$\nabla . (B \nabla A) = (\nabla B) . (\nabla A) + B \nabla^2 A$ . . .(ii)

Subtracting (ii) from (i) we get

$\nabla . (A \nabla B) - \nabla . (B \nabla A) = [(\nabla A . (\nabla B) + A\nabla^2 B] - [(\nabla B . (\nabla A) + B \nabla^2 A]$

or $\nabla . (A \nabla B - B \nabla A) = A \nabla^2 B . (\nabla A) + B \nabla^2 A$;

$[\because a . b = b . a]$

18. *Evaluate* $\nabla . (a \times r) r^n$, *where 'a' is a constant vector.*

Solution : $\nabla . (a \times r) r^n = \Sigma i . \frac{\partial}{\partial x}\{(a \times r) r^n\}$, by definition.

$$= \Sigma\, i \,.\left[\left\{\frac{\partial}{\partial x}(a \times r)\right\} r^n + (a \times r)\frac{\partial}{\partial x}(r^n)\right]$$

$$= \Sigma\, i \,.\left[\left(a \times \frac{\partial a}{\partial x}\right) r^n + (a \times r)\, n r^{n-1} \frac{\partial r}{\partial x}\right], \qquad \left[\because \frac{\partial a}{\partial x} = 0\right]$$

$$= \Sigma\, i \,.\, [(a \times i)\, r^n + (a \times r)\, n r^{n-1} (x/r)], \qquad \left[\because \frac{\partial r}{\partial x} = i, \frac{\partial r}{\partial x} = \frac{x}{r}\right]$$

$$= \Sigma\, i \,.\, [(a \times i)\, r^n + n\, x\, r^{n-2} (x \times r)]$$

$$= \Sigma\, [(i \,.\, a \times i)\, r^n + n\, r^{n-2}\, x\, i \,.\, (a \times r)]$$

$$= \Sigma\, n\, r^{n-2} (x\, i) \,.\, (a \times r), \qquad [\because (i \,.\, a \times i) = [i\ a\ i] = 0]$$

$$= n r^{n-2} (a \times r) \,.\, (\Sigma\, x\, i)$$

$$= n r^{n-2} (a \times r) \,.\, r = n r^{n-2} [a\ r\ r]$$

$$= 0, \text{ since } [a\ b\ c] = 0, \text{ if } b = c$$

19. *Verify divergence theorm for $F = (x^2 - yz)\, i + (y^2 - z\,x)\, j + (z^2 - xy)\, k$ taken over the rectangular parallelopiped $0 \leq x \leq a$, $0 \leq y \leq b$, $0 \leq z \leq c$.*

Solution : (consider it as rectangular parallelopiped whose edges are OB = a, OC = b and OF = c).

Now $\qquad F = (x^2 - yz)\, i + (y^2 - zx)\, j + (z^2 - xy)\, k$

$$\therefore \quad F = \frac{\partial}{\partial x}(x^2 - yz) + \frac{\partial}{\partial y}(y^2 - zx) + \frac{\partial}{\partial z}(z^2 - xy)$$

$$= 2x + 2y + 2z = 2\,(x + y + z)$$

$\therefore \int_V \text{div } F\, dV$, where V is the volume of the rectangular parallelopiped.

$$= \int_{x=0}^{a} \int_{y=0}^{b} \int_{z=0}^{c} 2(x + y + z)\, dx\, dy\, dz$$

$$= 2 \int_{x=0}^{a} \int_{y=0}^{b} \left[xz + yz + \frac{1}{2} z^2\right]_{z=0}^{c} dx\, dy$$

$$= 2 \int_{x=0}^{a} \int_{y=0}^{b} \left(xc + yc + \frac{1}{2} c^2\right) dx\, dy$$

$$= 2 \int_{0}^{a} \left(cxy + \frac{1}{2} cy^2 + \frac{1}{2} c^2 y\right)_{y=0}^{b} dx$$

$$= 2 \int_0^a \left(cbx + \frac{1}{2} cb^2 + \frac{1}{2} c^2 b \right) dx$$

$$= 2 \left[\frac{1}{2} cbx^2 + \frac{1}{2} cb^2 x + \frac{1}{2} c^2 bx \right]_0^a$$

$$= cba^2 + cb^2 a + c^2 ba = abc\,[a + b + c] \quad \ldots \text{(i)}$$

Now, we shall calculate $\int F \cdot n\, dS$ over the six faces of the rectangular parallelopiped separately and then add them to get the value of $\int_S F \cdot n\, dS$, where S is the entire surface of the rectangular parallelopiped.

For the face ABED, we have n = 1 x = a, dS = dy dz

$\therefore \int F \cdot n\, dS$ over the face ABED

$$= \int_{z=0}^{c} \int_{y=0}^{b} [(x^2 - yz)\, i + (y^2 - zx)\, j + (z^2 - xy)\, k] \cdot i\, dy\, dz$$

$$= \int_{z=0}^{c} \int_{y=0}^{b} (x^2 - yz)\, dy\, dz, \text{ over the face ABED for which } x = a$$

$$= \int_{z=0}^{c} \int_{y=0}^{b} (a^2 - yz)\, dy\, dz = \int_{z=0}^{c} \left(a^2 y - \frac{1}{2} y^2 z \right)_{y=0}^{b} dz$$

$$= \int_0^c \left(a^2 b - \frac{1}{2} b^2 z \right) dz = \left(a^2 bz - \frac{1}{4} b^2 z^2 \right)_0^a = a^2 bc - \frac{1}{4} b^2 c^2$$

$$= \frac{1}{4} bc\,(4a^2 - bc) \quad \ldots \text{(ii)}$$

For the face OCKF, we have n = – i, x = 0, dS = dy dz

$\therefore \int F \cdot n\, dS$ over the face OCKF

$$= \int_{z=0}^{c} \int_{y=0}^{b} [(x^2 - yz)\, i + (y^2 - zx)\, j + (z^2 - xy)\, k] \cdot (-i)\, dy\, dz$$

$$= \int_{z=0}^{c} \int_{y=0}^{b} (x^2 - yz)\, dy\, dz, \text{ for } x = 0$$

$$= \int_{z=0}^{c} \int_{y=0}^{b} yz\, dy\, dz = \left(\frac{1}{2} y^2 \right)_0^b \left(\frac{1}{2} z^2 \right)_0^a = \frac{1}{4} b^2 c^2 \quad \ldots \text{(iii)}$$

For the face ABOC, we have n = – k, dS = dx dy and z = 0

$\therefore \int F \cdot dS$ over the face ABOC

$$= \int_{x=0}^{a} \int_{y=0}^{b} [(x^2 - yz)\, i + (y^2 - zx)\, j + (z^2 - xy)\, k] \cdot (-k)\, dx\, dy$$

$$= - \int_{x=0}^{a} \int_{y=0}^{b} (z^2 - xy)\, dx\, dy, \text{ for } z = 0$$

$$= \int_{x=0}^{a} \int_{y=0}^{b} xy\, dx\, dy = \left(\frac{1}{2}x^2\right)_0^a \left(\frac{1}{2}y^2\right)_0^b = \frac{1}{4}a^2 b^2 \qquad \text{. . . (iv)}$$

For the face DEFK, we have n = k, dS = dx dy and z = c

$\therefore \int$ F . n dS over the face DEFK

$$= \int_{x=0}^{a} \int_{y=0}^{b} (z^2 - xy)\, dx\, dy, \text{ for } z = c, \text{ as above}$$

$$= \int_{x=0}^{a} \int_{y=0}^{b} (c^2 - xy)\, dx\, dy = \frac{1}{4} ab\,(4c^2 - ab) \qquad \text{. . . (v)}$$

For the face OADK we have n = j, dS = dx dz and y = b

$\therefore \int$ F . n dS over the face OADK

$$= \int_{x=0}^{a} \int_{z=0}^{c} (b^2 - xy)\, dx\, dz, \text{ proceeding as for face ABED}$$

$$= \frac{1}{4} ac\,(4b^2 - ac) \qquad \text{. . . (vi)}$$

And for the face OBEF, n = = j dS = dx dz and y = 0

$\therefore \int$ F . n dS over the face OBEF

$$= - \int_{x=0}^{a} \int_{z=0}^{c} (y^2 - xz)\, dx\, dz, \text{ for } y = 0 \text{ as for face OCKF}$$

$$= \int_{x=0}^{a} \int_{z=0}^{c} xz\, dx\, dz = \frac{1}{4} a^2 c^2 \qquad \text{. . . (vii)}$$

Adding (ii) to (vii) we get for the whole surface S of the rectangular parallelopiped.

$$\int_S F . n\, dS = \frac{1}{4} bc\,(4a^2 + bc) + \frac{1}{4} b^2c^2 + \frac{1}{4} a^2b^2 + \frac{1}{4} ab\,(4c^2 - ab) + \frac{1}{4} ac\,(4b^2 - ac) + \frac{1}{4} a^2c^2$$

$$= a^2 bc + abc^2 + acb^2 = abc\,(a + b + c) \qquad \text{. . . (viii)}$$

Hence from (i) and (vii) we find that

$$\int_S F . n\, dS = \int_V \text{div } F\, dV$$

Hence divergence theorem is verified.

20. (i) *Interpret the symbol a .* ∇. (ii) *Show that (a .* ∇*)* ϕ = *a .* ∇ ϕ. (iii) *Show that (a .* ∇*) r = a.*

Solution:

(i) Let a = a_1 i + a_2 j + a_3 k. Then

$$a \cdot \nabla = (a_1\, i + a_2\, j + a_3\, k) \cdot \left(i\frac{\partial}{\partial x} + j\frac{\partial}{\partial y} + k\frac{\partial}{\partial z} \right)$$

$$= a_1 \frac{\partial}{\partial x} + a_2 \frac{\partial}{\partial y} + a_3 \frac{\partial}{\partial z}.$$

(ii) $(a \cdot \nabla)\, \phi = \left(a_1 \frac{\partial}{\partial x} + a_2 \frac{\partial}{\partial y} + a_3 \frac{\partial}{\partial z} \right) f.$

Also $\quad a \cdot \nabla\, \phi = (a_1\, i + a_2\, j + a_3\, k) \cdot \left(\frac{\partial \phi}{\partial x} i + \frac{\partial \phi}{\partial y} j + \frac{\partial \phi}{\partial z} k \right)$

$$= \left(\frac{\partial \phi}{\partial x} i + \frac{\partial \phi}{\partial y} j + \frac{\partial \phi}{\partial z} k \right).$$

Hence $\quad (a \cdot \nabla)\, \phi = a \cdot \nabla\, \phi.$

(iii) $(a \cdot \nabla)\, r = \left(a_1 \frac{\partial}{\partial x} + a_2 \frac{\partial}{\partial y} + a_3 \frac{\partial}{\partial z} \right) r = a_1 \frac{\partial r}{\partial x} + a_2 \frac{\partial r}{\partial y} + a_3 \frac{\partial r}{\partial z}.$

But $r = xi + yj + zk.$

$$\therefore \quad \frac{\partial r}{\partial x} = i, \quad \frac{\partial r}{\partial y} = j, \quad \frac{\partial r}{\partial z} = k.$$

$$\therefore \quad (a \cdot \nabla)\, r = a_1\, i + a_2\, j + a_3\, k = a.$$

21. *If $F = xy^2\, i = 2\, x^2\, y\, z\, j - 3y\, z^2\, k$, find curl F at the point (1, –1, 1).*

Solution : Curl F = $\nabla \times F$

$$= \left(i \frac{\partial}{\partial x} + j \frac{\partial}{\partial y} + k \frac{\partial}{\partial z} \right) \times (xy^2\, i + 2x^2\, yz\, j - 3yz^2\, k)$$

$$= \begin{vmatrix} i & j & k \\ \frac{\partial}{\partial x} & \frac{\partial}{\partial y} & \frac{\partial}{\partial z} \\ xy^2 & 2x^2yz & -3yz^2 \end{vmatrix}$$

$$= i \frac{\partial}{\partial y}(-3yz^2) - \frac{\partial}{\partial z}(2x^2\, yz) - j \left[\frac{\partial}{\partial x}(-3yz^2) - \frac{\partial}{\partial z}(xy^2) \right]$$

$$+ k \left[\frac{\partial}{\partial y}(2x^2yz) - \frac{\partial}{\partial z}(xy^2) \right]$$

$$= i[-3z^2 - 2x^2y] - j\,[0] + k\,[4xyz = 2xy]$$

$= i[-3 \cdot 1^2 - 2 \cdot 1^2(-1)] + k[4 \cdot 1 \cdot (-1) \cdot 1 - 2 \cdot 1 \cdot (-1)]$, at $(1, -1, 1)$

$= -i = 2k$ **Ans.**

22. *If $f = (x + y + 1)\, i + j + (-x - y)\, k$, prove that $f \cdot \text{curl } f = 0$.*

Solution : Curl

$$f = \left(i\frac{\partial}{\partial x} + j\frac{\partial}{\partial y} + k\frac{\partial}{\partial z}\right) \times [(x + y + 1)\, i + j + (-x - y)\, k]$$

$$= \begin{vmatrix} i & j & k \\ \frac{\partial}{\partial x} & \frac{\partial}{\partial y} & \frac{\partial}{\partial z} \\ x+y+1 & 1 & -x-y \end{vmatrix}$$

$$i = \left[\frac{\partial}{\partial y}(-x - y) - \frac{\partial}{\partial z}(1)\right] - j\left[\frac{\partial}{\partial x}(-x - y) - \frac{\partial}{\partial z}(x + y + 1)\right]$$

$$+ k\left[\frac{\partial}{\partial x}(1) - \frac{\partial}{\partial y}(x + y + 1)\right]$$

$= i(-1 - 0) - j(-1 - 0) + k(0 - 1) = -i + j = k$

$\therefore \; f \cdot \text{curl } f = [(x + y + 1)\, i + j + (-x - y)\, k] \cdot (-i + j = k)$

$= -(x + y + 1) + 1(x + y)$, $\quad [\because i \cdot i = 1, i \cdot j = 0$ etc.$]$

$= -x - y - 1 + 1 + x + y = 0$ Hence proved.

23. *Prove curl $F = 0$, where $F = zi + xj + yk$.*

Solution : $\text{Curl } F = \nabla \times F = \left(i\frac{\partial}{\partial x} + j\frac{\partial}{\partial y} + k\frac{\partial}{\partial z}\right) \times (zi + xj + yk)$

$$= \begin{vmatrix} i & j & k \\ \frac{\partial}{\partial x} & \frac{\partial}{\partial y} & \frac{\partial}{\partial z} \\ z & x & y \end{vmatrix}$$

$$= i\left[\frac{\partial}{\partial y}(y) - \frac{\partial}{\partial z}(x)\right] - j\left[\frac{\partial}{\partial x}(y) - \frac{\partial}{\partial z}(z)\right] + \left[\frac{\partial}{\partial x}(x) - \frac{\partial}{\partial y}(z)\right]$$

$= i[1 - 0] = j[0 - 1] + k[1 - 0] = i + j + k$

$\therefore$ Curl F = curl $(i + j + k)$

$$= \nabla \times (i + j + k) = \begin{vmatrix} i & j & k \\ \dfrac{\partial}{\partial x} & \dfrac{\partial}{\partial y} & \dfrac{\partial}{\partial z} \\ 1 & 1 & 1 \end{vmatrix}$$

$$= i\left[\frac{\partial}{\partial y}(1) - \frac{\partial}{\partial z}(1)\right] - j\left[\frac{\partial}{\partial x}(1) - \frac{\partial}{\partial z}(1)\right] + k\,[\ldots\ldots]$$

$$= i\,(0) + j\,(0) + k\,(0) = 0$$

Hence proved.

24. *Show that* $\nabla \times (\phi\, V) = \phi\,(\nabla \times V) + (\nabla\phi) \times V$.

Solution : $\nabla \times (\phi\, V) = \nabla \times \{\phi\,(V_1 i + V_2\, j + V_3\, k)\}$

$$= \nabla \times \{\phi\,(V_1 i + V_2\, j + V_3\, k)\}$$

$$= \begin{vmatrix} i & j & k \\ \dfrac{\partial}{\partial x} & \dfrac{\partial}{\partial y} & \dfrac{\partial}{\partial z} \\ \phi\, V_1 & \phi V_2 & \phi V_3 \end{vmatrix}$$

$$= i\left[\frac{\partial}{\partial y}(\phi\, V_3) - \frac{\partial}{\partial z}(\phi\, V_2)\right] - j\left[\frac{\partial}{\partial z}(\phi\, V_1) - \frac{\partial}{\partial x}(\phi\, V_3)\right]$$

$$+ k\left[\frac{\partial}{\partial x}(\phi\, V_2) - \frac{\partial}{\partial y}(\phi\, V_1)\right]$$

$$= \left(\phi\,\frac{\partial V_3}{\partial y} + V_3\,\frac{\partial \phi}{\partial z} - V_2\,\frac{\partial \phi}{\partial z}\right) i + \left(\phi\,\frac{\partial V_1}{\partial z} + V_1\,\frac{\partial \phi}{\partial z} - \phi\,\frac{\partial V_3}{\partial z} - V_3\,\frac{\partial \phi}{\partial x}\right) j$$

$$+ \left(\phi\,\frac{\partial V_2}{\partial x} + V_2\,\frac{\partial \phi}{\partial z} - \phi\,\frac{\partial V_1}{\partial y} - V_1\,\frac{\partial \phi}{\partial y}\right) k$$

$$= \phi\left[\left(\frac{\partial V_3}{\partial y} - \frac{\partial V_2}{\partial z}\right) i + \left(\frac{\partial V_1}{\partial z} - \frac{\partial V_3}{\partial x}\right) j + \left(\frac{\partial V_2}{\partial x} - \frac{\partial V_1}{\partial x}\right) k\right]$$

$$+ \left[\left(\frac{\partial \phi}{\partial y}\,V_3 - \frac{\partial \phi}{\partial z}\,V_2\right) i + \left(\frac{\partial \phi}{\partial z}\,V_1 - \frac{\partial \phi}{\partial x}\,V_3\right)\right] j$$

$$+ \left(\frac{\partial \phi}{\partial x}\,V_2 - \frac{\partial \phi}{\partial y}\,V_1\right) k\Bigg]$$

$$= \phi (\nabla \times V) + \begin{vmatrix} i & j & k \\ \frac{\partial \phi}{\partial x} & \frac{\partial \phi}{\partial y} & \frac{\partial \phi}{\partial z} \\ V_1 & V_2 & V_3 \end{vmatrix}$$

$$= \phi (\nabla \times V) + \left(\frac{\partial \phi}{\partial x} i + \frac{\partial \phi}{\partial y} j + \frac{\partial \phi}{\partial z} k \right) \times (V_1 i + V_2 j + V_3 k)$$

or $\nabla \times (\phi V) = \phi (\nabla \times V) + (\nabla \phi) \times V.$

or Curl (ϕ V) = ϕ curl V + (grad ϕ) × V.

25. *Find curl (r × a), where 'a' is a constant vector.*

Solution : Let $a = a_1 i + a_2 j + a_3 k$. Also $r = xi + yj + zk$.

Then

$$r \times a = \begin{vmatrix} i & j & k \\ x & y & z \\ a_1 & a_2 & a_3 \end{vmatrix} = (a_3 y - a_2 z)i + (a_1 z - a_3 x) j + (a_2 x - a_1 y)k$$

$\therefore$ curl (r × a)

$$= \nabla \times [(a_3 y - a_2 z) i + (a_1 z - a_3 x) j + (a_2 x - a_1 y) k]$$

$$= \begin{vmatrix} i & j & k \\ \frac{\partial}{\partial x} & \frac{\partial}{\partial y} & \frac{\partial}{\partial z} \\ a_3 y - a_2 z & a_1 z - a_3 x & a_2 x - a_1 y \end{vmatrix}$$

$= [(2a_1) i + (2a_2) j + (2a_3) k]$, as in Ex. 13(a) above.

$= -2 (a_1 i + a_2 j + a_3 k) = -2a$ **Ans.**

26. *Prove that the vector $a = 3y^4z^2 i + 4x^2z^2 j - 3x^3y^2 k$ is not irrotational.*

Solution : Curl a = $\nabla \times a$ $= \begin{vmatrix} i & j & k \\ \frac{\partial}{\partial x} & \frac{\partial}{\partial y} & \frac{\partial}{\partial z} \\ 3y^4z^2 & 4x^3z^2 & -3x^2y^2 \end{vmatrix}$

$$= i \left[\frac{\partial}{\partial y} (-3x^2y^2) - \frac{\partial}{\partial z} (4x^3z^2) \right] + j \left[\frac{\partial}{\partial z} (3y^3z^2) - \frac{\partial}{\partial x} (-3x^2 y^2) \right]$$

$$+ k\left[\frac{\partial}{\partial x}(4x^3z^2) - \frac{\partial}{\partial y}(3y^4z^2)\right]$$

$$= i\left[-6x^2y - 8x^3z\right] + j\left[6y^4z + 6xy^2\right] + k\left[12x^2z^2 - 12y^3z^2\right]$$

$\neq 0.$ Hence a is not irrotational.

27. *When a rigid body is in motion, the curl of its linear velocity at any point gives twice its angular velocity in magnitude and direction.*

Solution : Let the linear and angular velocities of the body be v and ω respectively. If r be the position vector of the point whose linear velocity is r, then

$$v = \omega \times r = \begin{vmatrix} i & j & k \\ \omega_1 & \omega_2 & \omega_3 \\ x & y & z \end{vmatrix}$$

where $\omega = \omega_1\, i + \omega_2\, j + \omega_3\, k$

or $v = (\omega_2 z = \omega_3 y)\, i + (\omega_3 x - \omega_1 z)\, j + (\omega_1 y - \omega_2 x)\, k$

$$\therefore\ \text{curl } v = \nabla \times v = \begin{vmatrix} i & j & k \\ \dfrac{\partial}{\partial x} & \dfrac{\partial}{\partial y} & \dfrac{\partial}{\partial z} \\ \omega_2 z - \omega_3 y & \omega_3 x - \omega_1 z & \omega_1 y - \omega_2 x \end{vmatrix}$$

$= i\,[\omega_1 + \omega_1] + j\,[\omega_2 + \omega_2] + k\,[\omega_3 + \omega_3]$

$= 2\,[\omega_1 i + \omega_1 j + \omega_3 k = 2\omega$ Hence proved.

28. *If u and v are irrotationasl vectors then show that u × v is a solenoidal vector.*

Solution : If u and v are irrotational vectors, then by definiton

$$\nabla \times u = 0 \text{ and } \nabla \times v = 0 \qquad \ldots \text{(i)}$$

Now $\text{div}\,(u \times v) = \nabla \,.\, (u \times v)$

$= v \,.\, (\nabla \times u) = u \,.\, (\nabla \times v).$

$= v \,.\, 0 = u \,.\, 0$, from (i)

i.e., $\text{div}\,(u \times v) = 0$

Hence by definiton u × v is a solenodial vector.

29. *Show that div grad $r^m = m\,(m + 1)\, r^{m-2}$*

Solution : Grad $r^m = \left(i \frac{\partial}{\partial x} + j \frac{\partial}{\partial y} + k \frac{\partial}{\partial z}\right)$

$$= i\, mr^{m-1} \frac{\partial r}{\partial x} + j\, nr^{m-1} \frac{\partial r}{\partial x} + k\, mr^{m-1} \frac{\partial r}{\partial z}$$

$$= mr^{m-1}\left[i \frac{\partial r}{\partial x} + j \frac{\partial r}{\partial y} + k \frac{\partial r}{\partial z}\right]$$

$$= mr^{m-i}\left[i\left(\frac{x}{r}\right) + j\left(\frac{y}{r}\right) + k\left(\frac{z}{r}\right)\right] \qquad \left[\because \frac{\partial r}{\partial x} = \frac{x}{r} \text{ etc.}\right]$$

$$= mr^{m-2}\,[xi + yj + zk]$$

$\therefore$ div grad rm $= \nabla \,.\, [nrm^{-2}\,(xi + yj + zk)]$

$$= \left(i \frac{\partial}{\partial x} + j \frac{\partial}{\partial y} + k \frac{\partial}{\partial z}\right) . \left[mr^{m-2}\,(xi + yj + zk)\right]$$

$$= m\left[\frac{\partial}{\partial x}(x r^{m2})\, \frac{\partial}{\partial y}(y r^{m2})\, \frac{\partial}{\partial z}(z r^{m2})\right]$$

$$= m\left[\left\{r^{m-2} + (m-2)\, x r^{m-3} \frac{\partial r}{\partial x}\right\} + \left\{r^{m-2} + (m-2)\, y r^{m-3} \frac{\partial r}{\partial y}\right\}\right.$$

$$\left. + \left\{r^{m-2} + (m-2)\, x r^{n-3} \frac{\partial r}{\partial z}\right\}\right]$$

$$= 3mr^{m-2} + m\,(m-2)\, r^{m-2}\left[x \frac{\partial r}{\partial x} + y \frac{\partial r}{\partial y} + z \frac{\partial r}{\partial z}\right]$$

$$= 3mr^{m-2} + m\,(m-2)\, r^{m-2}\left[x\left(\frac{x}{r}\right) + y\left(\frac{y}{r}\right) + z\left(\frac{z}{r}\right)\right]$$

$$\left[\because \frac{\partial r}{\partial x} = \frac{x}{r}\right]$$

$$= 3\, mr^{m-2} + m\,(m-2)\, r^{m-4}\,(x^2 + y^2 + z^2)$$

$$= 3\, mr^{m-2} + m\,(m-2)\, r^{m-4}\,(r^2) \qquad [\because x^2 + y^2 + z^2 = r^2]$$

$$= [3m + m\,(m-2)\,]r^{m-2} = m\,(m+1)\, r^{m-2}$$

Hence proved.

30. *Evaluate curl grad r^n, where $r = |\, r \,| = |\, xi + yj + zk \,|$.*

Solution : Grad $r^n = \left(i\frac{\partial}{\partial x} + j\frac{\partial}{\partial y} + k\frac{\partial}{\partial z}\right) r^n$

$$= i\, nr^{n-1}\frac{\partial r}{\partial x} + j nr^{n-1}\frac{\partial r}{\partial y} + k\, nr^{n-1}\frac{\partial r}{\partial z}$$

$$= nr^{n-1}\left[i\frac{\partial r}{\partial x} + j\frac{\partial r}{\partial y} + k\frac{\partial r}{\partial z}\right]$$

$$= nr^{n-1}\left[i\frac{r}{x} + j\frac{r}{y} + k\frac{r}{z}\right], \qquad \left[\because \frac{\partial r}{\partial x} = \frac{x}{r} \text{ etc.}\right]$$

$$= nr^{n-2}\,(xi + yj + zk) \qquad \ldots \text{(i)}$$

$\therefore$ curl grad $r^n = \nabla \times (\text{grad } r^n)$

$$= \left(i\frac{\partial}{\partial x} + j\frac{\partial}{\partial y} + k\frac{\partial}{\partial z}\right) \times [nr^{n-2}\,(xi + yi + zk]$$

$$= \begin{vmatrix} i & j & k \\ \dfrac{\partial}{\partial x} & \dfrac{\partial}{\partial y} & \dfrac{\partial}{\partial z} \\ nr^{n-2}x & nr^{n-2}y & nr^{n-2}z \end{vmatrix}$$

$$= n\,\Sigma i\left[\frac{\partial}{\partial y}(r^{n-2}z) - \frac{\partial}{\partial z}(r^{n-2}y)\right]$$

$$= n\,\Sigma i\left[z\,(n-2)\,r^{n-3}\frac{\partial r}{\partial y} - y\,(n-2)\,r^{n-2}\frac{\partial r}{\partial z}\right]$$

$$= n\,(n-2)r^{n-3}\,\Sigma i\left[z\left(\frac{y}{r}\right) - y\left(\frac{z}{r}\right)\right],$$

$$\left[\because \frac{\partial r}{\partial y} = \frac{y}{r}\right]$$

$= n\,(n-2)\,r^{n-4}\,\Sigma\, i\,(zy = yz) = 0$ **Ans.**

31. *Prove that $\nabla^2\,(r\,r) = (4/r)\,r$.*

Solution : $\nabla^2\,(r r) = \dfrac{\partial^2}{\partial x^2}(r r) + \dfrac{\partial^2}{\partial y^2}(r r) + \dfrac{\partial^2}{\partial z^2}(r r)$

$$= \Sigma \frac{\partial^2}{\partial x^2}(r\,r) = \Sigma \frac{\partial}{\partial x}\left[\frac{\partial}{\partial x}(r\,r)\right] = \Sigma \frac{\partial}{\partial x}\left[\frac{\partial r}{\partial x} r + r\frac{\partial r}{\partial x}\right]$$

$$= \Sigma \frac{\partial}{\partial x}\left[\frac{x}{r} r + r i\right], \qquad \left[\because \frac{\partial r}{\partial x} = \frac{\partial}{\partial x}(xi + yj + zk) = i, \frac{\partial r}{\partial x} = \frac{x}{r}\right]$$

$$= \Sigma\left[\left\{\frac{r}{r} + \frac{x}{r}\frac{\partial r}{\partial x} - \frac{x}{r^2} r \frac{\partial r}{\partial x}\right\} + i\frac{\partial r}{\partial x}\right]$$

$$= \Sigma\left[\left\{\frac{r}{r} + \frac{x}{r}(i) - \frac{x}{r^2} r\left(\frac{x}{r}\right)\right\} + i\frac{x}{r}\right] \quad \left[\because \frac{\partial r}{\partial x} = \frac{x}{r}, \frac{\partial r}{\partial x} = i\right]$$

$$= \frac{3r}{r} + \frac{1}{r}(xi + yj + zk) - \frac{r}{r^3}(x^2 + y^2 + z^2) + \frac{1}{r}(xi + yj + zk)$$

$$= \frac{4}{r} r.$$ Hence proved.

32. *If* $\phi(x, y, z) = x^2y + y^2x + z^2$, *find* $\nabla\phi$ *at the point (1, 1, 1).*

Solution: We have

$$\nabla\phi = \frac{\partial\phi}{\partial x} i + \frac{\partial\phi}{\partial y} j + \frac{\partial\phi}{\partial z} k$$

$$= (2xy + y^2)\, i + (x^2 + 2xy)\, j + 2z\, k.$$

Putting $x = 1, y = 1, z = 1$, we get

$\nabla\phi$ at the point $(1, 1, 1) = 3i + 3j + 2k$.

33. *Find grad f, where f is given by* $f = x^3 - y^3 + xz^2$, *at the point (1, –1, 2).*

Solution: We have

$$\text{grad } f = \nabla f = \frac{\partial f}{\partial x} i + \frac{\partial f}{\partial y} j + \frac{\partial f}{\partial z} k$$

$$= \left[\frac{\partial f}{\partial x}(x^3 - y^3 + xz^2)\right]i + \left[\frac{\partial}{\partial y}(x^3 - y^3 + xz^2\right]j + \left[\frac{\partial}{\partial z}(x^3 - y^3 + xz^2)\right]k$$

$= (3x^2 + z^2)\ i + (-3y^2)\ j + 2xz\ k.$

Putting $x = 1, y = -1, z = 2$, we get

∇f at the point $(1,-1, 2) = 7i - 3j + 4k$.

34. *If $u = x + y + z$, $v = x^2 + y^2 + z^2$, $w = yz + zx + xy$, prove that (grd u) . [(grad v) × (grad w)] = 0.*

Solution: We have

$$\text{grad } u = \frac{\partial u}{\partial x}i + \frac{\partial u}{\partial y}j + \frac{\partial u}{\partial z}k$$

$= 1\ i + 1\ j + 1\ k = i + j + k,$

$$\text{grad } v = \frac{\partial v}{\partial x}i + \frac{\partial v}{\partial y}j + \frac{\partial v}{\partial z}k = 2x\ i + 2y\ j + 2z\ k$$

$$\text{and grad } w = \frac{\partial w}{\partial x}i + \frac{\partial w}{\partial y}j + \frac{\partial w}{\partial z}k = (y + z)\ i + (z + x)\ j + (x + y)\ k.$$

∴ grad u . [(grad v) × (grad w)] = scalar triple product of the vectors grad u. grad v and grad w

$$= \begin{vmatrix} 1 & 1 & 1 \\ 2x & 2y & 2z \\ y+z & z+x & x+y \end{vmatrix} = 2 \begin{vmatrix} 1 & 1 & 1 \\ x & y & z \\ y+z & z+x & x+y \end{vmatrix}$$

$$= 2 \begin{vmatrix} 1 & 1 & 1 \\ x+y+z & x+y+z & x+y+z \\ y+z & z+x & x+y \end{vmatrix}, \text{ by } R_2 + R_3$$

$$= 2\ (x + y + z) \begin{vmatrix} 1 & 1 & 1 \\ 1 & 1 & 1 \\ y+z & z+x & x+y \end{vmatrix}$$

$= 2\ (x + y + z)$, 0, the first two rows of the determinant being identical

$= 0.$

35. *If* $F = \left(y\frac{\partial f}{\partial z} - z\frac{\partial f}{\partial y}\right)i + \left(z\frac{\partial f}{\partial x} - x\frac{\partial f}{\partial z}\right)j + \left(x\frac{\partial f}{\partial y} - y\frac{\partial f}{\partial x}\right)k$, *prove that*

(i) $F = r \times \nabla f$,

(ii) $F \cdot r = 0$,

(iii) $F \cdot \nabla f = 0$.

Solution: We have

$$\nabla f = \frac{\partial f}{\partial x} i + \frac{\partial f}{\partial y} j + \frac{\partial f}{\partial z} k$$

and $r = xi + yj + zk$.

(i) $$r \times \nabla f = \begin{vmatrix} i & j & k \\ x & y & z \\ \partial f/\partial x & \partial f/\partial y & \partial f/\partial z \end{vmatrix}$$

$$= \left(y\frac{\partial f}{\partial z} - z\frac{\partial f}{\partial y}\right)i + \left(z\frac{\partial f}{\partial x} - x\frac{\partial f}{\partial z}\right)j + \left(x\frac{\partial f}{\partial y} - y\frac{\partial f}{\partial x}\right)k = F.$$

(ii) $F \cdot r = (r \times \nabla f) \cdot r$ $[\because F = r \times \nabla f]$

$= 0$,

because the value of a scalar triple product having two vectors equal is zero.

(iii) $F \cdot \nabla f = (r \times \nabla f) \cdot \nabla f = [r, \nabla f, \nabla f]$

$= 0$,

because the value of a scalar triple product having two vectors equal is zero.

36. *Prove that* $V \times \text{curl } V = \frac{1}{2}\nabla V^2 - (V \cdot \nabla) V$

Solution : Let $V = V_1 i + V_2 j + V_3 k$, then

$$\text{curl } V = \begin{vmatrix} i & j & k \\ \frac{\partial}{\partial x} & \frac{\partial}{\partial y} & \frac{\partial}{\partial z} \\ V_1 & V_2 & V_3 \end{vmatrix}$$

$$= \left(\frac{\partial V_3}{\partial y} - \frac{\partial V_2}{\partial z}\right) i + \left(\frac{\partial V_1}{\partial z} - \frac{\partial V_3}{\partial x}\right) j + \left(\frac{\partial V_2}{\partial x} - \frac{\partial V_1}{\partial y}\right) k$$

$$\therefore \quad V \times \text{curl } V = \begin{vmatrix} i & j & k \\ V_1 & V_2 & V_3 \\ \frac{\partial V_3}{\partial y} - \frac{\partial V_2}{\partial z} & \frac{\partial V_1}{\partial z} - \frac{\partial V_3}{\partial x} & \frac{\partial V_2}{\partial x} - \frac{\partial V_1}{\partial y} \end{vmatrix}$$

$$= \Sigma\, i \left[V_2 \left(\frac{\partial V_2}{\partial x} - \frac{\partial V_1}{\partial y} \right) - V_3 \left(\frac{\partial V_1}{\partial z} - \frac{\partial V_3}{\partial x} \right) \right]$$

$$= \Sigma\, i \left[\left(V_2 \frac{\partial V_2}{\partial x} + V_3 \frac{\partial V_3}{\partial x} \right) - \left(V_2 \frac{\partial V_1}{\partial y} + V_3 \frac{\partial V_1}{\partial z} \right) \right]$$

$$= \Sigma\, i \left[\left(V_1 \frac{\partial V_1}{\partial x} + V_2 \frac{\partial V_2}{\partial x} + V_3 \frac{\partial V_3}{\partial x} \right) \right.$$

$$\left. - \left(V_1 \frac{\partial V_1}{\partial x} + V_2 \frac{\partial V_1}{\partial y} + V_3 \frac{\partial V_1}{\partial z} \right) \right],$$

adding and subtracting $V_1 \dfrac{\partial V_1}{\partial x}$

$$= \Sigma\, i \left[\frac{1}{2} \frac{\partial}{\partial x} \left(V_1^2 + V_2^2 + V_3^2 \right) \right]$$

$$- \Sigma\, i \left(V_1 \frac{\partial}{\partial x} + V_2 \frac{\partial}{\partial y} + V_3 \frac{\partial}{\partial z} \right) V_1$$

$$= \frac{1}{2} \Sigma\, i \frac{\partial}{\partial x} (V)^2 - \left(V_1 \frac{\partial}{\partial x} + V_2 \frac{\partial}{\partial y} + V_3 \frac{\partial}{\partial z} \right) \Sigma\, V_1\, i,$$

$$\left[\because V^2 = V_1^2 + V_2^2 + V_3^2 \right]$$

$$= \frac{1}{2} \left(i \frac{\partial}{\partial x} + j \frac{\partial}{\partial y} + k \frac{\partial}{\partial z} \right) V^2 - (V \cdot \nabla)\, V,$$

$$\left[\because V \cdot \nabla = V_1 \frac{\partial}{\partial x} + V_2 \frac{\partial}{\partial y} + V_3 \frac{\partial}{\partial z} \right]$$

$= 1/2\ \nabla\ V^2 - (V \cdot \nabla)$ Hence proved.

37. *Find the directional derivative of the function $xy^2 + yz^2 + zx^2$ along the tangent to the curve $x = t$, $y = t^2$, $z = t^3$ at the point (2, 1, 1).*

Solution : Let $\phi = xy^2 + yz^2 + zx^2$

$$\therefore \text{ grad } \phi = i\frac{\partial \phi}{\partial x} + j\frac{\partial \phi}{\partial y} + k\frac{\partial \phi}{\partial z}$$

$$= i\,(y^2 + 2zx) + j\,(2xy + z^2) + k\,(2yz + x^2)$$

$$= i\,(1 + 2) + j\,(2 + 1) + k\,(2 + 1), \text{ at } (1, 1, 1)$$

$$= 3\,(i + j + k)$$

Also for the curve $x = t$, $y = t^2$, $z = t^3$ we have

$dx/dt = 1$, $dy/dt = 2t$, $dz/dt = 3t^2$

Also we know that direction cosines of the tangent to the curve at (x, y, z) are proporational to dx, dy, dz

i.e., proportional to dx/dt, dy/dt, dz/dt

∴ The direction cosines of the tangent to the given curve at (1, 1, 1) are proportional to 1, 2, 3.

i.e., the directional vector is $i + 2j + 3k$

∴ The unit vector in this direction

$$= \frac{i + 2j + 3k}{\sqrt{(1^2 + 2^2 + 3^2)}} = \frac{i + 2j + 3k}{\sqrt{(14)}} = \hat{a}, \text{ say}$$

∴ The required directional derivative

$$= \hat{a} \,.\, \text{grad } \phi \text{ at } (1, 1, 1)$$

$$= \frac{i + 2j + 3k}{\sqrt{(14)}} \,.\, 3\,(i + j + k) = \frac{3}{\sqrt{(14)}}\,(1 + 2 + 3) = \frac{18}{\sqrt{(18)}} \quad \textbf{Ans.}$$

38. *Find the values of constants a, b, c so that the directional derivative of $f = axy^2 + byz + cz^2x^3$ at (1, 2, –1) has maximum magnitude 64 in the direction paralled to z-axis.*

Solution : Grad $\phi = i\dfrac{\partial f}{\partial x} + j\dfrac{\partial f}{\partial y} + k\dfrac{\partial f}{\partial z}$

$$= i\,(ay2 + 3cz^2 x^2) + j\,(2axy + bz) + k\,(by + 2czx^3)$$

$\therefore$ At (1, 2, –1), grad $f = i\,(4a + 3c) + j\,(4a - b) + k\,(2b - 2c)$

$$= (4a + 3c)\, i + (4a - b)\, j + 2\,(b - c)\, k \quad \ldots(i)$$

Since the directional derivative of f at (1, 2, –1) is maximum along grad f and we are given that the directional derivative is maximum in the direction parallel to z-axis *i.e.*, parallel to the vector k, so from (i) we have

$$4a + 3c = 0,\ 4a - b = 0 \quad \ldots \text{(ii)}$$

equating the coefficients of i and j to zero.

i.e., $b = 4a$ and $3c = -4a \Rightarrow b = -3c$

$\therefore$ From (i) at (1, 2, –1), grad $f = 2(-3c - c)\,k$

or grad f at $(1, 2, -1) = (8c)\,k$

Also the maximum value of directional derivative

$$= |\text{grad } f| = |-8\,ck| = 8c = 64 \text{ (given)}$$

$\therefore$ $c = 8$

from (ii) we have $4a = -3c = -3(8) = -24$ or $a = -6$

and $b = -3c = -24$

Hence $a = -6,\ b = -24,\ c = 8$ **Ans.**

39. *Find the directional derivative of $\phi = x^2y^3z^4$ at the point (2, 3, –1) in the direction making equal angles with the positive x, y, and z axes.*

Solution : Here grad $\phi_1 = i\dfrac{\partial\phi}{\partial x} + j\dfrac{\partial\phi}{\partial y} + k\dfrac{\partial\phi}{\partial z}$

$$= i\,(2xy^3z^4) + j\,(3x^2y^2z^4) + k\,(4x^2y^3z^4)$$

$\therefore$ At (2, 3, –1), grad $\phi = i\,(2 \,.\, 2 \,.\, 3^3 \,.\, 1) + j\,(3 \,.\, 3^3 \,.\, 3^2 \,.\, 1) + k\,(-4 \,.\, 2^2 \,.\, 3^3 \,.\, 1)$

$$= 108\,(i + j - 4k)$$

Also the direction which makes equal angles with the positive x, y and z-axis is given by $i + j + k$.

40. *If $F = e^{xy}\,i + (x - 2y)\,j = + x \sin y\,k$, calculate*

(i) $\dfrac{\partial F}{\partial x}$, (ii) $\dfrac{\partial F}{\partial y}$, (iii) $\dfrac{\partial^2 F}{\partial x^2}$,

(iv) $\dfrac{\partial^2 F}{\partial x\,\partial y}$, (v) $\dfrac{\partial^2 F}{\partial y^2}$.

Solution:

(i) $$\frac{\partial F}{\partial x} = \left[\frac{\partial}{\partial x}(e^{xy})\right]i + \left[\frac{\partial}{\partial x}(x - 2y)\right]j + \left[\frac{\partial}{\partial x}(x \sin y)\right]k$$

$$= (ye^{xy})\,i + (1)\,j + (\sin y)\,k = ye^{xy}\,i + j + \sin y\,k.$$

(ii) $$\frac{\partial F}{\partial y} = \left[\frac{\partial}{\partial y}(e^{xy})\right]i + \left[\frac{\partial}{\partial y}(x - 2y)\right]j + \left[\frac{\partial}{\partial y}(x \sin y)\right]k$$

$= xe^{xy}\, i - 2\, j + x \cos y\, k.$

(iii) $$\frac{\partial^2 F}{\partial x^2} = \frac{\partial}{\partial x}\left(\frac{\partial F}{\partial x}\right) = \frac{\partial}{\partial x}[ye^{xy} i + j + \sin y k]$$

$$= \left[\frac{\partial}{\partial x}(ye^{xy}\right] i + \frac{\partial}{\partial x} j + \left[\frac{\partial}{\partial x}(\sin y)\right] k = y^2 e^{xy}\, i + 0 + k = y^2 e^{xy}\, i.$$

(iv) $$\frac{\partial^2 F}{\partial x \partial y} = \frac{\partial}{\partial y}\left(\frac{\partial F}{\partial x}\right) = \frac{\partial}{\partial y}[ye^{xy} i + j + \sin y\, k]$$

$$= \left[\frac{\partial}{\partial y}(ye^{xy})\right] i + \frac{\partial}{\partial y} j + \left(\frac{\partial}{\partial y} \sin y\right) k$$

$$= (e^{xy} + xye^{xy})\, i + 0 + \cos y\, k = e^{xy}(xy + 1)\, i + \cos y\, k.$$

(v) $$\frac{\partial^2 F}{\partial y^2} = \frac{\partial}{\partial y}\left(\frac{\partial F}{\partial y}\right) = \frac{\partial}{\partial y}[xe^{xy} i - 2j + x \cos y\, k]$$

$$= \left[\frac{\partial}{\partial y}(ye^{xy})\right] i + \frac{\partial}{\partial y}(-2j) + \left[\frac{\partial}{\partial y}(x \cos y)\right] k$$

$$= x^2 e^{xy}\, i + 0 - x \sin y\, k = x^2 e^{xy}\, i + x \sin y\, k.$$

41. *Evaluate $\int_c$ [yzdx + (xz + 1) dy + xy dz], where C is any path from (1, 0, 0) to 2, 1, 4).*

Solution : Since C is any path joining the points (1, 0, 0) and (2, 1, 40 therefore if yz dx + (xz + 1) + xy dz = F . dr, then F must be a conservative field.

Here $\quad F \cdot dr = yzdx + (xz + 1)\, dy + xydz$

$$= [yz\, i + (xz + 1)\, j + xy\, k] \cdot (i\, dx + j\, dy + k\, dz)$$

$$= [yz\, i + (xz + 1)\, j + xy\, k] \cdot dr$$

$$\therefore \quad F = yz\, i + (xz + 1)\, j + xy\, k$$

Hence curl F $$= \begin{vmatrix} i & j & k \\ \frac{\partial}{\partial x} & \frac{\partial}{\partial y} & \frac{\partial}{\partial z} \\ yz & xz+1 & xy \end{vmatrix}$$

$$= i\left[\frac{\partial}{\partial y}(xy) - \frac{\partial}{\partial z}(xz + 1)\right] + j\left[\frac{\partial}{\partial z}(yz) - \frac{\partial}{\partial x}(xy)\right]$$

$$+ k\left[\frac{\partial}{\partial x}(xz + 1) - \frac{\partial}{\partial y}(yz)\right]$$

$$= i\ n[x - x] + j\ [y - y] + k\ [z - z] = 0$$

$\therefore$ F is a conservative field conservative field and hence the above integral is independent of the path C joining (1, 0, 0) and (2, 1, 4).

Now $\quad \int_c [yz\,dx + (xz + 1)\,dy + xy\,dz]$

$$= \int_c [(yz\,dx + xz\,dy + xy\,dz) + dy]$$

$$= \int_c d\,(xyz + y)$$

$$= (XYZ + Y)\Big|_{(1,\,0,\,0)}^{(2,\,1,\,4)} = [(2.1.4 + 1) - (1.0.0 + 0)]$$

$= 9$ **Ans.**

42. *Show that $F = xi + yj + zk$ is conservative and find ϕ such that $F = \nabla\phi$.*

Solution : Curl $F = \nabla \times F = \nabla \times (xi + yj + zk)$

$$= \begin{vmatrix} i & j & k \\ \frac{\partial}{\partial x} & \frac{\partial}{\partial y} & \frac{\partial}{\partial z} \\ x & y & z \end{vmatrix}$$

$$= \Sigma\, i \left[\frac{\partial}{\partial y}(z) - \frac{\partial}{\partial z}(y)\right] = (0)\,i + (0)\,j + (0)\,k$$

$= 0$

As curl F = 0, so F is conservative.

Also $\quad F = \nabla\phi$, so $F\,.\,dr = (\Delta\phi)\,.\,dr$

or $\quad F\,.\,dr = \left(i\frac{\partial\phi}{\partial x} + j\frac{\partial\phi}{\partial y} + k\frac{\partial\phi}{\partial z}\right).\,(i\,dx + j\,dy + k\,dz)$

$$= \frac{\partial\phi}{\partial x}dx + \frac{\partial\phi}{\partial y}dy + \frac{\partial\phi}{\partial z}dz = d\phi$$

$\therefore \quad d\phi = F\,.\,dr = (xi + yj + zk).\,(i\,dx + j\,dy + k\,dz)$

$$= x\,dx + y\,dy + z\,dz$$

$$= d\left(\frac{1}{2}x^2\right) + f\left(\frac{1}{2}y^2\right) + d\left(\frac{1}{2}z^2\right) = \frac{1}{2}d\,(x^2 + y^2 + z^2)$$

or $\quad f = 1/2\,(x^2 + y^2 + z^2) + \text{constant}$ **Ans.**

43. *Verify Stokes' theorem for the function $F = x\,(ix + jy)$, integrated round the square, in the plane $z = 0$, whose sides are along the lines $x = 0$, $y = 0$, $x = a$, $y = a$,*

Solution: We clearly have

$$\oint_C F.dl + \int_{OA} F.dl + \int_{AB} F.dl + \int_{BC} F.dl + \int_{CO} F.dl,$$

where, $\int_{OA} F.dl = \int_o^a x(ix + jy).idx = \int_o^a x^2 dx = \frac{1}{3}a^3,$

$$\int_{AB} F.dl = \int_o^a x(ix + jy).jdy = \int_o^a aydy = \frac{1}{2}a^3,$$

$$\int_{BC} F.dl = \int_a^o x(ix + jy).idx = -\int_o^a x^2 dx = -\frac{1}{3}a^3 \text{ and}$$

$$\int_{CO} F.dl = \int_a^o x(ix + jy).jdx = 0$$

So that, $\oint_c F.dl = \frac{1}{3}a^3 + \frac{1}{2}a^3 - \frac{1}{3}a^3 + 0 = \frac{1}{2}a^3.$

Now, in accordance with Stokes' theorem, $\oint c\ F.d\ l = \iint S$ curl FdS.

Since curl x (ix + jy) = ky, we have

$$\iint_S \text{curl FdS} = \iint_S \text{curl}\, x(ix + jy)dS = \int_o^a\int_o^a ky.kdxdy$$

$$= \int_o^a\int_o^a ydxdy = \frac{1}{2}a^3.$$

Stokes' theorem thus stands verified.

44. *Show that*

$$V = \iint_S x\ dy\ dz = \iint_S z\ dx\ dy$$

$$= \frac{1}{3}\iint_S (x\ dy\ dz + y\ dz\ dx + z\ dx\ dy),$$

where V is the volume of a region T bounded by a surface S.

Solution : By Gauss' divergence theorem, we have

$$\iint_S x\ dy\ dz = \iiint_V \frac{\partial}{\partial x}(x)\ dx\ dy\ dz$$

$$\iiint_V dV, \qquad [\because dV = dx\ dy\ dz]$$

$$= V$$

Similarly we can prove that

$$\iint_S y\ dy\ dz = \text{and} \iint_S z\ dx\ dy = V$$

Adding these results, we get

$$3\ V = \iint_S x\ dy\ dz + \iint_S y\ dz\ dx + \iint_S z\ dx\ dy$$

or $$V = \frac{1}{3}\iint_S (x\ dy\ dz + y\ dz\ dx\ | + z\ dx\ dy)$$

45. *Evaluate* $\iint_S (x\,dy\,dz + y\,dz\,dx + z\,dx\,dy)$, *where S is the surface of the sphere* $x^2 + y^2 + z^2 = a^2$.

Solution : We know

$\iint_S (F_1\,dy\,dz + F_2\,dz\,dx + F_3\,dx\,dy)$

$$= \iiint_V \left(\frac{\partial F_1}{\partial x} + \frac{\partial F_2}{\partial y} + \frac{\partial F_3}{\partial z}\right) dx\,dy\,dz \qquad \text{. . . (i)}$$

Here $F_1 = x$, $F_2 = y$
and $F_3 = z$

$$\therefore \left(\frac{\partial F_1}{\partial x} = 1, \frac{\partial F_2}{\partial y} = 1, \frac{\partial F_3}{\partial z}\right) = 1$$

$\therefore$ From (i) we have

$\iint_S (x\,dy\,dz + y\,dz\,dx + z\,dx\,dy)$

$= \iint_V (1 + 1 + 1)\,dx\,dy\,dz$, where V is the volume enclosed by the surface S.

$$= 3 \iiint_V dx\,dy\,dz,$$

$$= 3 \text{ (volume of the given sphere)}$$

$$= 3\left(\frac{4}{3}\pi a^3\right) = 4\pi a^3 \quad \textbf{Ans.}$$

46. *By transforming into a triple integral, evaluate* $\iint_s (x^3\,dy\,dz + x^2 y\,dz\,dx + x^2 z\,dx\,dy)$, *where S is the closed surface consisting of the cylinder* $x^2 + y^2 = a^2$ $(0 \le z \le b)$ *and the circular dises* $z = 0$ *and* $z = b$ $(x^2 + y^2 \le a)$.

Solution : Here $F_1 = x^3$, $F_2 = x^2y$, $F_3 = x^2z$

$$\therefore \frac{\partial F_1}{\partial x} = 3x^2, \frac{\partial F_2}{\partial y} = x^2, \frac{\partial F_3}{\partial z} = x^2$$

$\therefore$ From cartesian form of Gauss' divergence theorem viz.

$\int_S (F_1\,dy\,dz + F_2\,dz\,dx + F_3\,dx\,dy)$

$= \iiint_V \left(\frac{\partial F_1}{\partial x} + \frac{\partial F_2}{\partial y} + \frac{\partial F_3}{\partial z}\right) dx\,dy\,dz$, we have

$\iint_s (x^3\,dy\,dz + x^2 y\,dz\,dx + x^2 z\,dx\,dy)$

$= \iiint_V (3x^2 + x^2 + x^2)\, dx\, dy\, dz,$

where V is the volume enclosed by S.

$= 5 \int_{x=-a}^{a} \int_{y=-\sqrt{(a^2-x^2)}}^{\sqrt{(a^2-x^2)}} \int_{z=0}^{b} x^2\, dx\, dy\, dz$

$= 5 \times 2 \times 2 \int_{x=0}^{a} x^2 \int_0^{\sqrt{(a^2-x^2)}} (z)_0^b\, dx\, dy$

$= 20\, b \int_{x=0}^{a} x^2\, (y)_0^{\sqrt{(a^2-x^2)}}\, dx = 20\, b \int_0^a x^2 \sqrt{(a^2 - x^2)}\, dx$

$= 20\, b \int_{\phi=0}^{\pi/2} a^2 \sin^2\phi \sqrt{(a^2 - a^2 \sin^2\phi)}\, a \cos\phi\, d\phi,$

putting $x = a \sin\phi$

$$= 20\, a^1\, b \int_0^{\pi/2} \sin^2\phi \cos\phi\, d\phi = 20\, a^2\, b \,.\, \frac{\Gamma\left(\frac{3}{2}\right)\Gamma\left(\frac{3}{2}\right)}{2\,\Gamma\,(3)}$$

$$= 20\, a^4\, v\, \frac{\frac{1}{2}\sqrt{\pi}\,.\,\frac{1}{2}\sqrt{\pi}}{2\,.\,2\,.\,1} = \frac{5}{4}\, a^4\, b.$$ **Ans.**

47. *Verify Stoke's theorm when $F = (2x - y)\, i = y\, z^2\, j - y^2\, z\, k$ where S is the upper half of the sphere $x^2 + y^2 + z^2 = i$ and C is the boundary.*

Solution : Given

$F = (2x - y)\, i - yz^2 j - y^2 zk$

Also we know

$r = xi + yj + zk$

$\therefore\ F \,.\, dr = [(2x - y)\, i - yz^2 j - y^2 j - y^2 zk] \,.\, [i\, dx + j\, dy + k\, dz]$

or $F \,.\, dr = (2r = y)\, dx - yz2\, dy - y^2 - y^2 zdz$ . . . (i)

Also here C is the circle $x^2 + y^2 = 1;\ z = 0$ *i.e.*, a circle of unit radius on x y-plane.

$\therefore$ For C we have $z = 0,\ dz = 0$. consequently from (i) we have

$F \,.\, dr = (2x - y)\, dx$ . . . (ii)

Also on C we have $x = \cos\phi$ and $y = \sin\phi$ and ϕ varies from 0 to 2π.

$\therefore$ From (ii) we have

$\int_C F \,.\, dr = \int_C (2x - y)\, dx$

$$= \int_{\phi=0}^{2\pi} (2\cos\phi - \sin\phi)(-\sin\phi)\, d\phi$$

$$= \int_{\phi=0}^{2\pi} (\sin^2\phi - \sin 2\phi)\, d\phi$$

$$= 4\int_0^{\pi/2} \sin^2\phi\, d\phi + \left[\frac{1}{2}\cos 2\phi\right]_0^{2\pi}$$

$$= 4 \cdot \frac{1}{2} \cdot \frac{1}{2}\pi + \frac{1}{2}[1-1] = \pi$$

Hence $\int_C F \,.\, dr = \pi$ . . . (iii)

Also the direction cosines of the line OP are

$$\sin\theta \cos\phi,\ \sin\theta \sin\phi,\ \cos\theta$$

$\therefore$ n = $(\sin\theta \cos\phi)$ i + $(\sin\theta \sin\phi)$ j + $(\cos\theta)$ k, where n is the unit outward drawn normal vector at P.

$$\text{Again curl } F = \begin{vmatrix} i & j & k \\ \dfrac{\partial}{\partial x} & \dfrac{\partial}{\partial y} & \dfrac{\partial}{\partial z} \\ 2x-y & -yz^2 & -y^2z \end{vmatrix}$$

$$= i\left[\frac{\partial}{\partial y}(-y^2x) - \frac{\partial}{\partial z}(-yz^2)\right] + j\left[\frac{\partial}{\partial z}(2x-y) - \frac{\partial}{\partial x}(-y^2z)\right]$$

$$+ k\left[\frac{\partial}{\partial x}(-yz^2) - \frac{\partial}{\partial y}(2x-y)\right]$$

$$= 0i + 0j + k = k$$

$\therefore$ n . curl F = $[(\sin\theta \cos\phi)\, i + (\sin\theta \sin\phi)\, j + (\cos\theta)\, k] \,.\, k$

$$= \cos\theta$$

$\therefore \int_S$ n . curl F dS

$= \int_S \cos\theta\, dS$, where $dS = \sin\theta\, d\theta\, d\phi$

$$= \int_{\theta=0}^{\pi/2} \int_{\phi=0}^{2\pi} \cos\theta \sin\theta\, d\theta\, d\phi$$

$$= \int_{\theta=0}^{\pi/2} \int_{\phi=0}^{2\pi} \frac{1}{2}\sin 2\theta\, d\theta\, d\phi = \frac{1}{2}\left[-\frac{1}{2}\cos 2\theta\right]_0^{\pi/2} [\phi]_0^{2\pi}$$

$$= \frac{1}{2}\left[\frac{1}{2}\cos\theta + \frac{1}{2}\cos\theta\right] 2\pi = \frac{1}{2}\left(\frac{1}{2} + \frac{1}{2}\right) . \ 2\pi = \pi \qquad . . . \text{(iv)}$$

Hence from (iii) and (iv) we have

$$\int_C F \cdot dr = \pi = \int_S n \cdot \text{curl } F \, dS$$

which verifies Stoke's theorem.

48. *Verify Stocke's theorem when $F = x^2 i + x y j$, where C is the perimeter of the square in xy-plane whose sides are along the lines $x = 0$, $y = 0$, $x = a$, $y = a$.*

Solution : Here $F = x^2 i + xyj$ and $dr = (dx)\, i + (dy)\, j$

$\therefore F \cdot dr = (x^2 i + xyj) \cdot (idx + jdy) = x^2 dx + xy^2 dy$ . . . (A)

$\therefore \int_C F \cdot dr$

$= \int_{C1} F \cdot dr + \int_{C2} F \cdot dr + \int_{C3} F \cdot dr + \int_{C4} F \cdot dr$ . . . (i)

where C_1, C_2, C_3, and C_4 are the lines from a to O, O to D, D to B and B to A respectively

Now for C_1 we have $x = 0$, $dx = 0$ and y varies from a to 0, so from (A)

we get $\int_{C1} F \cdot dr = 0$ . . . (ii)

For C_2 we have $y = 0$, $dy = 0$ and x varies form 0 to a, so from (A) we get

$$\int_{C2} F \cdot dr = \int_{x=0}^{a} x^2 dx = \left(\frac{1}{3} x^3\right)_0^a = \frac{1}{3} a^3 \quad \text{. . . (iii)}$$

For C_3 we have $x = a$, $dx = 0$ and x varies form 0 to a, so form (A)

$$\text{we get } \int_{C3} F \cdot dr = \int_{y=0}^{a} ay\, dy = \frac{1}{2} a \left(y^2\right)_0^1 = \frac{1}{2} a^3 \quad \text{. . . (iv)}$$

For C_4 we have $y = a$, $dy = 0$ and x varies forom a to 0 so, from (A)

$$\text{we get } \int_{C4} F \cdot dr = \int_{x=0}^{a} x^2\, dx = \frac{1}{2} \left(x^3\right)_a^0 = -\frac{1}{3} a^3 \quad \text{. . . (v)}$$

$\therefore$ From (i), (ii), (iii), (iv) and (v) we have

$$\int_C F \cdot dr = 0 + \frac{1}{3} a^3 + \frac{1}{2} a^3 - \frac{1}{3} a^3 = \frac{1}{2} a^3 \quad \text{. . . (vi)}$$

Again for the square ABDO, the unit outward drawn normal vector $n = k$.

$$\text{Also curl } F = \begin{vmatrix} i & j & k \\ \frac{\partial}{\partial x} & \frac{\partial}{\partial y} & \frac{\partial}{\partial z} \\ x^2 & xy & 0 \end{vmatrix} = i\,(0) + j(0) + k(y)$$

or $\quad \text{Curl } F = yk$

$\therefore \quad n \,.\, \text{curl } F = k \,.\, yk = y$

Also $\quad dS = dx\, dy$

$\therefore \quad \int_S n \,.\, \text{curl } F \, dS = \int_{x=0}^{a} \int_{y=0}^{a} y \, dx\, dy$

$$=(x)_0^1 \left(\frac{1}{2} y^2\right)_0^1 = \frac{1}{2} a^3 \qquad \ldots \text{(vii)}$$

From (vi) and (vii) we get $\int_C F \,.\, dr = \int_S n \,.\, \text{curl } F \, dS$

49. *Evaluate $\iint_s (x^3 \, dy\, dz + y^2 \, dz\, dx + z^2 \, dx\, dy)$, over the surface S of a cube bounded by the coordinate planes and the planes*

$$x = y = z = a.$$

Solution : Here $F_1 = x^3$,

$$F_2 = y^3, \; F_3 = z^3.$$

$$\therefore \quad \frac{\partial F_1}{\partial x} = 3x^2, \frac{\partial F_2}{\partial y} = 3y^2 \text{ and } \frac{\partial F_3}{\partial z} = 3z^2$$

$\therefore$ By cartesian form of Gauss' divergence theorem, we have

$\iint_S (xh3 \; dy\, dz + y^3 \, dz\, dx + z^3 \, dx\, dy)$

$= \iiint_V (3x^2 + 3y^2 + 3z^2) \, dx\, dy\, dz,$

where V is the volume enclosed by S,

$= 3 \int_{x=0}^{a} \int_{y=0}^{a} \int_{z=0}^{a} (x^2 + y^2) \, dx\, dy\, dz$

$= 3 \int_{x=0}^{a} \int_{y=0}^{a} \left(x^2 z + y^2 z + \frac{1}{3} z^3 \right)_{z=0}^{a} dx\, dy$

$= 3 \int_{x=0}^{a} \int_{y=0}^{a} \left(x^2 a + y^2 a + \frac{1}{3} a^3 \right) dx\, dy$

$= 3 \int_{x=0}^{a} \left(x^2 ay + \frac{1}{3} ay^3 + \frac{1}{3} a^3 y \right)_0^a dx = 3 \int_0^a \left(a^2 x^2 + \frac{2}{3} a^4 \right) dx$

50. *Evaluate $\iint_s (x^3 \, dy\, dz + y^2 \, dz\, dx)$, where S is the surface of the sphere $x^2 + y^2 + z^2 = a^2$.*

Solution : Here $F_1 = x^3$,

$$F_2 = y^3 \text{ and } F_3 = 0$$

$\therefore \quad \dfrac{\partial F_1}{\partial x} = 3x^2, \dfrac{\partial F_2}{\partial y} = 3y^2$ and $\dfrac{\partial F_3}{\partial z} = 0$

$\therefore$ By cartesian form of Gauss' theorem, we have

$$\iint_S (x^3\, dy\, dz + y^3\, dz\, dx)$$

$= \iiint_V (3x^2 + 3y^2) dx\, dy\, dz$, where V is the volume enclosed by the sphere $x^2 + y^2 + z^2 = a^2$.

$= 3 \iiint_V (x^2 + y^2) dx\, dy\, dz$,

$= \dfrac{3}{\rho} \iiint_V (x^2 + y^2)\, \rho\, dx\, dy\, dz$, where ρ is the density of the material of the sphere assmed constant

$= \dfrac{3}{\rho} \iiint_V (x^2 + y^2)\, dm$, where $dm = \rho\, dx\, dy\, dz$

$= \dfrac{3}{\rho}$ [Moment of inertia of the given sphere about z-axis]

$= \dfrac{3}{\rho}$ [$(2/5)\, Ma^2$], where M is the mass of the sphere

$= \dfrac{3}{\rho}$ [$(2/5) \cdot (4/3)\, \pi\, a^3 \rho \cdot a^2$], $\qquad$ [$\because M = (4/3)\, \pi\, a^3 \rho$.]

$= (8/5)\, \pi\, a^5$.

51. *Prove that*

$$\iint_S [lu + mv + nw]\, dS = \iiint_V \left(\frac{\partial u}{\partial x} + \frac{\partial v}{\partial y} + \frac{\partial w}{\partial z}\right) dx\, dy\, dz$$

where l, m, n, are the direction cosines of the outward drawn normal at a point P (x, y, z) of the closed surface S enclosing hte volume V.

Solution : $\because$ (l, m, n) are the direction cosines of the outward drawn normal vector at any point P (x, y, z) on the surface S, so we have

$$n = l\, i + mj + nk \qquad \text{. . . (i)}$$

Also let
$$F = u\, i + vj + wk \qquad \text{. . . (ii)}$$

$\therefore$ From (i) and (ii) we have

$$F \cdot n = (ui + yj + wk) \cdot (li + mj + nk)$$

or
$$F \cdot n = lu + mv + nw \qquad \text{. . . (iii)}$$

And $\text{div } F = \nabla . F = \left(i\frac{\partial}{\partial x} + j\frac{\partial}{\partial y} + k\frac{\partial}{\partial z}\right).(ui + vj + wk)$

$$\text{div } F = i\frac{\partial u}{\partial x} + \frac{\partial v}{\partial y} + \frac{\partial w}{\partial z} \quad \text{. . . (iv)}$$

∴ Substituting these values of F . n and div F in Gauss' theorem $\int_S$ F . n dS = $\int_V$ div F dV, where dV = dx dy dz

we have $\int_S (lu + mv + nw)\, dS = \left(\frac{\partial u}{\partial x} + \frac{\partial v}{\partial y} + \frac{\partial w}{\partial z}\right) dx\, dy\, dz$

Hence proved.

52. *Prove that* $\int_C \phi \nabla \psi . dr = \int_S (\nabla\phi \times \nabla\psi) . n\, dS$

Solution : By Stoke's theroem, we have

$$\int_S F . dr = \int_S \text{curl } F . n\, dS \quad \text{. . (i)}$$

Let $F = \phi \nabla \psi$. then

curl $F = \text{curl } (\phi \nabla \psi)$

$= \nabla\phi \times \nabla\psi + \phi \text{ curl } (\nabla\psi).$

∴ $\nabla \times (\psi\nabla) = \nabla\phi \times V + \phi (\nabla \times V).$

or curl $F = \nabla\phi \times \nabla\psi$, [∵ curl (grad ψ) = 0]

Subsituting these values in (i) we get

$$\int_S \phi \nabla \psi . dr = \int_S (\nabla \phi \times \nabla \psi) . n\, dS$$

Hence proved.

53. *Evaluate* $\int_s F . n\, dS$, *where* $F = a x i + b y j + c z k$ *and S is the surface of sphere* $x^2 + y^2 + z^2 = 1$.

Solution : Here div F = ∇ . F

$$= \left(i\frac{\partial}{\partial x} + j\frac{\partial}{\partial y} + \frac{\partial}{\partial z}\right).(axi + byj + czk)$$

$$= \frac{\partial}{\partial x}(ax) + \frac{\partial}{\partial y}(by) + \frac{\partial}{\partial z}(cz) = a + b + c$$

∴ By Gauss' theorem we have

$$\int_S F . n\, dS = \int_V \text{div } F\, dV,$$

where V is the volume enclosed by the surface S

$= \int_V (a + b + c)\, dV = (a + b + c) \int_V dV$

$= (a + b + c) V$, where V = volume enclosed by S = $4/3\, \pi\, (1)^3$

$= 4/3\ \pi\ (a + b + c).$

54. *Verify the divergence theorem for $F = 4x\,i - 2y^2\,j + z^2\,k$ taken over the region S bounded by $x^2 + y^2 = 4$, $z = 0$ and $z = 3$.*

Solution : Here $\nabla F = \dfrac{\partial}{\partial x}(4x) + \dfrac{\partial}{\partial y}(-2y^2) + \dfrac{\partial}{\partial z}(z^2)$

$$= 4 - 4y + 2z = 2\,(2 - 2y + z)$$

∴ By Gauss' Divergence theorem we have

$\int_S F \,.\, n\, dS = \int_V \nabla \,.\, F\, dV$, where V is the volume enclosed by S.

Now

$$\int_t \nabla \,.\, F dV = \int_{x=-2}^{2} \int_{y=-\sqrt{(4-x^2)}}^{\sqrt{(4-x^2)}} \int_{z=0}^{3} (2(2 - 2y + z)\, dx\, dy\, dz$$

$$= 2 \int_{x=-2}^{2} \int_{y=-\sqrt{(4-x^2)}}^{\sqrt{(4-x^2)}} \left[2z - 2yz + \frac{1}{2} z^2\right]_0^3 dx\, dy$$

$$= 2 \int_{x=-2}^{2} \int_{y=-\sqrt{(4-x^2)}}^{\sqrt{(4-x^2)}} [6 - 6y + (9/2)]\, dx\, dy$$

$$= 21 \int_{x=-2}^{2} \int_{y=-\sqrt{(4-x^2)}}^{\sqrt{(4-x^2)}} dx\, dy \;\; - 12 \int_{x=-2}^{2} \int_{y=-\sqrt{(4-x^2)}}^{\sqrt{(4-x^2)}} dx\, dy$$

$$= 21 \times 4 \int_{x=-2}^{2} \int_{y=0}^{\sqrt{(4-x^2)}} dx\, dy - 1\,[0],$$

as y is an odd function in the second integral

$$= 84 \int_{x=0}^{2} \int_{y=0}^{\sqrt{(4-x^2)}} dx = 84 \int_0^2 \sqrt{(2^2 - x^2)}\, dx$$

$$= 84 \left[\frac{1}{2}\, x\, \sqrt{(2^2 - x^2)} + \frac{1}{2} \,.\, 2^2 \,.\, \sin^{-1}\left(\frac{1}{2} x\right)\right]_0^2$$

$= 84\ [2 \sin^{-1} 1] = 84\ [2\ (\pi/2] = 84\ \pi$ . . . (i)

Again the surface S is the surface of the given cylinder consisting of the curved surface S_1 (say) and two equal circular ends S_2 and S_3 (say0 on the planes $z = 0$ and $z = 3$ respectively.

$$\therefore\ \int_S F \,.\, n\, dS = \int_{S_1} F \,.\, n\, dS + \int_{S_2} F \,.\, n\, dS + \int_{S_3} F \,.\, n\, dS \quad ...(ii)$$

Now for the surface S_2 *i.e.*, the lower circular end of the cylinder on the plane $z = 0$, the outward drawn normal is downwards and parallel to z-axis so n = – k.

$\therefore \int_{S_2} F . n\, dS = \int_{S_2} (4xi - 2y^2 j + z^2 k) . (-k)\, dS$

$= \int_{S_2} z^2\, dS = 0,$ . . . (iii)

as z = 0 for S_2.

Also for the surface S_3 *i.e.,* the upper curcular end of the cylinde on the plane z = 3, the outward drawn normal is upwards and parallel to z-axis, so n = k

$\therefore \int_{S_3} F . n\, dS = \int_{S_3} (4xi - 2y^2 j + z^2 k) . k\, dS$

$= 9 \int_{S_3} dS,$ [$\because$ z = 3 for S_3]

= 9 S_3, where S_3 is the area of the upper end of the cylinder $x^2 + y^2 = 4$ *i.e.,* a circle of radius 2

$= 9 [\pi (2)^2] = 36\pi$ . . . (iv)

And for the curved surface S_1 of the cylinder, we have

$$n = \frac{\nabla \phi}{|\nabla \phi|}, \text{ where } \phi \equiv x^2 + y^2 = 4$$

$$= \frac{2(xi + yj)}{|2(xi + yj)|},$$

$$\left[\because \nabla \phi = i \frac{\partial \phi}{\partial x} + j \frac{\partial \phi}{\partial y} + k \frac{\partial \phi}{\partial z} = 2xi + 2yj\right]$$

$$= \frac{si + yj}{\sqrt{(x^2 + y^2)}} = \frac{xi + yj}{2}, \qquad [\because x^2 + y^2 = 4]$$

Also for S_1, z varies from z = 0 to z = 3

and take x = 2 cos θ y = 2 sin θ so that θ varies from 0 to 2π.

$\therefore \int_{S_1} F . n\, dS = \int_{S_1} (4xi - 2y^2 j + z^2 k) . \frac{1}{2} (xi + j)\, dS$

$= \int_{S_1} (2x^2 - y^2)\, dS$

$= \int_{z=0}^{3} \int_{\theta=0}^{2\pi} [2(2 \cos\theta)^2 - (2 \sin \theta)^3]\, 2\, d\theta\, dz,$

$\because$ dS = element of the area on the curved surface

= (2 d θ) dz

$= \int_{z=0}^{3} \int_{\theta=0}^{2\pi} 8 (\cos^2 \theta - \sin^3 \theta)\, d\theta\, dz$

$= 16 [z]_0^3 \int_0^{2\pi} (\cos^2 \theta - \sin^3 \theta)\, d\theta$

$= 16 \times 3 \times 2 \int_0^{\pi} (\cos^2 \theta\, d\theta,$ $\qquad [\because \{\sin (2\pi - \theta)\}^3 = - \sin^3\theta,]$

so by the property of definite integral value of the second integral is zero

$= 96 \times 2 \int_0^{\pi/2} \cos^2 \theta\, d\theta = 192 \left[\frac{1}{2} \cdot \frac{\pi}{2}\right] = 48\,\pi$ . . . (v)

$\therefore$ From (ii), (iii), (iv), and (v) we get

$\int_S F \cdot n = 0 + (36\,\pi) + (48\,\pi) = 84\,\pi$ . . . (iv)

$\therefore$ From (i) and (vi) we have

$\int_S F \cdot n\, dS = \int_V \nabla \cdot F\, dV,$

and thus divergence theorem is verifed.

55. *Verify the divergence theorem for $F = 2x^2 i - y^2 j + 4xz\, k$ taken over the region S in the first octant bounded by $y^2 + z^2 = 9$, $x = 0$ and $x = 2$.*

Solution : Here $\nabla \cdot F = \frac{\partial}{\partial x}(2x^2) + \frac{\partial}{\partial y}(-y^2) + k \frac{\partial}{\partial z}(4xz)$

$= 4x - 2y + 4x = 8x - 2y$

$\therefore \int_V \nabla \cdot F\, dV = \int_V (8x - 2y)\, dV,$

where V is the colume enclosed by S

$= \int_{x=0}^{2} \int_{y=0}^{3} \int_{z=0}^{\sqrt{(9-y^2)}} (8x - 2y)\, dx\, dy\, dz,$

as in last example.

$= \int_{x=0}^{2} \int_{y=0}^{3} (8x - 2y) \sqrt{(9 - y^2)}\, dx\, dy$

$= 8 \int_{x=0}^{2} \int_{y=0}^{3} x \sqrt{(9 - y^2)}\, dx\, dy - 2 \int_{x=0}^{2} \int_{y=0}^{3} y \sqrt{(9 - y^2)}\, dx\, dy$

$= 8 \left[\frac{1}{2} x^2\right]_0^2 \left[\frac{1}{2} y \sqrt{(9 - y^2)} + \frac{1}{2}\, 9 \sin^{-1}\left(\frac{1}{3} y\right)\right]_0^3$

$- 2 [x]_0^2 \int_0^2 y \sqrt{(9 - y^2)}\, dy$

$$= 8\left[9\left(\frac{1}{2}\pi\right) + 4\int_3^0 t^2\,dt,\right] \text{ Putting } 9 - y^2 = t^2$$

$$= 36\pi - 4\left[\frac{1}{3}t^3\right]_0^3 = 36\pi - 36 \qquad \ldots \text{(i)}$$

Again the surface S is the surface consisting of two equal quadrants of circular ends of the cylinder given by x = 0 and x = 2 and let us call them as S_2 and S_3 respectively; the curved surface S_1 (say of the cylinder lying between the planes y = 0 and y = 0 respectively

For the surface S_1 as defined above, we have

$\phi \equiv y^2 + z^2 = 9$, [$\because$ the equation of the cylinder is $y^2 + z^2 = 9$]

$$\therefore \quad \nabla\phi = \left(i\frac{\partial}{\partial x} + j\frac{\partial}{\partial y} + k\frac{\partial}{\partial z}\right)\phi$$

$$= i\frac{\partial}{\partial x}(y^2 + z^2 - 9) + j\frac{\partial}{\partial y}(y^2 + z^2 - 9) + k\frac{\partial}{\partial z}(y^2 + z^2 - 9)$$

or $$\nabla\phi = j\,(2y) + (2z) = 2\,(yj + zk)$$

$$\therefore \quad 'n' = \frac{\nabla\phi}{|\nabla\phi|} = \frac{2(yj + zk)}{\sqrt{[(2y)^2 + (2z)^2]}} = \frac{2\,(yj + zk)}{2\sqrt{(y^2 + z^2)}}$$

$$= \frac{1}{3}(yj + zk), \text{ since } y^2 + z^2 = 9$$

$$\therefore \quad F \cdot n = (2x^2\,i - y^2\,j + 4\,xzk) \cdot 1/3\,(yj + zk)$$

$$= \frac{1}{3}(-y^3 + 4xz^2)$$

Also as in take $y = r\cos\phi$, $z = r\sin\phi$ and on the cylinder $y^2 + z^2 = 9$ $r = 3$ ϕ varies from 0 to $1/2\,\pi$.

Also dS = elementary area on the surface of the cylinder

$$= 3\,d\phi\,dx \qquad \therefore$$

$$\int_{S_1} F \cdot n\,dS = \int_{x=0}^{2}\int_{\phi=0}^{\pi/2}\frac{1}{3}(-y^3 + 4xz^2)\,3\,d\phi\,dx$$

$$= \int_{x=0}^{2}\int_{\phi=0}^{\pi/2}(-27\cos^3\phi + 36x\sin^2\phi)\,d\phi\,dx,$$

Putting $y = 3\cos\phi$, $z = 3\sin\phi$

$$= -27(x)_0^2 \int_0^{\pi/2} \cos^3 \phi d\phi + 36\left(\frac{1}{3}x^2\right)_0^2 \int_0^{\pi/2} \sin^2 \phi \, d\phi$$

$$= -(2)\left(\frac{2}{3}\right) + 36\,(2)\left(\frac{1}{2}\cdot\frac{1}{2}\pi\right) = 18\pi - 36. \qquad \ldots \text{(ii)}$$

On the area S_2, which is a quadrant of the circular cross section of the cylinder $y^2 + z^2 = 9$ by the plane $x = 0$, we have n = outwad drawn unit normal vector to $S_2 = -i$, as the outward drawn normal to this surface will be in the negative side of x-axis.

$$\therefore \int_{S_2} F \cdot n \, dS = \int_{S_2} (2x^2 i - y^2 j + 4xzk) \cdot (-i)\, dS$$

$$= -2 \int_{S_2} x^2 \, dS = 0, \text{ as } x = 0, \text{ on } S_2 \qquad \ldots \text{(iii)}$$

Similarly, on the area S_3, which is a quadrant of the circular cross-section of the cylinder $y^2 + z^2 = 9$, by the plane $x = 2$, we have 'n' = i, as the outward drawn normal to the area S_3 will be in the positive direction of x-axis.

$$\therefore \int_{S_3} F \cdot n \, dS = \int_{S_3} (2x^2 i + y^2 j + 4xzk) \cdot i\, dS$$

$$= 2 \int_{S_3} x^2 \, dS, \text{ on the plane } x = 2$$

$$= 8 \int_{S_3} dS = 8\, S_3, \text{ where } S_3 \text{ is as defined above}$$

$$= 8\left[\frac{1}{4}\pi\,(3)^2\right] = 18\,\pi \qquad \ldots \text{(iv)}$$

Again for S_4 which is a rectangle on the plane $z = 0$, we have $n = k$, as the outward drawn unit normal vector to S_1 will be in the negative direction of z-axis.

$$\therefore \int_{S_4} F \cdot n \, dS = \int_{S_4} (2x^2 i - y^2 j + 4xzl) \cdot (-k)\, dS$$

$$= \int_{S_4} (-4xz)\, dS, \text{ where } S_4 \text{ lies on } z = 0$$

$$= 0, \text{ as } z = 0 \qquad \ldots \text{(v)}$$

And for S_5, which is a rectangle on the plane $y = 0$, we have $n = -j$, as the outward drawn normal vector to S_5 will be in the negative direction of y-axis.

$$\therefore \int_{S_5} F \cdot n \, dS = \int_{S_5} (2x^2 i - y^2 j + 4xzk) \cdot (-j)\, dS$$

$$= \int_{S_5} y^2 \, dS, \text{ on the plane } y = 0$$

$$= 0 \qquad \ldots \text{(vi)}$$

$\therefore$ Adding (ii), (iii), (iv), (v) and (vi) we have for the entire surface S of the quadrant of the cylinder

$$\int_S \mathbf{F} \cdot d\mathbf{S} = (18\pi - 36) + 0 + 18\pi + 0 + 0 = 36\pi - 36$$

$$= \int_V \nabla \cdot \mathbf{F}\, dV, \text{ from (i).}$$

Hence Gauss' Theorem is verified.

56. *Evaluate $\int_s F \cdot n\, dS$ where S denotes the surface of the ellipsoid $(x^2/a^2) + (y^2/b^2) + (z^2/c^2) = 1$ and $F = x^2i + y^2j + z^2k$.*

Solution : Given $F = x^2i + y^2j + z^2k$

$$\therefore \text{div } F = \nabla \cdot F = \left(i\frac{\partial}{\partial x} + j\frac{\partial}{\partial y} + k\frac{\partial}{\partial z}\right) \cdot (x^2i + y^2j + z^2k)$$

$$= \frac{\partial}{\partial x}(x^2) + \frac{\partial}{\partial y}(y^2) + \frac{\partial}{\partial z}(z^2) = 2(x + y + z) \qquad \text{. . . (i)}$$

Now by Gauss' divergence theorem, we have

$$\int_S F \cdot n\, dS = \int_V \text{div } F\, dV,$$

where V is the volume enelosed by the given ellipsoid.

Also we know that if $(\bar{x}, \bar{y}, \bar{z})$ be the corrdinates of the centroid of a solid then.

$$\bar{x} = \frac{\int_V x\,dV}{\int_V dV};\ \bar{y} = \frac{\int_V y\,dV}{\int_V dV};\ \bar{z}\ \frac{\int_V z\,dV}{\int_V dV}$$

In the case of the given ellipsoid, its centroid is the origin

$\therefore\ \bar{x} = 0 = \bar{y} = \bar{z}$

Consequently, $\int_V dV = 0 = \int_V y\, dV = \int_V z\, dV$ . . . (ii)

Now $\int_S F \cdot n\, dS = \int_V \text{div } F\, dV = \int_V 2(x + y + z)\, dV$, from (i)

$= 2\int_V (x + y + z)\, dV = 2\int_V x\, dV + 2\int_V y\, dV + 2\int_V z\, dV$

$= 2(0) + 2(0) + 2(0)$, from (ii)

$= 0$ **Ans.**

57. *Verify divergence theorm for $F = (x^2 - yz)\, i + (y^2 - z\,x)\, j + (z^2 - xy)\, k$ taken over the rectangular parallelopiped $0 \leq x \leq a$, $0 \leq y \leq b$, $0 \leq z \leq c$.*

Solution : (consider it as rectangular parallelopiped whose edges are OB = a, OC = b and OF = c).

Now $\qquad F = (x^2 - yz)\, i + (y^2 - zx)\, j + (z^2 - xy)\, k$

$$\therefore \; F = \frac{\partial}{\partial x}(x^2 - yz) + \frac{\partial}{\partial y}(y^2 - zx) + \frac{\partial}{\partial z}(z^2 - xy)$$

$$= 2x + 2y + 2z = 2(x + y + z)$$

$\therefore \int_V \text{div } F \, dV$, where V is the volume of the rectangular parallelopiped.

$$= \int_{x=0}^{a} \int_{y=0}^{b} \int_{z=0}^{c} 2(x + y + z)\, dx\, dy\, dz$$

$$= 2 \int_{x=0}^{a} \int_{y=0}^{b} \left[xz + yz + \frac{1}{2} z^2 \right]_{z=0}^{c} dx\, dy$$

$$= 2 \int_{x=0}^{a} \int_{y=0}^{b} \left(xc + yc + \frac{1}{2} c^2 \right) dx\, dy$$

$$= 2 \int_{0}^{a} \left(cxy + \frac{1}{2} cy^2 + \frac{1}{2} c^2 y \right)_{y=0}^{b} dx$$

$$= 2 \int_{0}^{a} \left(cbx + \frac{1}{2} cb^2 + \frac{1}{2} c^2 b \right) dx$$

$$= 2 \left[\frac{1}{2} cbx^2 + \frac{1}{2} cb^2 x + \frac{1}{2} c^2 bx \right]_0^a$$

$$= cba^2 + cb^2 a + c^2 ba = abc\,[a + b + c] \qquad \ldots \text{(i)}$$

Now, we shall calculate $\int F \,.\, n \, dS$ over the six faces of the rectangular parallelopiped separately and then add them to get the value of $\int_S F \,.\, n \, dS$, where S is the entire surface of the rectangular parallelopiped.

For the face ABED, we have $n = 1$ $x = a$, $dS = dy\, dz$

$\therefore \int F \,.\, n \, dS$ over the face ABED

$$= \int_{z=0}^{c} \int_{y=0}^{b} [(x^2 - yz)\, i + (y^2 - zx)\, j + (z^2 - xy)\, k] \,.\, i \, dy\, dz$$

$$= \int_{z=0}^{c} \int_{y=0}^{b} (x^2 - yz)\, dy\, dz, \text{ over the face ABED for which } x = a$$

$$= \int_{z=0}^{c} \int_{y=0}^{b} (a^2 - yz)\, dy\, dz = \int_{z=0}^{c} \left(a^2 y - \frac{1}{2} y^2 z \right)_{y=0}^{b} dz$$

$$= \int_{0}^{c} \left(a^2 b - \frac{1}{2} b^2 z \right) dz = \left(a^2 bz - \frac{1}{4} b^2 z^2 \right)_0^a = a^2 bc - \frac{1}{4} b^2 c^2$$

$$= \frac{1}{4} bc\,(4a^2 - bc) \qquad \ldots \text{(ii)}$$

For the face OCKF, we have $n = -i$, $x = 0$, $dS = dy\, dz$

$\therefore \int$ F . n dS over the face OCKF

$$= \int_{z=0}^{c} \int_{y=0}^{b} [(x^2 - yz)\, i + (y^2 - zx)\, j + (z^2 - xy)\, k] . (-i)\, dy\, dz$$

$$= \int_{z=0}^{c} \int_{y=0}^{b} (x^2 - yz)\, dy\, dz, \quad \text{for } x = 0$$

$$= \int_{z=0}^{c} \int_{y=0}^{b} yz\, dy\, dz = \left(\frac{1}{2}y^2\right)_0^b \left(\frac{1}{2}z^2\right)_0^a = \frac{1}{4}\, b^2c^2 \qquad \text{. . . (iii)}$$

For the face ABOC, we have n = – k, dS = dx dy and z = 0

$\therefore \int$ F . dS over the face ABOC

$$= \int_{x=0}^{a} \int_{y=0}^{b} [(x^2 - yz)\, i + (y^2 - zx)\, j + (z^2 - xy)\, k] . (-k)\, dx\, dy$$

$$= - \int_{x=0}^{a} \int_{y=0}^{b} (z^2 - xy)\, dx\, dy, \quad \text{for } z = 0$$

$$= \int_{x=0}^{a} \int_{y=0}^{b} xy\, dx\, dy = \left(\frac{1}{2}x^2\right)_0^a \left(\frac{1}{2}y^2\right)_0^b = \frac{1}{4}\, a^2 b^2 \qquad \text{. . . (iv)}$$

For the face DEFK, we have n = k, dS = dx dy and z = c

$\therefore \int$ F . n dS over the face DEFK

$$= \int_{x=0}^{a} \int_{y=0}^{b} (z^2 - xy)\, dx\, dy, \text{ for } z = c, \text{ as above}$$

$$= \int_{x=0}^{a} \int_{y=0}^{b} (c^2 - xy)\, dx\, dy = \frac{1}{4}\, ab\, (4c^2 - ab) \qquad \text{. . . (v)}$$

For the face OADK we have n = j, dS = dx dz and y = b

$\therefore \int$ F . n dS over the face OADK

$$= \int_{x=0}^{a} \int_{z=0}^{c} (b^2 - xy)\, dx\, dz, \text{ proceeding as for face ABED}$$

$$= \frac{1}{4}\, ac\, (4b^2 - ac) \qquad \text{. . . (vi)}$$

And for the face OBEF, n = = j dS = dx dz and y = 0

$\therefore \int$ F . n dS over the face OBEF

$$= - \int_{x=0}^{a} \int_{z=0}^{c} (y^2 - xz)\, dx\, dz, \quad \text{for } y = 0 \text{ as for face OCKF}$$

$$= \int_{x=0}^{a} \int_{z=0}^{c} xz\, dx\, dz = \frac{1}{4}\, a^2c^2 \qquad \text{. . . (vii)}$$

Adding (ii) to (vii) we get for the whole surface S of the rectangular parallelopiped.

$$\int_S \mathbf{F} \cdot \mathbf{n}\, dS = \frac{1}{4} bc\,(4a^2 + bc) + \frac{1}{4} b^2c^2 + \frac{1}{4} a^2b^2 + \frac{1}{4} ab\,(4c^2 - ab)$$

$$+ \frac{1}{4} ac\,(4b^2 - ac) + \frac{1}{4} a^2c^2$$

$$= a^2 bc + abc^2 + acb^2 = abc\,(a + b + c) \qquad \ldots \text{(viii)}$$

Hence from (i) and (vii) we find that

$$\int_S \mathbf{F} \cdot \mathbf{n}\, dS = \int_V \operatorname{div} \mathbf{F}\, dV$$

Hence divergence theorem is verified.

58. *From the equation* $\nabla \cdot D = \rho$, $\nabla \cdot H = 0$,

$\nabla \times D = \frac{1}{c}\frac{\partial H}{\partial t}$, $\nabla \times H = \frac{1}{c}\left(\frac{\partial D}{\partial t} + \rho v\right)$ *in which t is the time variable, c a constant and the other symbols denote point-function, prove that*

(i) $\nabla^2 D - \frac{1}{c^2}\frac{\partial^2 D}{\partial t^2} = \nabla \rho + \frac{1}{c^2}\frac{\partial}{\partial t}(\rho v)$

and (ii) $\nabla^2 H - \frac{1}{c^2}\frac{\partial^2 H}{\partial t^2} = -\frac{1}{c^2}\nabla \times (\rho v)$

Solution : (i) Given that $\nabla \times D = -\frac{1}{c}\frac{\partial H}{\partial t}$

Taking curl of both sides, we get

$$\nabla \times (\nabla \times D) = \nabla \times \left(-\frac{1}{c}\frac{\partial H}{\partial t}\right) = -\frac{1}{c}\frac{\partial}{\partial t}(\nabla \times H)$$

$$= -\frac{1}{c}\frac{\partial}{\partial t}\left[\frac{1}{c}\frac{\partial D}{\partial t} + \rho v\right], \text{ putting value of } \nabla \times H$$

or $$\nabla(\nabla \cdot D) - \nabla^2 D = -\frac{1}{c^2}\frac{\partial^2 D}{\partial t^2} - \frac{1}{c^2}\frac{\partial}{\partial t}(\rho v)$$

or $$\nabla(\rho) - \nabla^2 D = -\frac{1}{c^2}\frac{\partial^2 D}{\partial t^2} - \frac{1}{c^2}\frac{\partial}{\partial t}(\rho v), \qquad [\because \nabla \cdot D = \rho]$$

or $$\nabla^2 D - \frac{1}{c^2}\frac{\partial^2 D}{\partial t^2} = \nabla \rho + \frac{1}{c^2}\frac{\partial}{\partial t}(\rho v).$$ Hence proved.

(ii) Given that $\nabla \times H = \frac{1}{c}\left(\frac{\partial D}{\partial t} + \rho v\right)$

Taking curl of both sides, we get

$$\nabla \times (\nabla \times H) = \nabla \times \left[\frac{1}{c}\left(\frac{\partial D}{\partial t} + \rho\ v\right)\right]$$

$$= \frac{1}{c}\frac{\partial}{\partial t}(\nabla \times D) + \frac{1}{c}\nabla \times (\rho\ v)$$

or $\nabla(\nabla . H) - \nabla^2 H = \frac{1}{c}\frac{\partial}{\partial t}\left[-\frac{1}{c}\frac{\partial H}{\partial t}\right] + \frac{1}{c}\nabla \times (\rho\ v)$

substituting value of $\nabla \times D$

or $\nabla(0) - \nabla^2 H = -\frac{1}{c^2}\frac{\partial^2 H}{\partial t^2} + \frac{1}{c}\nabla \times (\rho\ v)$ $[\because \nabla . H = 0]$

or $\nabla^2 H - \frac{1}{c^2}\frac{\partial^2 H}{\partial t^2} + -\frac{1}{c}\nabla \times (\rho\ v)$ Hence proved.

59. *Find the directional directive of the function in the direction of the vector 2i + j – k at the point (2, 3, 1).*

Solution : If $\hat{a}$ be the unit vector in the given direction, then

$$\hat{a} = \frac{2i + j - k}{\sqrt{[2^2 + 1^2 + (-1)^2]}} = \frac{2i + j - k}{\sqrt{6}}$$

and grad $\phi = i\frac{\partial \phi}{\partial x} + j\frac{\partial \phi}{\partial y} + k\frac{\partial \phi}{\partial z}$

$$= i\ (yz^2) + j\ (xz^2) + k\ (2x\ y\ z)$$

$\therefore$ grad f at (2, 3, 1)

$= i\ [3\ .\ 1^2] + j\ [2\ .\ 1^2] + k\ [2\ .\ 2.\ 3.\ 1] = 3i + 2j + 12k$

$\therefore$ The required directional derivative at (2, 3, 1)

$= \frac{"d\phi"}{ds} = \hat{a}$. grad ϕ at (2, 3, 1)

$= \frac{2i + j - k}{\sqrt{6}} . (3i + 2j + 12k)$

$= (1/\sqrt{6})\ (2i + j - k)\ .\ (3i + 2j + 12k) = (1/\sqrt{6})\ (6 + 2 - 12)$

$= -4/\sqrt{6} = -(2/\sqrt{6})/3$ **Ans.**

60. *Find the directional derivative of $\phi = x^2yz + 4xz^2$ at (1, –2, –1) in the direction 2i –j – 2k.*

Solution : Here $\hat{a}$ = the unit vector in the direction of 2i = j = 2k

$$= \frac{2i - j - 2k}{\sqrt{[2^2 + (-1)^2 + (-2)^2]}} = \frac{2i - j - 2k}{3}$$

And grad ϕ at (1, –2, –1)

$$= i\frac{\partial \phi}{\partial x} + j\frac{\partial \phi}{\partial y} + k\frac{\partial \phi}{\partial z} \text{ at } (1, -2, -1)$$

$$= i\,(2xyz + 4z^2) + j\,(x^2z) + k\,(x^2y + 8xz)$$

at (1, –2, –1), $[\because = x^2\,yz + 4xz^2]$

$$= i\,[8] + j\,[-1] + k\,[-10] = 8i = j - 10k$$

∴ The required directional derivative at (1, –2, –1)

$$= \frac{"d\phi"}{ds} = \hat{a}\ .\ \text{grad } f \text{ at } (1, -2, -1)$$

$$= \frac{1}{3}\,(2i - j - 2k)\ .\ (8i - j - 10k)$$

$$= \frac{1}{3}\,(16 + 1 + 20) = \frac{1}{3}\,(37)$$ **Ans.**

61. *Find the directional derivative of $\phi = xy + yz + zx$ in the direction of the vector 2i + j – k at the point (1, 0, 2). Also find the maximum value of directional derivative at the point.*

Solution : Here $\hat{a}$ = the unit vector in the direction of 2i + j – k

$$= \frac{2i + j - k}{\sqrt{[2^2 + 1^2 + (-1)^2]}} = \frac{2i + j - k}{\sqrt{6}}$$

And grad $\phi = i\frac{\partial \phi}{\partial x} + j\frac{\partial \phi}{\partial y} + k\frac{\partial \phi}{\partial z}$ at (1, 0, 2)

$= i\,(z + z) + j\,(z + x) + k\,(x + y)$ at (1, 0, 2) $[\because \phi = xy + yz + zx]$

$= i\,(0 + 2) + j\,(2 + 1) + k\,(1 + 0)$

$= 2i + 3j + k$ . . . (i)

∴ The required directional derivative at (1, 0, 2)

$$= \frac{"d\phi"}{ds} = \hat{a}\ .\ \text{grad } \phi \text{ at } (1, 0, 2)$$

$$= \frac{2i + j + k}{\sqrt{6}} . (2i + 3j + k) = \frac{1}{\sqrt{6}} (4 + 3 + 1) = \sqrt{6}$$ **Ans.**

Also maximum value of directional derivative at (1, 0, 2)

= modulus of grad ϕ at (1, 0, 2)

= modulus of 2i + 3j + k, from (i)

$= \sqrt{(2^2 + 3^2 + 1^2)} = \sqrt{(14)}$ **Ans.**

62. *Find the unit normal to the surface $x^4 - 3xyz + z^2 + 1 = 0$ at the point (1, 1, 1).*

Solution : Let $\phi = x^4 - 3x\,y\,z + z^2 + 1$

$$\therefore \text{ grad } \phi = i \frac{\partial \phi}{\partial x} + j \frac{\partial \phi}{\partial y} + k \frac{\partial \phi}{\partial z}$$

$= i\,(4x^3 - 3yz) + j\,(-3xz) + (-3xy + 2z)$ at (1, 1, 1)

$= i\,(4 - 3) + j\,(-3) + k\,(-3 + 2)$

$= i - 3j - k$

We know grad ϕ is a vector normal to the surface ϕ (x, y, z) = constant at the point (x, y, z)

$$\therefore \text{ Required unit normal } = \frac{\text{grad } \phi}{|\text{grad } \phi|} = \frac{i - 3j - k}{|i - 3j - k|}$$

$$= \frac{\text{grad } \phi}{|\text{grad } \phi|} = \frac{i - 3j - k}{|i - 3j - k|}$$ **Ans.**

63. *If ϕ (x, y, z) = xy^2z and $f = xz\,i - xy\,j + yz^2\,k$, show that* $\frac{\partial^2}{\partial x^2 \partial z}(\phi f)$ *at (2, – 1, 1) is 4i + 2j.*

Solution:

We have $\phi f = xy^2z\,(xz\,i - xy\,j + yz^2 k)$

$= x^2y^2z^2\,i - x^3y^3z\,j + xy^3z^3\,k.$

$$\therefore \frac{\partial}{\partial x}(\phi f) = 2xy^2z^2\,i - 2xy^3z\,j + y^3z^3\,k.$$

$$\frac{\partial^2}{\partial x^2}(\phi f) = \frac{\partial}{\partial x}\left[\frac{\partial}{\partial x}(\phi f)\right]$$

$$= \frac{\partial}{\partial x}[2xy^2z^2\,i - 2xy^3z\,j + y^3z^3\,k]$$

$$= 2y^2z^2\ i - 2y^3z\ j + 0\ k = 2y^2z^2\ i - 2y^3z\ j$$

and $\frac{\partial^3}{\partial x^2 \partial z}(\phi f) = \frac{\partial}{\partial z}\left[\frac{\partial^2}{\partial x^2}(\phi f)\right]$

$$= \frac{\partial}{\partial z}(2y^2z^2\,i - 2y^3 - j)$$

$$= 4y^2z\ i - 2y^3\ j.$$

$\therefore$ at the point (2, – 1, 1), $\frac{\partial^3}{\partial x^2 \partial z}(\phi f)$

$$= 4\,.\,(-1)^2\,.\,1\ i - 2\,.\,(-1)^3\ j$$

$$= 4\ i + 2\ j.$$

64. *If $f = (2x^2y - x^4)\ i + (e^{xy} - y \sin x)\ j + x^2 \cos y\ k$, erify that*

$$\frac{\partial^2 f}{\partial y\, \partial x} = \frac{\partial^2 f}{\partial x\, \partial y}.$$

Solution:

We have $\frac{\partial f}{\partial x}$

$$= \left[\frac{\partial}{\partial x}(2x^2y - x^4)\right]i + \left[\frac{\partial}{\partial x}(e^{xy} - y\sin x)\right]j + \left[\frac{\partial}{\partial x}(x^2\cos y)\right]k$$

$$= (4xy - 4x^3)\ i + (ye^{xy} - y\cos x)\ j + 2x\cos y\ k.$$

$$\therefore \quad \frac{\partial^2 f}{\partial x\, \partial y} = \frac{\partial}{\partial y}\left(\frac{\partial f}{\partial x}\right)$$

$$= \frac{\partial}{\partial y}[(xxy - 4x^3)i + (ye^{xy} - y\cos x)j + 2x\cos y\ k]$$

$$= \left[\frac{\partial}{\partial y}(4xy - 4x^3)\right]i + \left[\frac{\partial}{\partial y}(ye^{xy} - y\cos x\right]j + \left[\frac{\partial}{\partial y}(2x\cos y)\right]k$$

$$= (4x)\ i + (e^{xy} + xy\,e^{xy} - \cos x)\ j + (-2x\sin y)\ k$$

$$= 4x\ i + (e^{xy} + xy\,e^{xy} - \cos x)\ j - 2x\sin y\ k. \qquad \text{...(i)}$$

Again $\quad \frac{\partial f}{\partial y} = \left[\frac{\partial}{\partial y}(2x^2y - x^4)\right]j + \left[\frac{\partial}{\partial y}(e^{xy} - y\sin x)\right]$

$$j+\left[\frac{\partial}{\partial x}(x^2 \sin y)\right]k$$

$= 2x^2\, i + (xe^{xy} - \sin x)\, j - x^2 \sin y\, k.$

$$\therefore \quad \frac{\partial^2 f}{\partial y\, \partial x} = \frac{\partial}{\partial x}\left(\frac{\partial f}{\partial y}\right) = \left[\frac{\partial}{\partial x}(2x^2)\right]i+\left[\frac{\partial}{\partial x}(xe^{xy} - \sin x)\right]$$

$$j-\left[\frac{\partial}{\partial x}(x^2 \sin y)\right]k$$

$= 4x\, i + (e^{xy} + xye^{xy} - \cos x)\, j - 2x \sin y\, k.$ ...(2)

From (1) and (2), we have

$$\frac{\partial^2 f}{\partial x\, \partial y} = \frac{\partial^2 f}{\partial y\, \partial x}.$$

65. *Prove that $f(u)\, \nabla u = \nabla^* f(u)\, du$.*

Solution:

We have $\nabla^* f(u)\, du = \sum i \frac{\partial}{\partial x}\left\{\int f(u)du\right\}$ [By def. of gradient]

$$= \sum i\left\{\frac{d}{du}\int f(u)du\right\}\frac{\partial u}{\partial x}$$

$$= \sum i f(u)\frac{\partial u}{\partial x} = f(u) \sum i\frac{\partial u}{\partial x} = f(u)\, \nabla u.$$

66. *Show that*

(i) *grad (r.a) = a.*

(ii) *grad [r, a, b] = a × b,*

where a and be are constant vectors.

Solution:

(i) Let $a = a_1\, i + a_2 j + a_3 k$.

Then a_1, a_2, a_3 are constants.

Also $r = xi + yj + zk$.

$\therefore$ $r.a = a_1 x + a_2 y + a_3 z$.

$\therefore$ grad (r.a) $= \nabla$ (r.a) $= \nabla\, (a_1 x + a_2 y + a_3 z)$

$$= i\frac{\partial}{\partial x}(a_1x+a_2y+a_3z)+j\frac{\partial}{\partial y}(a_1x+a_2y+a_3z)+k\frac{\partial}{\partial z}(a_1x+a_2y+a_3z)$$

$= a_1 \, i + a_2 \, j + a_3 \, k = a.$

(ii) grad [r, a, b] = grad {r.(a × b)}, where a × b is a constant vector = a × b, as in part (i).

67. *Prove that* $A \cdot \left(\nabla \frac{1}{r}\right) = -\frac{A.r}{r^3}$.

Solution:

We have

$$\nabla\left(\frac{1}{r}\right) = i\frac{\partial}{\partial x}\left(\frac{1}{r}\right) + j\frac{\partial}{\partial y}\left(\frac{1}{r}\right) + k\frac{\partial}{\partial}\left(\frac{1}{r}\right)$$

$$= i\left(-\frac{1}{r^2}\frac{\partial r}{\partial x}\right) + j\left(-\frac{1}{r^2}\frac{\partial r}{\partial y}\right) + k\left(-\frac{\partial}{r^2}\frac{\partial r}{\partial z}\right)$$

$$= -\frac{1}{r^2}\left(\frac{\partial r}{\partial x}i + \frac{\partial r}{\partial y}j + \frac{\partial r}{\partial z}k\right) = -\frac{1}{r^2}\left(\frac{x}{r}i + \frac{y}{r}j + \frac{z}{r}k\right)$$

$$= -\frac{1}{r^3}(xi + yj + zk) = -\frac{1}{r^3}r.$$

$$\therefore \; A \cdot \nabla\left(\frac{1}{r}\right) = A \cdot \left(-\frac{r}{r^3}\right) = -\frac{A.r}{r^3}.$$

68. *Prove that* $\nabla r^{-3} = -3r^{-5}\, r$.

Solution:

We have r = x i + y j + z k

and $\quad r = |r| = \sqrt{(x^2 + y^2 + z^2)}$

so that $r^2 = x^2 + y^2 + z^2$.

Now $\quad \nabla r^{-3} = \left(\frac{\partial}{\partial x}r^{-3}\right)i + \left(\frac{\partial}{\partial y}r^{-3}\right)j + \left(\frac{\partial}{\partial z}r^{-3}\right)k$

$$= -3r^{-4}\frac{\partial r}{\partial x}i - r^{-4}\frac{\partial r}{\partial y} - 3r^{-4}\frac{\partial r}{\partial z}k$$

$$= -3r^{-4}\left(\frac{\partial r}{\partial x}i + \frac{\partial r}{\partial y}j + \frac{\partial r}{\partial z}k\right). \qquad ...(1)$$

Differentiating both sides of

$r^2 = x^2 + y^2 + z^2$

partially w.r.t. w, we have

$$2r\frac{\partial r}{\partial x} = 2x \text{ or } \frac{\partial r}{\partial x} = \frac{x}{r}.$$

Similarly $\frac{\partial r}{\partial y} = \frac{y}{r}$ and $\frac{\partial r}{\partial z} = \frac{z}{r}$.

$\therefore$ from (1), we have

$$\nabla r^{-3} = -3r^{-4}\left(\frac{x}{r}i+\frac{y}{r}j+\frac{z}{r}k\right)$$

$$= -3r^{-5}(xi + yj + zk) = -3r^{-5}\,r.$$

69. *Establish the results :*

(i) div (ϕ A) = ϕ div A + A. grad ϕ and

(ii) div (A × B) = B. curl A – A. curl B.

Solution:

(i) We have
$$\text{div}(\phi A) = \sum i.\frac{\partial(\phi A)}{\partial x} = \sum i\left(\phi\frac{\partial A}{\partial x}+\frac{\partial \phi}{\partial x}A\right)$$

$$= \sum i.\phi\frac{\partial A}{\partial x}+\sum i.\frac{\partial \phi}{\partial x}A = \phi\sum i.\frac{\partial A}{\partial x}+\left(\sum i\frac{\partial \phi}{\partial x}\right)A.$$

$= \phi$ div A + A. grad. ϕ.

(ii)
$$\text{div}(A\times B) = \sum i.\frac{\partial(A\times B)}{\partial x} = \sum i\left(\frac{\partial A}{\partial x}\times B + A\times\frac{\partial B}{\partial x}\right)$$

$$= \sum i.\frac{\partial A}{\partial x}\times B+\sum i.A\times\frac{\partial B}{\partial x} = \sum i\times\frac{\partial A}{\partial x}B-\sum i\times\frac{\partial B}{\partial x}A$$

$$= \left(\sum i\times\frac{\partial A}{\partial x}\right).B-\left(\sum i\times\frac{\partial B}{\partial x}\right).A \quad \text{curl } A.B-\text{curl } B.A$$

= B. curl A – A. curl B.

70. *If $\phi = 1/r$, where $r^2 = x^2 + y^2 + z^2$ then find $\Delta\phi$.*

Solution : $\nabla\phi = [1/\sqrt{(x^2 + y^2 + z^2)}]$

$$= \left(i\frac{\partial}{\partial x}+j\frac{\partial}{\partial y}+k\frac{\partial}{\partial z}\right)\left[\frac{1}{\sqrt{(x^2+y^2+z^2}}\right]$$

$$= \sum i\frac{\partial}{\partial x}\left[\frac{1}{\sqrt{(x^2+y^2+z^2}}\right]$$

$$= i\left[-\frac{1}{2}\left(x^2+y^2+z^2\right)^{-3/2}2x\right]+j\left[-\frac{1}{2}\left(x^2+y^2+z^2\right)^{-3/2}2y\right]$$

$$+ j\left[-\frac{1}{2}\left(x^2 + y^2 + z^2\right)^{-3/2} 2z\right]$$

$$= \frac{-xi - yj - zk}{\left(x^2 + y^2 + z^2\right)^{3/2}} - \frac{r}{r^3}$$, where r = xi + yi + zk **Ans.**

71. *Prove that a . grad* $\left(\frac{a}{r}\right) = \frac{a \cdot r}{r^3}$ *where a is a constant vector and r = xi + yi + zk.*

Solution:

(a) above we can prove (to be proved) that

grad (1/r) = ∇ (1/r) = –r/r³

$$\therefore \quad a \cdot \text{grad}\left(\frac{1}{r}\right) = a \cdot \left(-\frac{r}{r^3}\right) = -\frac{a \cdot r}{r^3}.$$ Hence proved.

72. *If* $\phi(x, y) = \log \sqrt{(x^2 + y^2)}$, *show that grad* $\phi = \frac{r - (k \cdot r)\, k}{\{r - (k \cdot r)\, k\} \cdot (k \cdot r)\, k\}}$

Solution : grad $\phi = \nabla\phi = \nabla \log \sqrt{(x^2 + y^2)}$

$$= \frac{1}{2} \nabla \log (x^2 + y^2)$$

$$= \frac{1}{2}\left(i\frac{\partial}{\partial y} + j\frac{\partial}{\partial y} + k\frac{\partial}{\partial z}\right) \log (x^2 + y^2)$$

$$= \frac{1}{2}\left[i\frac{\partial}{\partial y}\log (x^2 + y^2) + j\frac{\partial}{\partial y}\log (x^2 + y^2) + k\frac{\partial}{\partial z}\log (x^2 + y^2)\right]$$

$$= \frac{1}{2}\left[\left\{\frac{1}{x^2 + y^2} 2x\right\} j + \left\{\frac{1}{x^2 + y^2} 2y\right\} j + \{0\}\, k\right]$$

$$= (xi + yj)/(x^2 + y^2)$$

$$= \frac{(xi + yj)}{(xi + yj) \cdot (xi + yj)},$$ [∵ i . i = 1, i . j = 0]

$$= \frac{r - zk}{(r - zk) \cdot (r - zk)},$$ [∵ r = xi + yj + zk]

$$= \frac{r - (k \cdot r)\, k}{\{r - (k \cdot r)\, k\} \cdot \{r - (k \cdot r)\, k\}},$$ [∵ k . r = k . {xi + yj + zk) = z]

73. *If r = xi + yi + zk , find* Δr^n.

Solution : r = | r | = $\sqrt{(x^2 + y^2 + z^2)}$

or $$r^2 = x^2 + y^2 + z^2$$

$\therefore$ $$2r\frac{\partial r}{\partial x} = 2x \text{ or } \frac{\partial r}{\partial x} = \frac{x}{r} \quad \ldots \text{(i)}$$

Similarly $$\frac{\partial r}{\partial y} = \frac{y}{r} \text{ and } \frac{\partial r}{\partial z} = \frac{z}{r} \quad \ldots \text{(ii)}$$

$$\therefore \quad \nabla r^n = \left(i\frac{\partial}{\partial x} + j\frac{\partial}{\partial y} + k\frac{\partial}{\partial z}\right) r^n$$

$$= i\frac{\partial}{\partial x}(r^n) + j\frac{\partial}{\partial y}(r^n) + k\frac{\partial}{\partial z}(r^n)$$

$$= i\left(nr^{n-1}\frac{\partial r}{\partial x}\right) + j\left(nr^{n-1}\frac{\partial r}{\partial y}\right) + k\left(nr^{n-1}\frac{\partial r}{\partial z}\right)$$

$$= nr^{n-1}\left[i\left(\frac{x}{r}\right) + j\left(\frac{y}{r}\right) + \left(\frac{z}{r}\right)\right], \text{ from (i) and (ii)}$$

$$= nr^{n-2}(x\,i + y\,j + z\,k) = nr^{n-2}\,r. \textbf{ Ans.}$$

74. *Prove that* $\nabla\phi \,.\, dr = d\phi$.

Solution: We have

$$\nabla\phi = \frac{\partial\phi}{\partial x}i + \frac{\partial\phi}{\partial y}j + \frac{\partial\phi}{\partial z}k. \quad \ldots(1)$$

Also $$r = xi + yj + zk$$

$\Rightarrow dr = dx\,i + dy\,j + dz\,k.$ ...(2)

From (1) and (2), $\nabla\phi \,.\, dr$

$$= \left(\frac{\partial\phi}{\partial x}i + \frac{\partial\phi}{\partial y}j + \frac{\partial\phi}{\partial z}k\right).(dx\,i + dy\,j + dz\,k)$$

$$= \frac{\partial\phi}{\partial x}dx + \frac{\partial\phi}{\partial y}dy + \frac{\partial\phi}{\partial z}dz = d\phi.$$

75. *Show that* $\frac{d\phi}{ds} = \nabla\phi \,.\, \frac{dr}{ds}$, *where* $r = xi + yj + zk$ *and* ϕ *is a function of x, y and z.*

Solution: We have

$$\nabla\phi = \frac{\partial\phi}{\partial x}i + \frac{\partial\phi}{\partial y}j + \frac{\partial\phi}{\partial z}k \quad \ldots(1)$$

Also $\quad r = xi + yj + zk$

$$\Rightarrow \quad \frac{dr}{ds} = \frac{dx}{ds}i + \frac{dy}{ds}j + \frac{dz}{ds}k. \qquad ...(2)$$

From (1) and (2), we have

$$\nabla\phi \cdot \frac{dr}{ds} = \left(\frac{\partial\phi}{\partial x}i + \frac{\partial\phi}{\partial y}j + \frac{\partial\phi}{\partial z}k\right) \cdot \left(\frac{dx}{ds}i + \frac{dy}{ds}j + \frac{dz}{ds}k\right)$$

$$= \frac{\partial\phi}{\partial x}\frac{dx}{ds} + \frac{\partial\phi}{\partial y}\frac{dy}{ds} + \frac{\partial\phi}{\partial z}\frac{dz}{ds} = \frac{d\phi}{ds}.$$

76. *ρ and p are two scalar point functions such that ρ is a function of p: show that*

$$\nabla\rho = \frac{d\rho}{dp}\nabla p.$$

Solution: We have

$$\nabla\rho = \frac{\partial\rho}{\partial x}i + \frac{\partial\rho}{\partial y}j + \frac{\partial\rho}{\partial z}k. \qquad ...(1)$$

Since ρ is a function of p, therefore

$$\frac{\partial\rho}{\partial x} = \frac{d\rho}{dp}\frac{\partial p}{\partial x}, \quad \frac{\partial\rho}{\partial y} = \frac{d\rho}{dp}\frac{\partial p}{\partial y}, \quad \frac{\partial\rho}{\partial z} = \frac{d\rho}{dp}\frac{\partial p}{\partial z}.$$

∴ from (1), we have

$$\nabla\rho = \frac{d\rho}{dp}\frac{\partial p}{dx}i + \frac{\partial\rho}{dp}\frac{\partial p}{\partial y}j + \frac{d\rho}{dp}\frac{\partial p}{\partial z}k$$

$$= \frac{d\rho}{dp}\left(\frac{\partial p}{\partial x}i + \frac{\partial p}{\partial y}j + \frac{\partial p}{\partial z}k\right) = \frac{d\rho}{dp}\nabla p.$$

77. *That* $\nabla^2 (r^n r) = n(n+3) r^{n-2} r.$

Solution : $\nabla^2 (r^n r) = \left(\frac{\partial^2}{\partial x^2} + \frac{\partial^2}{\partial y^2} + \frac{\partial^2}{\partial z^2}\right)(r^n r) = \Sigma \frac{\partial^2}{\partial x^2}(r^n r)$

$$= \Sigma \frac{\partial}{\partial x}\left[\frac{\partial}{\partial x}(r^n r)\right] = \Sigma \frac{\partial}{\partial x}\left[n\, r^{n-1}\frac{\partial r}{\partial x} r + r^n \frac{\partial^2 r}{\partial x}\right]$$

$$= \Sigma \frac{\partial}{\partial x}\left[nr^{n-1}\left(\frac{x}{r}\right) r + r^n\, (i)\right], \quad \left[\because \frac{\partial r}{\partial x} = \frac{x}{r}, \frac{\partial r}{\partial x} = i, \text{ as above}\right]$$

$$= \Sigma \frac{\partial}{\partial x}\left[n r^{n-2}\, r + r^{n}\, i\right]$$

$$= \Sigma\left[n\left\{r^{n-2}\, r + x\,(n-2) r^{n-3}\frac{\partial r}{\partial x} r + x\, r^{n-2}\frac{\partial r}{\partial x}\right\} + n r^{n-1}\frac{\partial r}{\partial x} i\right]$$

$= n\,\Sigma\,[r^{n-2}\, r + (n-2)\, x^2\, r^{n-4}\, r + xi\, r^{n-2} + r^{n-2}\, xi]$,

$$\left[\because \frac{\partial r}{\partial x} = \frac{x}{r}, \frac{\partial r}{\partial x} = i\right]$$

$= n\,[3\, r^{n-2}\, r + (n-2)\, r^{n-4}\, r\,(x^2 + y^2 + z^2) + r^{n-2}\,(xi + yj + zk)$
$+ r^{m-2}\,(xi + yj + zk)]$

$= n\,[3\, r^{n-2}\, r + (n-2)\, r^{n-2}\, r + r^{n-2}\, r + r^{n-2}\, r^{n-2}]$, $[\because r = xi + yj + zk]$

$= n\,[3 + (n-2) + 1 + 1]\, r^{n-2}\, r = n\,(n+3)\, r^{n-2}\, r.$

Hence proved.

78. *Show that* $\nabla^2 f(r) = (2/r)\, f'(r)\, f''(r)$.

Solution : $\nabla^2 f(r) = \dfrac{\partial^2}{\partial x^2} f(r) + \dfrac{\partial^2}{\partial y^2} f(r)\, \dfrac{\partial^2}{\partial z^2} f(r)$

$$= \Sigma \frac{\partial^2}{\partial x^2} f(r) = \Sigma \frac{\partial}{\partial x}\left[\frac{\partial}{\partial x} f(r)\right]$$

$$= \Sigma \frac{\partial}{\partial x}\left[f'(r)\frac{\partial r}{\partial x}\right] = \Sigma \frac{\partial}{\partial x}\left[f'(r)\frac{x}{r}\right], \qquad \left[\because \frac{\partial r}{\partial x} = \frac{x}{r}\right]$$

$$= \Sigma\left[\frac{f'(r)}{r} + f''(r)\frac{x}{r}\frac{\partial r}{\partial x} + x f'(r)\left(-\frac{1}{r^2}\frac{\partial r}{\partial x}\right)\right]$$

$$= \Sigma\left[\frac{f'(r)}{r} + \frac{x^2}{r^2} f''(r) - \frac{x^2}{r^3} f'(r)\right], \qquad \left[\because \frac{\partial r}{\partial x} = \frac{x}{r}\right]$$

$$= \frac{3f'(r)}{r} + \frac{f''(r)}{r^2}(x^2 + y^2 + z^2) - \frac{f'(r)}{r}(x^2 + y^2 + z^2)$$

$$= \frac{3f'(r)}{r} + \frac{f''(r)}{r^2}(r^2) - \frac{f'(r)}{r^3}, \qquad \left[\because x^2 + y^2 + z^2 = r^2\right]$$

$$= \frac{2}{r} f'(r) + f''(r)$$ Hence proved.

79. *If a is constant vector, r = xi + yi + zk and r = √(x² + y² + z²), prove that* $\nabla\left(a\,.\,\nabla\frac{1}{r}\right) = \frac{a}{r^3} + \frac{3(a\,.\,r)\,r}{r^5}$

Solution : $\nabla\left(\frac{1}{r}\right) = \left(i\frac{\partial}{\partial x} + j\frac{\partial}{\partial y} + k\frac{\partial}{\partial z}\right)\left[\frac{1}{\sqrt{(x^2+y^2+z^2)}}\right]$

$$= \Sigma i\left[-\frac{1}{2}\left(x^2+y^2+z^2\right)^{-3/2}.\,2x\right] = -\,\Sigma\, i\left[x\left(x^2+y^2+z^2\right)^{-3/2}\right]$$

$$= -\,\Sigma\, i\,[x\,(r^2)^{-3/2}] = -\,r^{-3}\,\Sigma\,(xi) = -\,r^{-3}\,r$$

$\therefore$ $a\,.\,\nabla\,(1/r) = a\,.\,(-r^{-3}r) = -r|\,(a\,.\,r)$

$= -r^{-3}\,(a_1x + z_2y + a_3z)$, if $a = a_1i + a_2j + a_3k$

$\therefore$ $[a\,.\,\nabla\,(1/r)]$

$$= \left(i\frac{\partial}{\partial x} + j\frac{\partial}{\partial y} + k\frac{\partial}{\partial z}\right)\left[-\frac{a_1}{r^3}x - \frac{a_2}{r^3}y - \frac{a_3}{r^3}z\right]$$

$$\nabla\left[a\,.\,\nabla\left(\frac{1}{r}\right)\right] = -\,\Sigma i\left[-\frac{\partial}{\partial x}\left(\frac{x a_1}{r^3}\right) + \frac{\partial}{\partial x}\left(\frac{a_2y}{r^3}\right) + \frac{\partial}{\partial x}\left(\frac{a_3z}{r^3}\right)\right] \quad ...(i)$$

Now $\frac{\partial}{\partial x}\left[\frac{x a_1}{r^3}\right] = \left[\frac{a_1}{r^3} - \frac{3}{r^4}\,xa_1\,\frac{\partial}{\partial x}\right]$, where $\frac{\partial r}{\partial x} = \frac{x}{r}$

or $$\frac{\partial}{\partial x}\left[\frac{x a_1}{r^3}\right] = \left[\frac{a_1}{r^3} - \frac{3a_1x^2}{r^5}\right] \quad ...\,(ii)$$

$$\frac{\partial}{\partial x}\left[\frac{a_1y}{r^3}\right] = a_2y\left[-\frac{3}{r^4}\,\frac{\partial r}{\partial x}\right] = \frac{3a_2y}{r^4} = -\,\frac{3a_2xy}{r^5} \quad ...\,(iii)$$

and $$\frac{\partial}{\partial x}\left[\frac{a_1z}{r^4}\right] = a_3z\left[-\frac{3}{r^4}\,\frac{\partial r}{\partial x}\right] = -\,\frac{3a_3xz}{r^5} \quad ...\,(iv)$$

From (i), (ii), (iii), and (iv) we get

$$\nabla\,[a\,.\,\nabla\,(1/r) = -\,\Sigma\, i\left[\frac{a_1}{r^3} - \frac{3a_1x^2}{r^5} - \frac{3a_2xy}{r^5} - \frac{3a_2zx}{r^5}\right]$$

$$= -\,\Sigma\, i\left[\frac{1}{r^3}a_1 - \frac{3x}{r^5}\,(a_1x \;\; a_2y + a_3z)\right]$$

$$= -\,\frac{1}{r^3}\,\Sigma\, i\, a_1 + \frac{3}{r^5}\,(a_1x + a_2y + a_3z)\,\Sigma\, x i$$

$$= -\frac{a}{r^3} + \frac{3}{r^5}(a \cdot r)\, r \text{ Hence proved.}$$

80. *Verify Stoke's theorm when $F = (2x - y)\, i = y\, z^2\, j - y^2\, z\, k$ where S is the upper half of the sphere $x^2 + y^2 + z^2 = i$ and C is the boundary.*

Solution : Given

$$F = (2x - y)\, i - yz^2 j - y^2 zk$$

Also we know

$$r = xi + yj + zk$$

$\therefore$ $F \cdot dr = [(2x - y)\, i - yz^2 j - y^2 j - y^2 zk] \cdot [i\, dx + j\, dy + k\, dz]$

or $F \cdot dr = (2r = y)\, dx - yz2\, dy - y^2 - y^2 zdz$ . . . (i)

Also here C is the circle $x^2 + y^2 = 1$; $z = 0$ *i.e.*, a circle of unit radius on x y-plane.

$\therefore$ For C we have $z = 0$, $dz = 0$. consequently from (i) we have

$$F \cdot dr = (2x - y)\, dx \qquad \text{. . . (ii)}$$

Also on C we have $x = \cos\phi$ and $y = \sin\phi$ and ϕ varies from 0 to 2π.

$\therefore$ From (ii) we have

$$\int_C F \cdot dr = \int_C (2x - y)\, dx$$

$$= \int_{\phi=0}^{2\pi} (2\cos\phi - \sin\phi)(-\sin\phi)\, d\phi$$

$$= \int_{\phi=0}^{2\pi} (\sin^2\phi - \sin 2\phi)\, d\phi$$

$$= 4\int_0^{\pi/2} \sin^2\phi\, d\phi + \left[\frac{1}{2}\cos 2\phi\right]_0^{2\pi}$$

$$= 4 \cdot \frac{1}{2} \cdot \frac{1}{2}\pi + \frac{1}{2}[1 - 1] = \pi$$

Hence $\int_C F \cdot dr = \pi$ . . . (iii)

Also the direction cosines of the line OP are

$$\sin\theta\cos\phi,\ \sin\theta\sin\phi,\ \cos\theta$$

$\therefore$ $n = (\sin\theta\cos\phi)\, i + (\sin\theta\sin\phi)\, j + (\cos\theta)\, k$, where n is the unit outward drawn normal vector at P.

$$\text{Again curl } F = \begin{vmatrix} i & j & k \\ \dfrac{\partial}{\partial x} & \dfrac{\partial}{\partial y} & \dfrac{\partial}{\partial z} \\ 2x - y & -yz^2 & -y^2z \end{vmatrix}$$

$$= i\left[\frac{\partial}{\partial y}(-y^2x) - \frac{\partial}{\partial z}(-yz^2)\right] + j\left[\frac{\partial}{\partial z}(2x - y) - \frac{\partial}{\partial x}(-y^2z)\right]$$

$$+ k\left[\frac{\partial}{\partial x}(-yz^2) - \frac{\partial}{\partial y}(2x - y)\right]$$

$$= 0i + 0j + k = k$$

$$\therefore\ n \cdot \text{curl } F = [(\sin\theta \cos\phi)\, i + (\sin\theta \sin\phi)\, j + (\cos\theta)\, k] \cdot k$$

$$= \cos\theta$$

$$\therefore \int_S n \cdot \text{curl } F\, dS$$

$$= \int_S \cos\theta\, dS, \text{ where } dS = \sin\theta\, d\theta\, d\phi$$

$$= \int_{\theta=0}^{\pi/2} \int_{\phi=0}^{2\pi} \cos\theta \sin\theta\, d\theta\, d\phi$$

$$= \int_{\theta=0}^{\pi/2} \int_{\phi=0}^{2\pi} \frac{1}{2} \sin 2\theta\, d\theta\, d\phi = \frac{1}{2}\left[-\frac{1}{2}\cos 2\theta\right]_0^{\pi/2} [\phi]_0^{2\pi}$$

$$= \frac{1}{2}\left[\frac{1}{2}\cos\theta + \frac{1}{2}\cos\theta\right] 2\pi = \frac{1}{2}\left(\frac{1}{2} + \frac{1}{2}\right) . 2\pi = \pi \qquad \ldots \text{(iv)}$$

Hence from (iii) and (iv) we have

$$\int_C F \cdot dr = \pi = \int_S n \cdot \text{curl } F\, dS$$

which verifies Stoke's theorem.

81. *If $= xyz\, i + xz^2 j - y^3 k$ and $v = x^3 i - xyz\, j + x^2z\, k$, calculate* $\dfrac{\partial^2 u}{\partial y^2} \times \dfrac{\partial^2 v}{\partial x^2}$ *at the point (1, 1, 0).*

Solution: We have

$$\frac{\partial u}{\partial y} = \left[\frac{\partial}{\partial y}(xyz)\right] i + \left[\frac{\partial}{\partial y}(xz^2)\right] j - \left[\frac{\partial}{\partial y}(y^3)\right] k$$

$$= xz\, i + 0\, j - 3y^2\, k = xz\, i - 3y^2\, k.$$

$$\text{and } \frac{\partial^2 u}{\partial y^2} = \frac{\partial}{\partial y}\left(\frac{\partial u}{\partial y}\right) = 0\, i - 6y\, k = -6y\, k.$$

Again $$\frac{\partial v}{\partial x} = \left[\frac{\partial}{\partial x}(x^3)\right]i - \left[\frac{\partial}{\partial x}(xyz)\right]j + \left[\frac{\partial}{\partial x}(x^2z)\right]k$$

$$= 3x^2\ i - yz\ j + 2xz\ k$$

and $$\frac{\partial^2 v}{\partial x^2} = \frac{\partial}{\partial x}\left(\frac{\partial v}{\partial x}\right) = 6x\ i - 0\ j + 2z\ k = 6x\ i + 2z\ k.$$

$$\therefore \quad \frac{\partial^2 u}{\partial y^2} \times \frac{\partial^2 v}{\partial x^2} = (-6y\ k) \times (6x\ i + 2z\ k)$$

$$= -36xy\ k \times i - 12yz\ k \times k$$

$\therefore$ at the point (1, 1, 0), we have [$\because$ $k \times i = j$ and $k \times k = 0$]

$$\frac{\partial^2 u}{\partial y^2} \times \frac{\partial^2 v}{\partial x^2} = -36\ j.$$

82. *If $A = x^2yz\ i - 2xz^3\ j + xz^2\ k$, $B = 2z\ i + y\ j - x^2\ k$, find the value of $\frac{\partial^2}{\partial x\,\partial y}(A \times B)$ at (1, 0 – 2).*

Solution: We have

$$A \times B = \begin{vmatrix} i & j & k \\ x^2yz & -2xz^3 & xz^2 \\ 2z & y & -x^2 \end{vmatrix}$$

$$= (2x^3z^3 - xyz^2)\ i + (2xz^3 + x^4yz)\ j + (x^2y^2z + 4xz^4)\ k.$$

$$\therefore \quad \frac{\partial}{\partial y}(A \times B) = -xz^2\ i + x^4z\ j + 2x^2yz\ k.$$

Again $$\frac{\partial^2}{\partial x\,\partial y}(A \times B) = \frac{\partial}{\partial x}\left\{\frac{\partial}{\partial y}(A \times B)\right\}$$

$$= -z^2\ j + 4x^3z\ j + 4xyz\ k. \qquad ...(1)$$

Putting x = 1, y = 0 and z = – 2 in (1), we get the required derivative at the point (1, 0, – 2) = – 4 i – 8j.

83. *If $f(x, y, z) = 3x^2y - y^3z^2$, find grad f at the point (1, – 2, – 1).*

Solution: We have

$$\text{grad } f = \nabla f = \left(i\frac{\partial}{\partial x} + j\frac{\partial}{\partial y} + k\frac{\partial}{\partial z}\right)(3x^2y - y^3z^2)$$

$$= i\frac{\partial}{\partial x}(3x^2y - y^3z^2) + j\frac{\partial}{\partial y}(3x^2y - y^3z^2) + k\frac{\partial}{\partial z}(3x^2y - y^3z^2)$$

$$= i\ (6xy) + j(3x2 - 3y2z2) + k\ (-2y3z)$$

$$= 6xy\ i + (3x^2 - 3y^2z^2)\ j - 2y^3z\ k.$$

Putting $x = 1$, $y = -2$, $z = -1$, we get

$$\nabla f = 6\ (1)\ (-2)\ i + \{3(1)^2 - 3\ (-2)^2\ (-1)^2\}\ j - 2\ (-2)^3\ (-1)\ k$$

$$= -\ 121 - 9j - 16k.$$

84. *Find the curl of the vector $V = (x^2 + yz)\ i + (y^2 + zy\ j\ (z^2 + xy)\ k$ at the point (1, 2, 3).*

Solution : Curl V = $\nabla \times V$

$$= \left(i\frac{\partial}{\partial x} + j\frac{\partial}{\partial y} + \frac{\partial}{\partial z}\right) \times \left[(x^2 + yz)\ i + (y^2 + zx)\ j + (z^2 + xy)k\right]$$

$$= \begin{vmatrix} i & j & k \\ \frac{\partial}{\partial x} & \frac{\partial}{\partial y} & \frac{\partial}{\partial z} \\ x^2 + yz & y^2 + xz & z^2 + xy \end{vmatrix}$$

$$= i\left[\frac{\partial}{\partial y}(z^2 + xy) - \frac{\partial}{\partial z}(y^2 + zx)\right] + j\left[\frac{\partial}{\partial z}(x^2 + yz) - \frac{\partial}{\partial x}(z^2 + xy)\right]$$

$$+ k\left[\frac{\partial}{\partial x}(y^2 + zx) - \frac{\partial}{\partial y}(x^2 + yz)\right]$$

$$= i\ (x = x) + j\ (y - y) + k\ (z - z) = 0 \qquad \forall\ x, y, z$$

$\therefore$ Curl V at (1, 2, 30 is 0 **Ans.**

85. *If $r = |\,r\,|$ where $r = xi + yj + zk$, prove that*

(i) $\nabla f(r) = f'(r)\ \nabla r,$

(ii) $\nabla r = \frac{1}{r}r,$

(iii) $\nabla f(r) \times r = 0,$

(iv) $\nabla\left(\frac{1}{r}\right) = -\frac{r}{r^3},$

(v) $\nabla \log |r| = \frac{r}{r^2}$,

(vi) $\nabla r^n = n r^{n-2} r$.

Solution: If $r = xi + yj + zk$, then $r = |r| = \sqrt{(x^2 + y^2 + z^2)}$.

$\therefore \quad r^2 = x^2 + y^2 + z^2$.

(i) $\nabla f(r) = \left(i\frac{\partial}{\partial x} + j\frac{\partial}{\partial y} + k\frac{\partial}{\partial z} \right) f(r) = i\frac{\partial}{\partial x} f(r) + j\frac{\partial}{\partial y} f(r) + k\frac{\partial}{\partial z} f(r)$

$$= i\, f'(r) \frac{\partial r}{\partial x} + j f'(r) \frac{\partial r}{\partial y} + k\, f'(r) \frac{\partial r}{\partial z}$$

$$= f'(r) \left(i\frac{\partial r}{\partial x} + j\frac{\partial r}{\partial y} + k\frac{\partial r}{\partial z} \right) = f'(r)\, \nabla r.$$

(ii) We have $\nabla r = i\frac{\partial r}{\partial x} + j\frac{\partial r}{\partial y} + k\frac{\partial r}{\partial z}$.

Now $\quad r^2 = x^2 + y^2 + z^2$.

$\therefore \quad 2r \frac{\partial r}{\partial x} = 2x$ *i.e.,* $\frac{\partial r}{\partial x} = \frac{x}{r}$.

Similarly, $\quad \frac{\partial r}{\partial y} = \frac{y}{r}$ and $\frac{\partial r}{\partial z} = \frac{z}{r}$.

$\therefore \quad \nabla r = \frac{x}{r} i + \frac{y}{r} j + \frac{z}{r} k = \frac{1}{r} = \frac{1}{r} r = \hat{r}$.

(iii) We have in part (i), $\nabla f(r) = f'(r)\, \nabla r$.

But as in part (ii), $\nabla r = \frac{1}{r} r$.

$\therefore \quad \nabla r(r) = f'(r) \frac{1}{r} r$.

$\therefore \quad \nabla f(r) \times r = \left\{ f'(r) \frac{1}{r} r \right\} \times r$

$$= \left\{ \frac{1}{r} f'(r) \right\} (r \times r) = 0. \quad [\because r \times r = 0]$$

(iv) We have $\nabla \left(\frac{1}{r} \right) = i\frac{\partial}{\partial x}\left(\frac{1}{r} \right) + j\frac{\partial}{\partial y}\left(\frac{1}{r} \right) + k\frac{\partial}{\partial z}\left(\frac{1}{r} \right)$

$$= i\left(-\frac{1}{r^2}\frac{\partial r}{\partial x}\right) + j\left(-\frac{1}{r^2}\frac{\partial r}{\partial y}\right) + k\left(-\frac{1}{r^2}\frac{\partial r}{\partial z}\right)$$

$$= -\frac{1}{r^2}\left(\frac{\partial r}{\partial x}i + \frac{\partial r}{\partial y}j + \frac{\partial r}{\partial z}k\right)$$

$$= -\frac{1}{r^2}\left(\frac{x}{r}i + \frac{y}{r}j + \frac{z}{r}k\right)$$

$$= -\frac{1}{r^3}(xi + yj + zk) = -\frac{1}{r^3}r. \qquad \text{[sec part (ii)]}$$

(v) We have $\nabla \log |r| = \nabla \log r$

$$= i\frac{\partial}{\partial x}\log r + j\frac{\partial}{\partial y}\log r + k\frac{\partial}{\partial z}\log r$$

$$= \frac{1}{r}\frac{\partial r}{\partial x}i + \frac{1}{r}\frac{\partial r}{\partial y}j + \frac{1}{r}\frac{\partial r}{\partial z}k$$

$$= \frac{1}{r}\left(\frac{x}{r}i + \frac{y}{r}j + \frac{z}{r}k\right)$$

$$= \frac{1}{r^2}(xi + yj + zk) = \frac{1}{r^2}r.$$

(vi) We have $\nabla r^n = i\frac{\partial}{\partial x}r^n + j\frac{\partial}{\partial y}r^n + k\frac{\partial}{\partial z}r^n$

$$= i\, nr^{n-1}\frac{\partial r}{\partial x} + j\, nr^{n-1}\frac{\partial r}{\partial y} + k\, nr^{n-1}\frac{\partial r}{\partial z}$$

$$= nr^{n-1}\left(i\frac{\partial r}{\partial x} + j\frac{\partial r}{\partial y} + k\frac{\partial r}{\partial z}\right) = nr^{n-1}\nabla r$$

$$= nr^{n-1}\frac{1}{r}r \qquad \left[\because \nabla r = \frac{r}{r} \text{ a sin part (ii)}\right]$$

$$= nr^{n-2}r.$$

86. *Of 'a' is a constant vector and r any vector, then show that $\Delta(a \cdot r) = a$*

Solution. Let $a = a_1 i + a_2 j + a_3 k$

Then $a \cdot r = (a_1 i + a_2 j + a_3 k) \cdot (x i + y j + z k)$

$= (z_1 x + a_2 y + a_3 z)$

$$\therefore \quad \nabla (a \cdot r) = \left(i \frac{\partial}{\partial x} + j \frac{\partial}{\partial y} + k \frac{\partial}{\partial z} \right) (a_1 x + a_2 y + a_3 z)$$

87. *If ϕ and ψ are two scalar point functions and ϕ is a function of ψ, then prove that* $\Delta\phi = \Delta\phi \dfrac{d\phi}{d\psi}$.

Solution : $\nabla\phi = \Sigma i \dfrac{\partial \phi}{\partial x}$, by definition of grad ϕ or $\nabla\phi$

$$= \Sigma i \frac{\partial \phi}{\partial \psi} \frac{\partial \psi}{\partial x}, \qquad [\because \phi \text{ is a function of } \psi]$$

$$= \frac{\partial \phi}{\partial \psi} \Sigma i \frac{\partial \psi}{\partial x} = \frac{d\phi}{dx} \nabla \psi \quad \text{Hence proved.}$$

EXERCISES

1. A proton of 1.6×10^{-19} curl charge and 1.7×10^{-27} kg mass has velocity of 10^7 met/sec in a direction inclined at 30° to the direction of a magnetic field of strength 1 weber/metre^2. Calculate Lorentz force.
2. If ϕ (x, y, z) = $3x^2y - y^3x^2$; calculate gradient ϕ at the point (1, –2. –1).
3. Evaluate $\iint_S \vec{F}.d\vec{S}$ where $\vec{F} = 4xz\hat{i}_1 - y^2\hat{i}_2 + yz\hat{i}_3$; S is the surface bounded by x = 0 , x = 1, y = 1, z = 0, z = 1, and $\hat{i}_1, \hat{i}_2, \hat{i}_3$ are unit vectors along x, y, z axes.
4. Verify the divergence theorem for $F = 2x^2 i - y^2 j + 4xz\, k$ taken over the region S in the first octant bounded by $y^2 + z^2 = 9$, x = 0 and x = 2.
5. Explain clearly what do you understand by divergence and curl of a vector field $\vec{F}$ obtain expressions in general orthogonal coordinates for

   (a) div $\vec{F}$ and (b) curl $\vec{F}$
6. What do you mean by the flux of a vector field ? Show that flux of a vector $\vec{F}$ through the surface is given by

   $\iint_S \vec{F}\hat{n}\, dS$ or $\int_S \vec{F}.\hat{n}\, dS$,

   where $\hat{n}$ is unit vector pointing outward at right angle to the surface.

7. The gradient of scalar field grad f or $\vec{\nabla}$ f is

   (a) a vector (b) a scalar

8. if $\vec{\nabla} \times \vec{A} = 0$, there is a f such that

   (a) $\vec{A} = \vec{\nabla} f$ (b) $\vec{A} = \vec{\nabla} \times f$

9. If $\vec{\nabla} \cdot \vec{D} = 0$; there is a $\vec{C}$ such that

   (a) $\vec{D} = \vec{\nabla} \cdot \vec{C}$ (b) $\vec{D} = \vec{\nabla} \times \vec{C}$

10. If $\vec{B}$ is a magnetic field and its divergence $\vec{B} = 0$, everywhere, it means

    (a) the lines of $\vec{B}$ could have a source in the field $\vec{B}$

    (b) an isolated magnetic pole is possible

    (c) an isolated magnetic pole is not possible.

11. Why the vector fields are classified on the basis of different association of curl and divergence ? What are these combinations ? Illustrate by giving example.

12. Show that curl of a gradient of scalar is zero and also that the div of a curl of vector is zero *i.e.*,

    $$\text{curl grad f} = 0, \text{ where f is scalar}$$

    div curl $\vec{F} = 0$ where $\vec{F}$ is vector

    What you conclude from these relations ? How the idea of vector potential is introduced ?

13. What do you understand by the field of any physical quantity? Given examples of scalar and vector fields.

14. What do you understand by gradient of a scalar field ? If $\vec{r}$ be the position vector of any particle find the value of grad (1/r).

5

DIVERGENCE THEOREM OF GAUSS

INTRODUCTION

Suppose V is the volume bounded by a closed piecewise smooth surface S Suppose F(x, y, z) is a vector function of position which is continuous and has continuous first partial derivatives in V. Then

$$\iiint_V \nabla . F dV = \iiint_V \nabla . F dV,$$

where n is the outwards drawn unit normal vector to S.

Since F . n is the normal component of vector F, therefore divergence theorem may also be stated as follows:

The surface integral of the normal component of a vector taken over a closed surface is equal to the integral of the divergence of F taken over the volume enclosed by the surface.

Cartesian Equivalent of Divergence Theorem

Let $F = F_1 i + F_2 j + F_3 k.$

Then $\nabla F = \text{div } F = \frac{\partial F_1}{\partial x} + \frac{\partial F_2}{\partial y} + \frac{\partial F_3}{\partial z}.$

If α, β, γ are the angles which outward drawn unit normal n makes with positive directions of x, y, z-axes, then cos α, cos β, cos γ are direction cosines of n and we have

$$n = \cos\alpha \, i + \cos\beta \, j + \cos\gamma \, k.$$

$$\therefore \quad F . n = (F_1 i + F_2 j + F_3 k) \cdot (\cos\alpha \, i + \cos\beta \, j + \cos\gamma \, k)$$

$$= F_1 \cos\alpha + F_2 \cos\beta + F_3 \cos\gamma.$$

Therefore the divergence theorem can be written as

$$\iiint_V \left(\frac{\partial F_1}{\partial x} + \frac{\partial F_2}{\partial y} + \frac{\partial F_3}{\partial z} \right) dx\,dy\,dz$$

$$= \iint_S (F_1 \cos\alpha + F_2 \cos\beta + F_3 \cos\gamma) dS$$

$$= \iint_S (F_1 \cos\alpha + F_2 \cos\beta + F_3 \cos\gamma) dS.$$

The significance of divergence theorem lies in the fact that a surface integral may be expressed as a volume integral and vice varies.

Proof of the divergence theorem. We shall first prove the theorem for a special region V which is bounded by a piecewise smooth closed surface S and has time property that any straight line parallel to any cane of the coordinate axes and intersecting V has only one segment (or a single point) in common with V. If R is the orthogonal projection of S own the xy-plane, then V can be represented in the forms $f(x, y) \le z \le g(x, y)$ where (x, y) varies in R.

Obviously $z = g(x, y)$ represents the upper portion S_1 of S, $z = f(x, y)$ represents the lower portion S_2 of S and there may the a remaining vertical portion S_3 of S.

We have

$$\iiint_V \frac{\partial F_3}{\partial z} dV = \iiint_V \frac{\partial F_3}{\partial z} dx\,dy\,dz$$

$$= \iiint_V \frac{\partial F_3}{\partial z} dx\,dy\,dz$$

$$= \iiint_V \frac{\partial F_3}{\partial z} dx\,dy\,dz$$

$$= \iint_R [F_3[x, y, g(x, y)] - F_3[x, y, f(x, y)]] dx\,dy$$

$$= \iint_R F_3[x, y, g(x, y)] dx\,dy - \iint_R F_3[x, y, f(x, y)] dx\,dy$$

Now for the vertical portion S_3 of S, the normal n_3 to S_3 makes a right angle γ with k. Therefore

$$\iint_{S_3} F_3 k.n_3 dS_3 = 0, \text{ since } k \,.\, n_3 = 0.$$

For the upper portion S_1 of S, the normal n_1 to S_1 makes an acute angle γ_1 with k. Therefore

$$k \cdot n_1\, dS_1 = \cos\gamma_1\, dS_1 = dx\, dy.$$

Hence $\iint_{S_1} F_3 k.n_1 dS_1 = \iint_R F_3[x, y, g(x, y)]dx\,dy$

$\therefore \quad \iint_S F_3 k.n_3 dS_3 + \iint_{S_1} F_3 k.n_1 dS_1 \iint_{S_2} F_3 k.n_2 dS_2$

$= 0 + \iint_R F_3[x, y, g(x, y)]dx\,dy - \iint_R F_3[x, y, f(x, y)]dx\,dy$

or with the help of (1), we get

$$\iint_S F_3 k.ndS = \iint_V \frac{\partial F_3}{\partial z} dV. \qquad ...(2)$$

Similarly, be projecting S on the other co-ordinate planes, we get

$$\iint_S F_2 j.ndS = \iiint_V \frac{\partial F_2}{\partial y} dV \qquad ...(3)$$

and $\iint_S F_1 i.ndS = \iiint_V \frac{\partial F_1}{\partial x} dV \qquad ...(4)$

Adding (2), (3) and (4), we get

$$\iint_S (F_1 i + F_2 j + F_3 k).ndS$$

$$= \iiint_V \left(\frac{\partial F_1}{\partial x} + \frac{\partial F_2}{\partial y} + \frac{\partial F_3}{\partial z} \right) dV$$

or $\iiint_V \nabla.FdV = \iint_S F.ndS.$

The proof of the theorem can now be extended to a region V which can be subdivided into finitely many special regions of the above type by drawing auxiliary surfaces. In this case we apply the theorem to each sub-region and then add the results. The sum of the volume integrals over parts of V will be equal to the volume integral over V. The surface integrals over auxiliary surfaces cancel in pairs, while the sum of the remaining surface integrals is equal to the surface integral over the whole boundary S of V.

Note: The divergence theorem is applicable for a region V if it is bounded by two closed surfaces S1 and S2 one of which lies within the other. Here outward drawn normals will have the directions.

SOME DEDUCTIONS FROM DIVERGENCE THEOREM

1. Green's theorem : *Let ϕ and ϕ be scalar point functions which together with their derivatives in any direction are uniform and continuous within the region V bounded by a closed surface S, then*

$$\iiint_V (\phi\nabla^2\psi - \psi\nabla^2\phi)dV = \iint_S (\phi\nabla\psi - \psi\nabla\phi).n\,dS.$$

Proof:

By divergence theorem, we have $\iiint_V \nabla.F\,dV = \iint_S F.n\,dS$

Putting $F = \phi\nabla\psi$, we get $\nabla . F = \nabla.(\phi\nabla\psi) = \phi(\nabla . \nabla\psi) + (\nabla\phi).(\nabla\psi)$ $= \phi\nabla^2\psi + (\nabla\phi).(\nabla\psi)$.

Also $\quad F . n = (\phi\nabla\psi) . n.$

$\therefore$ divergence theorem gives

$$\iiint_V [\phi\nabla^2\psi + (\nabla\phi).(\nabla\psi)]dV = \iint_S (\phi\nabla\psi).n\,dS \qquad ...(1)$$

This is called *Green's first identify or theorem.*

Interchanging ϕ and ψ in (1), we get

$$\iiint_V [\psi\nabla^2\phi + (\nabla\psi).(\nabla\phi)]dV = \iint_S [\psi\nabla\phi].n\,dS \qquad ...(2)$$

Subtracting (2) from (1), we get

$$\iiint_V (\phi\nabla\psi - \psi\nabla^2\phi)dV = \iint_S (\phi\nabla\psi - \psi\nabla\phi).n\,dS \qquad ...(3)$$

This is called *Green's second identity or Green's theorem in symmetrical form.*

Since $\quad \nabla\psi = \dfrac{\partial\psi}{\partial n}n$ and $\nabla\phi = \dfrac{\partial\phi}{\partial n}n$, therefore

$$(f\nabla\psi - \psi\nabla\phi) . n = \left(\phi\frac{\partial\psi}{\partial n}n - \psi\frac{\partial\phi}{\partial n}n\right).n = \phi\,\frac{\partial\psi}{\partial n} - \psi\frac{\partial\phi}{\partial n}.$$

Hence (3) can also be written as

$$\iiint_V (\phi\nabla^2\psi - \psi\nabla^2\phi)dV = \iint_S \left(\phi\frac{\partial\psi}{\partial n} - \psi\frac{\partial\phi}{\partial n}\right)dS.$$

Note : Harmonic function. *If axalar point functicn ϕ satisfies Laplac' seqaution $\nabla^2\phi = 0$, then ϕ is called harmonic function. If f and y are both harmonic functions, then $\nabla^2\phi = 0$, $\nabla^2\psi = 0$.*

Hence from Green's second entity, we get

$$\iint_S \left(\phi \frac{\partial \psi}{\partial n} - \psi \frac{\partial \phi}{\partial n} \right) dS = 0.$$

2. *Prove that* $\iiint_V \nabla \phi \, dV = \iint_S \phi n \, dS.$

Proof:

By divergence theorem, we have

$$\iiint_V \nabla . F dV = \iint_S F . n dS.$$

Taking $F = \phi C$ where C is an arbitrary constant non-zero vector, we get

$$\iiint_V \nabla . (\phi C) dV = \iint_S (\phi C) . n dS.$$

Now $\nabla . (\phi C) = (\nabla \phi) . C + \phi(\nabla . C) = (\nabla \phi), C,$

since $\nabla . C = 0.$

Also $\phi(C) . n = C . (\phi n).$

$\therefore$ (1) becomes

$$\iiint_V C.(\nabla \phi) dV = \iint_S C.(\phi n) dS \text{ or } C . \iiint_V \nabla \phi \, dV = C . \iint_S (\phi n) dS$$

or $$C . \left[\iiint_V \nabla \phi \, dV - \iint_S \phi n \, dS \right] = 0.$$

Since C is an arbitrary vector, therefore we must have

$$\iiint_V \nabla \phi \, dV = \iint_S \phi n \, dS.$$

3. *Prove that* $\iiint_V \nabla \times B dV = \iint_S n \times B dS.$

Proof:

In divergence theorem taking $F = B \times C$, where C is an arbitrary constant vector, we get

$$\iiint_V \nabla . (B \times C) dV = \iint_S (B \times C) . n dS.$$

Now $\nabla . (B \times C) = C . \text{curl } B - B . \text{curl } C = C \cdot \text{curl } B,$

since curl C = 0.

Also $(B \times C) . n = [B, C, n] = [C, n, B] = C . (n \times B)$.

∴ (1) becomes

$$\iiint_V (C.\text{curl}B)dV = \iint_S C.(n \times B)dS$$

or $$C . \iiint_V (\nabla \times B)dV = C \cdot \iint_S (n \times B)dS$$

or $$C . \left[\iiint_V (\nabla \times B)dV - \iint_S (n \times B)dS\right] = 0.$$

Since C is an arbitrary vector therefore we can take C as a non-zero vector which is not perpendicular to the vector

$$\iiint_V (\nabla \times B)dV - \iint_S (n \times B)dS.$$

Hence we must have

$$\iiint_V (\nabla \times B)dV - \iint_S (n \times B)dS$$

$$= 0 \text{ or } \iiint_V (\nabla \times B)dV - \iint_S (n \times B)dS.$$

The theorem states that the *normal surface integral of a function* F *over the boundary of a closed surface* S (*i.e., the flux across* S) *is equal to the volume integral of the divergence of the function over the volume* V *enclosed by the surface, i.e..,*

$$\text{flux} = \iint_S F.dS = \iiint \hat{V} \text{ div } FdV = \iiint \hat{V} (\nabla.F)dV.$$

This may be easily shown as follows:

Let the surface S enclosing a volume V be divided up into a very large number of elementary volumes in the form of cubes or rectangular parallelopipeds adjoining each other. Imagine one such cube (Fig. 4.4). With its edges along the three Cartesian coordinate axes and their lengths dx, dy and dz respectively. The flux outwards through the left face is obviously given by $-\int F_x dydz$, where F_x is the x-component of the vector field F and the integral extends over the axes of the face.

Since the cube considered is an infinitesimal one this integral-may be taken to be very nearly equal to the product of the x-component of F at the centre P_1 of the face and the area dydz of the face. So that, denoting the x-component of F at the centre of the face by F_x (P_1), we have

flux through left face of the cube $= - F_x (P_1)$ dydz

Similarly, *flux through right face of the cube* $= F_x (P_2)$ dydz, the component $F_x (P_2)$ now being positive.

Taking $F_x(P_2) = F_x(P_1) + \frac{\partial F_x}{\partial x} dx$, we have

flux through right face of the cube $= \left[F_x(P_1) + \frac{\partial F_x}{\partial x} dx\right]$ dydz.

∴ *flux through the left and right faces*

$$= \left[F_x(P_1) + \frac{\partial F_x}{\partial x} dx\right] dydz - F_x(P_1) dydz = \frac{\partial F_x}{\partial x} dxdydz.$$

Similarly, *flux through top and bottom faces of the cube*

$$= \frac{\partial F_y}{\partial y} dxdydz$$

and *flux through the remaining faces of the cube*

$$= \frac{\partial F_z}{\partial z} dxdydz.$$

Therefore, flux through all the faces of the cube is given by the sum of all these. So that,

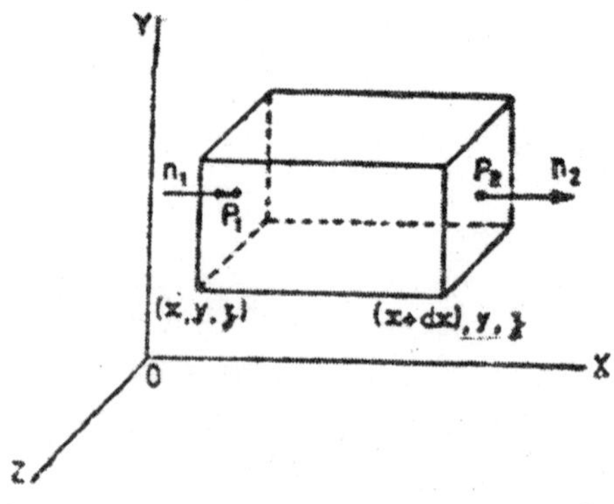

Fig. 1

$$\iint_{\substack{\text{Surface}\\ \text{of cube}}} F\, dS = \left(\frac{\partial F_x}{\partial x} + \frac{\partial F_y}{\partial y} + \frac{\partial F_z}{\partial z}\right) dxdydz.$$

where dS is the area of the surface of the cube.

Now $$\frac{\partial F_x}{\partial x}+\frac{\partial F_y}{\partial y}+\frac{\partial F_z}{\partial z} = \nabla.F \text{ or div } F$$

and $$dxdydz = dV,$$

volume of the cube.

We, therefore, have $$\iint_{\substack{\text{Surface}\\ \text{of cube}}} F\, dS = (\nabla.F)\, dV.$$

Therefore, summing up the fluxes through all the elemental cubes into which the surface is divided up, we have

$$\iint_S F dS = \iiint \hat{V}\, (\nabla.F) dV = \iiint \hat{V}\, (\text{div}\, F)\, dV,$$

which is Gauss's theorem.

Perhaps, more rigorously, it may be written as

$$\iint_S F.n\; dS = \iiint \hat{V}\, (\nabla.F) dV = \iiint \hat{V}\, (\text{div}\, F)\, dV,$$

where n is the *unit outward drawn normal vector.*

It may be noted that $\iint_S F dS$ gives the surface integral of F over the area of the bounding surface S enclosing volume V, because surfaces of the elemental cubes other than those forming part of surface S itself are common to two adjacent cubes and their surface integrals thus all cancel out in view of their oppositely directed normals.

Obviously, since $F.dS = F_x dS_x + F_y dS_y + F_z dS_z = F_x dydz + F_y dzdx + F_z dxdy$, we can express *Gauss's* theorem in terms of the Cartesian coordinates as (Fig. 2).

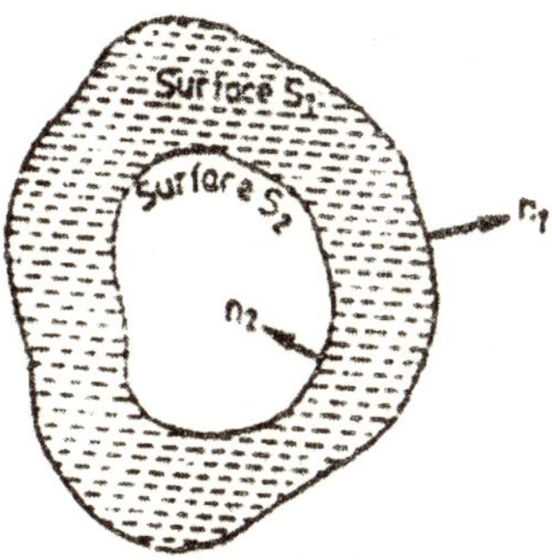

Fig. 2

$$\iint_S \left(F_x dS_x + F_y dS_y + F_z dS_z\right) = \iiint \hat{V} \left(\frac{\partial F_x}{\partial x}+\frac{\partial F_y}{\partial y}+\frac{\partial F_z}{\partial z}\right) dxdydz.$$

It may as well be mentioned here that Gauss's theorem applies also to a volume bounded by two closed surface (and, indeed by any number of closed surfaces). Only, it must be remembered that the outward drawn normals (n_1 and n_2) at any points on the surfaces concerned must point away from the enclosed volume, as shown.

STOKES' THEOREM

This theorem states that *the line integral (or the circulation) of a vector field* F *around any closed curve* C *is equal to the normal surface integral of the curl of* F *taken over any surface* S *of which curve* C *forms the contour or the boundary, i.e.,*

$$\oint_C F dl = \iint_S \text{curl } F.dS = \iint_S (\nabla \times F).dS.$$

This may be shown as follows:

Let a closed curve forming the boundary of a surface S of whatever shape (Fig. 3) be divided up into a very large number of small loops, infinitesimal in size and squarish in shape, all lying on the surface. These loops enclose small areas dS_1, dS_2 etc. inside them which are all flat, in view of the small size of the loops. Considering one such small squarish loop enclosing area dS_1, we have *line integral (or circulation) of vector* F *around it, i.e.,*

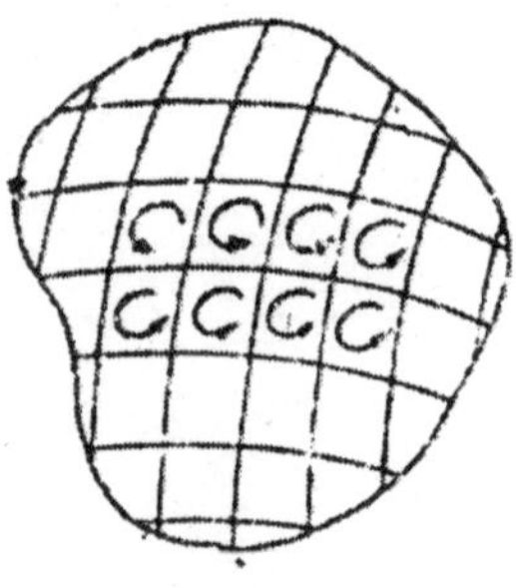

Fig. 3

$$\oint F dl = \text{curl } F dS_1 \, ,$$

The direction of the curl being perpendicular to the plane of area dS in accordance with the right-handed screw rule.

Now, if we sum up the left hand side, *i.e.*, the circulation around all these small loops into which the curve has been divided, we find that the contributions due to all the bounding lines of the loops, other than those that form part of the curve itself, cancel out since these contributions are all in the form of equal and oppositely directed pairs. So that, the summation simply gives the circulation or the line integral of the vector around the original closed curve alone.

And, summation on the right hand side of the expression obviously gives the surface integral of *curl* F over the entire surface S closed by the curve. So that, we have

$\oint$ Fdl = $\iint_S$curl F.dS, which is the required theorem.

APPLICATIONS OF GAUSS'S AND STOCKES' THEOREMS

(1) *To show that the curl of the gradient of a scalar function ϕ is equal to zero, i.e.,* $\nabla \times \Delta\ \phi = 0$. We know that the line integral of the gradient $\nabla\phi$ of the function along a closed loop is zero, *i.e.*,

$$\oint_{\text{loop}} \nabla\phi.\ dl = 0.$$

Now in accordance with Stokes' theorem, this line integral of the vector $\nabla\phi$ along a closed loop is equal to the surface integral of the curl of the vector over the surface B bounded by the loop, *i.e.*,

$$\oint \nabla\phi\ dl = \iint \hat{S}\ \nabla \times (\nabla\phi)\ dS.$$

It follows, therefore, that the integral

$$\iint \hat{S}\ \nabla \times (\nabla\phi) dS = 0$$

over any surface. And, since the integral is zero, the integrand must-necessarily be zero, *i.e.*,

$$\nabla \times \nabla\phi = 0,$$

a result we had obtained by vector algebra earlier.

(2) *To show that the divergence of the curl of a vector function f is zero. i.e.*, $\nabla . \nabla \times f = 0$. We know that according to Stocke' theorem, the normal surface integral of the curl of a vector f over a surface is equal to the line integral of the vector over the closed loop that forms the boundary of the surface. It follows, therefore that as the boundary, or the loop, decreases, *i.e.*, as the surface becomes more and more closed, (Fig. 4), the line integral of the vector around the loop, and hence the normal surface integral of the curl of the vector over the surface. Must progressively decrease until, when the loop decreases to a mere point, *i.e.*, the surface becomes closed, the line and the surface integrals become zero. *Thus, for a closed surface.*

$$\iint_S (\nabla \times f)\ dS = 0.$$

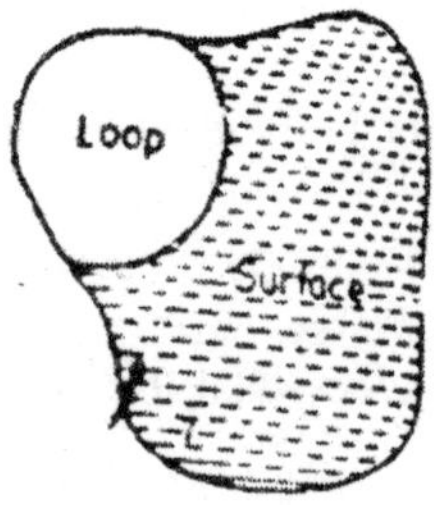

Fig. 4

But, in accordance with Gauss's theorem, this normal surface integral of a function over the boundary of a closed surface S is equal to the volume integral of the divergence of the function over the volume V enclosed by the surface. We therefore, have

$$\iint_S (\nabla \times f)\, dS \times \iiint v\ \Delta.(\nabla \times f)\, dV.$$

Obviously, therefore, $\iiint \hat{V}\, \nabla.(\nabla \times f)\, dV = 0$ for any vector field f. And, since this is so for any volume, it follows that at every point in space the

integrand $\nabla.(\nabla \times f) = 0$,

which, again, is a result we had arrived at earlier (iii) by vector algebra.

GREEN'S THEOREM IN THE PLANE

Let R be a closed bounded region in the x-y plane whose boundary C consists of finitely many smooth curves. Let M and N be continuous functions of x and y having continuous partial derivatives

$$\frac{\partial M}{\partial y} \ \textit{and}\ \frac{\partial N}{\partial x} \ \textit{in R. Then}$$

$$\iint_R \left(\frac{\partial N}{\partial x} - \frac{\partial M}{\partial y} \right) dx\, dy = \oint_C (M\, dx + N\, dy),$$

the line integral being taken along the entire boundary C of R such that R is on the left as one advances in the direction of integration.

Proof:

We shall first prove the theorem for a special region R bounded by a closed curve C and having the property that any straight line parallel to any one of the coordinate axes and intersecting R has only one segment (or a single point) in common with R. This means that R can be represented in both of the forms

$a \le x \le, f(x) \le y \le g(x)$

and $c \le y \le d$, $p(y) \le x \le q(y)$.

In the adjoining figure, the equations of the curves AEB and BFA are $y = f(x)$ and $y = g(x)$ respectively. Similarly, the equations of the curves FAE and EBF are $r = p(y)$ and $x = q(y)$ respectively.

We have

$$\iint_R \frac{\partial M}{\partial y} dx\, dy = \int_{x=d}^{b} \left\{ \int_{y=f(x)}^{f(t)} \frac{\partial M}{\partial y} dy \right\} dx$$

$$= \int_{x=d}^{b} [M(x,y)]_{y=f(t)}^{y=f(t)} dx$$

$$= \int_{x=d}^{b} [M|x, g(x)| - M|x, f(x)|] dx$$

$$= -\int_{d}^{b} M[x, f(x)] dx - \int_{b}^{d} M[r, g(r)] dx$$

$= -\oint_C M(x,y) dx$, since $y = f(x)$ represents the curve AEB and $y = g(r)$ represents the curve BFA.

If portions of C are segments parallel to y-axis such as GH and PQ in the adjoining figure, then the above result is not affected. The line integral $\int M\, dx$ over GH is zero because on GH, we have x = constant implies $dx = 0$. Similarly the line integral over PQ is zero. The equations of QG and HP are $y = f(x)$ and $y = g(x)$ respectively. Hence we have

$$-\iint_R \frac{\partial M}{\partial y} dx\, dy = \oint_C M(x,y) dx. \qquad ...(1)$$

Similarly, $\iint_R \frac{\partial N}{\partial x} dx\, dy$

$$= \int_{y=c}^{d} \left[\int_{x=p(y)}^{q(y)} \frac{\partial N}{\partial x} dx \right] dy$$

$$= \int_{y=c}^{d} [N(x,y)]_{r=p(y)}^{r=q(y)} dy$$

$$= \int_{y=c}^{d} [N|q(y), y| - N|P(y), y|] dy$$

$$= \int_c^d N[q(y), y]dy + \int_d^c N[p(y), y]dy$$

$$= \oint_C N(x, y)dy.$$

From (1) and (2), we get on adding

$$\iint_R \left(\frac{\partial N}{\partial x} - \frac{\partial M}{\partial y}\right)dx\,dy = \oint_C (M\,dx + N\,dy).$$

The proof of the theorem can now be extended to a region R which can be subdivided into finitely many special regions of the above type by drawing lines. In this case we apply the theorem to each subregion and then add the results. The sum of the left hand members will be equal to the integral over R. The sum of the right hand members will be equal to the integral over C plus the line integrals over the curves introduced for subdividing R. Each of the latter integrals comes twice, taken once in each direction. Therefore these two integrals cancel each other and thus the sum of the right hand members will be equal to the line integral over C.

Note: Extension of Green's theorem in plane to multiply connected regions.

Green's theorem in the plane is also valid for a multiply connected region R such as shown in the adjoining figure. Here the boundary C of R consists of two parts, the exterior boundary C_1 is traversed in the left, while the interior boundary C_2 is traversed in the clockwise sense so that R is on the left.

In order to establish the theorem, we construct a line such as AD (*called a cross cut*) connecting the exterior and interior boundaries. The region bounded by ADEFGDAPWLHA is simply connected and so Green's theorem is valid for it. Therefore

$$\oint_C M\,dx + N\,dy = \iint_R \left(\frac{\partial N}{\partial x} - \frac{\partial M}{\partial y}\right)dx\,dy$$

The integral on the left hand side leaving out the integrand is equal to

$$\int_{AD} + \int_{C_2} + \int_{DA} + \int_{C_1} = \int_{C_2} + \int_{C_1}, \text{ since } \int_{AD} = \int_{DA}$$

$$= \oint_C (M\,dx + N\,dy).$$

Hence the theorem.

GREEN'S THEOREM IN THE PLANE IN VECTOR NOTATION

We have r = xi + yj so that dr = dx i + dy j.

Let F = Mi + Nj.

Then $\quad M\,dx + N\,dy = (mi + Nj)\,.\,(dx\,i + dy\,j) = F\,.\,dr.$

Also curl F = ∇ F = $\begin{vmatrix} i & j & k \\ \dfrac{\partial}{\partial x} & \dfrac{\partial}{\partial y} & \dfrac{\partial}{\partial z} \\ M & N & 0 \end{vmatrix}$

$$= -\frac{\partial N}{\partial z}i + \frac{\partial M}{\partial z}j + \left(\frac{\partial N}{\partial x} - \frac{\partial M}{\partial y}\right)k.$$

$$\therefore \quad (\nabla \times F)\,.\,k = \frac{\partial N}{\partial x} - \frac{\partial M}{\partial y}.$$

Hence Green's theorem in plane can be written as

$$\iint_R (\nabla \times F).K = \oint_C F.dr$$

where dR = dx dy and k is unit vector perpendicular to the xy-plane.

If s denotes the are length of C and t is the unit tangent vector to C, then

$$dr = \frac{dr}{ds}ds = tds.$$

Therefore the above result can also be written as

$$\iint_R (\nabla \times F).k\,dR = \oint_C F.t\,ds.$$

LINE INTEGRALS

Any integral which is to be evaluated along a curve is called a line integral.

Suppose r (t) = x (t) i + y (t) j + z (t) k, where r (t) is the position vector of (x, y, z) *i.e.*, r (t) = xi + yj + jk, defines a piecewise smooth curve joining two points A and B. Let $t = t_1$ at A and $t = t_2$ at B. Suppose F (x, y, z) = $F_1i + F_2j + F_3k$ is a vector point function defined and continuous along C. If s denotes the are length of the curve C, then $\dfrac{dr}{ds}$ = t is a unit vector along the tangent to the curve C at the point r. The component of the vector F along this tangent is F . $\dfrac{dr}{ds}$. The integral of F .

$\frac{dr}{ds}$ along C from A to B written as

$$\int_A^B \left[F.\frac{dr}{ds}\right] ds = \int_A^B F.dr = \int_C F.dr$$

is an example of a *line integral.* It is called *the tangent line integral* of F along C.

Since r = xi + yj + zk, therefore, dr = dx i + dy j + dz k.

$\therefore$ F . dr = $(F_1 i + F_2 j + F_3 k) . (dx\ i + dy\ j + dz\ k)$

$= F_1\ dx + F_2\ dy + F_3\ dz.$

Therefore in components form the above line integral is written as

$$\int_C F.dr = \int_C (F_1 dx + F_2 dy + F_3 dz).$$

The parametric equations of the curve C are x = x (t) and z = z (t). Therefore we may write

$$\int_C F.dr = \int_{t_1}^{t_2} \left[F_1 \frac{dx}{dt} + F_2 \frac{dy}{dt} + F_3 \frac{dz}{dt}\right] dt.$$

Circulation : *If C is simple closed curve (i.e., a curve which does not intersect it sell anywhere), then the tangent line integral of F around C is called the circulation of F about C. It is often denoted by*

$$\oint_C F.dr = \oint_C (F_1 dx + F_2 dy + F_3 dz).$$

Work Done by a Force : Suppose a force F acts upon a particle. Let the particle be displaced along a given path C in space. If r denotes the position vector of a point on C, then $\frac{dr}{ds}$ is a unit vector along the tangent to C at the point r in the direction of s increasing. The component of force F along tangent to C is F . $\frac{dr}{ds}$. Therefore the work done by F during small displacement ds of the particle along C is $\left[F.\frac{dr}{ds}\right] ds$ *i.e.*, F . dr. The total work W done by F in this displacement along C, is given by the line integral

$$W = \int_C F.dr,$$

the integration being taken in the sense of the displacement.

SURFACE INTEGRALS

Any integral which is to be evaluated over a surface is called a surface integral.

Suppose S is a surface of finite area. Suppose f (x, y, z) is a single valued function of position defined over S. Subdivide the area S into n elements of areas $\delta S_1, \delta S_2, \ldots, \delta S_n$. In each part δS_k we choose an arbitrary point P_k whose coordinates are (x_k, y_k, z_k).

We define f(Pk) = f(xk, yk, zk).

Form the sum $\sum_{k=1}^{n} f(P_k)\delta S_k$.

Now take the limit of this sum as $n \to \infty$ in such a way the largest of the areas δS_k approaches zero. This limit if it exists, is called the *surface integral* of f(x, y, z) over S and is denoted by

$$\iint_S f(x,y,z)dS.$$

It can be shown that if the surface S is piecewise smooth and the function f(x, y, z) is continuous over S, then the above limit exists *i.e.*, is independent of the choice of sub-divisions and points P.

Flux : Suppose S is a piecewise smooth surface and F(x, y, z) is a vector function of position defined and continuous over S. Let P be any point on the surface S and let n be the unit vector at P in the direction of outward normal to the surface S at P. Then F . n is normal component of F at P. The integral of F . n over S is

$$\iint_S F.n dS.$$

It is called the *flux* of F over S.

Let us associate with the differential of surface area dS a vector dS (called vector area) whose magnitude is dS and whose direction is that of n. Then dS = n dS. Therefore we can write

$$\iint_S F.n dS = \iint_S F.dS.$$

Suppose the outward drawn normal to the surface S at P makes angles α, β, γ with the positive directions of x, y and z axes respectively, If *l*, m, n are the direction cosines of the outward drawn normal, then

$$l = \cos\alpha,\ m = \cos\beta,\ n = \cos\gamma$$

Also $\quad n = \cos\alpha\, i + \cos\beta\, j + \cos\gamma\, k = li + mj + nk.$

Let $F(x, y, z) = F_1 i + F_2 j + F_3 k$. Then

$$F \cdot n = F_1 \cos\alpha + F_2 \cos\beta + F_3 \cos\gamma = F_1 l + F_2 m + F_3 n.$$

Therefore we can write

$$\iint_S F.n\,dS = \iint_S (F_1 \cos\alpha + F_2 \cos\beta + F_3 \cos\gamma)\,dS$$

$$= \iint_S (F_1\,dy\,dz + F_2\,dz\,dx + F_3\,dx\,dy),$$

if we define $\iint_S F_1 \cos\alpha\,dS = \iint_S F_1\,dy\,dz,$

$$\iint_S F_2 \cos\beta\,dS = \iint_S F_2\,dz\,dx,\ \iint_S F_3 \cos\gamma\,dS = \iint_S F_3\,dx\,dy.$$

Notes:

1. Other examples of surface integrals are $\iint_S fn\,dS$, $\iint_S F\times dS$ where $f(x, y, z)$ is a scalar function of position.
2. In order to evaluate surface integrals it is convenient to express them as double integrals taken over the orthogonal projection of the surface S on one of the coordinate planes. But this is possible only if any line perpendicular to the co-ordinate plane chosen meets the surface S in no more than one point. If the surface S does not satisfy this condition, then it can be sub-divided into surfaces which do satisfy this condition.

   Suppose the surface S is such that any line perpendicular to the y-plane meets S in no more than one point. Then the equation of the surface S can be written in the form

   $$z = h(x, y).$$

Let R be the orthogonal projection of S on the xy-plane. If γ is the acute angle which the undirected normal n at P(x, y, z) to the surface S makes with z-axis, then it can be shown that

$$\cos\gamma\,dS = dr\,dy,$$

where dS is the small element of area of surface S at the point P.

Therefore $dS = \dfrac{dx\,dy}{\cos\gamma} = \dfrac{dx\,dy}{|n.k|}$, where k is the unit vector along z-axis.

Hence $\displaystyle\iint_S F.n\,dS = \iint_S F.n \frac{dx\,dy}{|n.k|}.$

Thus the surface integral on S can be evaluated with the help of a double integral integrated over R.

VOLUME INTEGRALS

Suppose V is a volume bounded by a surface S. Suppose f(x, y, z) is a single valued function of position defined over V. Subdivide the volume V into n elements of volumes $\delta V_1, \delta V_2, \ldots, \delta V_n$. In each part δV_k we choose an arbitrary point P_k whose co-ordinates are (x_k, y_k, z_k). We define

$$f(P_k) = f(x_k, y_k, z_k).$$

Form the sum $\sum_{k=1}^{n} f(P_k)\delta V_k$.

Now take the limit of this sum as $n \to \infty$ in such a way that the largest of the volumes δV_k approaches zero. This limit, if it exists, is called the volume integral of f(x, y, z) over V and is denoted by

$$\iiint_V f(x,y,z)dV.$$

It can be shown that if the surface is peccewise smooth and the function f(x, y, z) is continuous over V, then the above limit exists *i.e.*, is independent of the choice of sub-divisions and points P_k.

If we subdivide the volume V into small cuboids by drawing lines parallel to the three co-ordinates axes, then dV = dx dy dz and the above volume integral becomes

$$\iiint_V f(x,y,z)dx\,dy\,dz.$$

If F(x, y, z) is a vector function, then $\iiint_V F\,dV$ is also an example of a volume integral.

IMPORTANT DEFINITIONS AND FORMULAE

1. *Green's Theorem (Vector Form)* : If ϕ and ψ are two scalar point functions such that ϕ and ψ and their derivatives in any direction are uniform and continuous in a certain region which is bounded by a closed surface S, then

$$\iiint_V \left(\phi\nabla^2\psi - \psi\nabla^2\phi\right)dv = \iiint_S \left(\phi\nabla\psi - \psi\nabla\phi\right).\hat{n}ds.$$

2. *Green's Theorem (Tantestian Form)* : If c be a regular closed curve in the xy plane and s the region bounded by c, then

$$\int_c (F_1 dx + F_2 dy) = \iint_s \left(\frac{\partial F_2}{\partial x} \frac{\partial F_1}{\partial y}\right) dxdy.$$

Where F_1 and F_2 are continuously differentiable function inside and on c.

3. *Harmonic Function :* A function ϕ is said to be har monic if it satisfies the Laplace equation *i.e.* $\nabla^2\phi = 0$.

4. *Stoke's Theorem (Vector Form) :* If s is open, two sided surface bounded by a non-intersecting curve c, then for $\vec{F}$ having continuous derivatives

$$\int_c \vec{F}.d\vec{r} = \int_s \left(\nabla \times \vec{F}\right).\vec{n}\, ds$$

where $\hat{n}$ is the unit outward drawn normal vector at any point of s and c is traversed in the positive direction.

5. *Stoke's Theorem (Cartesian Form) :* If $\vec{F} = F_1\hat{i} + F_2\hat{j} + F_3\hat{k}$ and F_1, F_2 F_3 be functions of x, y, z then

$$\int_c (F_1 dx + F_2 dy + F_3 dz) = \iint_s \left[\frac{\partial F_3}{\partial y} - \frac{\partial F_2}{\partial z}\right] dydz + \int \left(\frac{\partial F_3}{\partial x} - \frac{\partial F_3}{\partial x}\right) dzdx + \left(\frac{\partial F_2}{\partial x} - \frac{\partial F_1}{\partial y}\right) dxdy$$

where the symbol have meaning on the result above.

6. *Gauss's Divergence Theorem (Vector Form) :* If V is the volume bonded by a closed surface S and F is a vector point function with continuous derivatives, then

$$\int_S \vec{F}.\hat{n}\, ds = \int_V \text{div}\, \vec{F}\, dY$$

$$\Rightarrow \qquad \int_S \vec{F}.\hat{n}\, ds = \int_V \text{div}\, \vec{F}\, dV$$

where $\hat{n}$ is the unit outward drawn normal vector surface S.

7. If ϕ is a scalar point function then

$$\int_S f\hat{n}\, ds = \int_V \nabla\phi\, dV$$

where $\hat{n}$, S and V are as defined as in result previous question.

SOME USEFUL RESULTS FROM GAUSS'S AND STOKE'S THEOREM

(1) *The curl of the gradient of a scalar ϕ is zero:*

$\vec{\nabla} \times \vec{\nabla} \phi = 0$. As we have discussed that the line integral of the gradient of a scalar function along a closed path is zero, *i.e.*, if

$$\vec{F} = \vec{\nabla} \phi$$

$$\oint_{\text{loop}} \vec{\nabla} \phi . d\vec{l} = 0$$

Hence from Stoke's theorem,

$$\oint_C \vec{\nabla} \phi . d\vec{l} = \iint_S \text{curl} \vec{\nabla} \phi . d\vec{S} = 0$$

Since the integral $\iint \text{curl} \vec{\nabla} \phi . d\vec{S}$ is zero, the integrand itself should be zero *i.e.*,

$$\text{curl } \vec{\nabla} \phi = 0 \text{ or } \nabla \times \nabla \phi = 0.$$

(2) *The divergence of the curl of a vector function* $\vec{F}$ *is zero* :

According to the Stoke's theorem the line integral around a curve is equal to the surface integral of the curl $\vec{F}$ taken over any surface S of which the curve C is a boundary edge.

It means, that as the boundary or the loop decreases which corresponds as the surface tends to close and the value of integral on both side of decreases, when the boundary or loop decreases to a mere point, the surface is closed and the value of two integrals is zero, so that for a closed surface,

$$\iint_S (\vec{\nabla} \times \vec{F}) . d\vec{S} = 0$$

Now, according to Gauss's theorem, we can transform surface integral to volume integral,

$$-\iint S\, (\vec{\nabla} \times \vec{F}) . d\vec{S} = \iiint_V \vec{\nabla} . (\vec{\nabla} \times \vec{F})\, dv$$

clearly, $\iiint_V \vec{\nabla} . (\vec{\nabla} \times \vec{F})\, dv = 0$

i.e., $\vec{\nabla} \cdot (\vec{\nabla} \times \vec{F}) = 0$, at every point

Stoke's Theorem. *Let S be a piecewise smooth open surface bounded by a piecewise smooth simple closed curve C. Let F (x, y, z) be a continuous*

sector function which has continuous first partial derivatives in a region of space which contains S in its interior. Then

$$\oint_C F.dr = \iint_S (\nabla \times F.)n\,dS = \iint_S (\text{curl}F).dS$$

where C is traversed in the positive direction. The direction of C is called positive if an observe, walking on the boundary of S in this direction, with hashed pointing in the direction of outward drawn normal n to S, has the surface on the left.

Note. $\oint_C F.dr = \oint_C \left(F.\frac{dr}{ds}\right)ds = \oint_C (F.t)ds,$

where t is unit tangent vector to C. Therefore F . t is the component of F in the direction of the tangent vector of C. Also $(\nabla \times F) . n$ is the component of curl F in the direction of outward drawn normal vector n of S. Therefore in words Stoke's theorem may be stated as follows:

The line integral of the tangential component of vector F taken around a simple closed curve C is equal to the surface integral of the normal component of the curl of F taken over any surface S having C as its boundary.

Charisma Equivalent of Stoke's Theorem

Let $F = F_1 i + F_2 j + F_3 k$. Let outward drawn normal vector n of S make angles α, β, γ with positive directions x, y, z axes. Then $n = \cos\alpha\, i + \cos\beta\, j + \cos\gamma\, k$.

Also $$\nabla \times F = \begin{vmatrix} i & j & k \\ \frac{\partial}{\partial x} & \frac{\partial}{\partial y} & \frac{\partial}{\partial z} \\ F_1 & F_2 & F_3 \end{vmatrix}$$

$$= \left(\frac{\partial F_3}{\partial y} - \frac{\partial F_2}{\partial z}\right)i + \left(\frac{\partial F_1}{\partial z} - \frac{\partial F_3}{\partial x}\right)j + \left(\frac{\partial F_2}{\partial x} - \frac{\partial F_1}{\partial y}\right)k.$$

$\therefore \quad (\nabla \times F) \cdot n$

$$= \left(\frac{\partial F_3}{\partial y} - \frac{\partial F_2}{\partial z}\right)\cos\alpha + \left(\frac{\partial F_1}{\partial z} - \frac{\partial F_3}{\partial x}\right)\cos\beta + \left(\frac{\partial F_2}{\partial x} - \frac{\partial F_1}{\partial y}\right)\cos\gamma.$$

Also $$F \cdot dr = (F_1 i + F_2 j + F_3 k) \cdot (dx\, i + dy\, j + dz\, k)$$
$$= F_1\, dx + F_2\, dy + F_3\, dz.$$

$\therefore$ Stoke's theorem can be written as

$$\oint_C F_1 dx + F_2 dy + F_3 dz$$

Proof of Stoke's Theorem : Let S be a surface which is such that its projection on the xy, yz and zx planes are regions bounded by simple closed curves. Suppose S can be represented simultaneous in the forms z = f (x, y), = g (x, z), x = h (z, y) where f, g, h are continuous function and have continuous firs partial derivatives.

Consider the integral $\iint_S [\nabla \times (F_1 i)].n\,dS$.

We have $\nabla \times (F_1 i) = \begin{vmatrix} i & j & k \\ \frac{\partial}{\partial x} & \frac{\partial}{\partial y} & \frac{\partial}{\partial z} \\ F_1 & 0 & 0 \end{vmatrix} = \frac{\partial F_1}{\partial y} k.$

$$\therefore \quad [\nabla \times (F_1 i)] \cdot n = \left(\frac{\partial F_1}{\partial z} j.n - \frac{\partial F_1}{\partial y} k.n\right)$$

$$= \frac{\partial F_1}{\partial z}\cos\beta - \frac{\partial F_1}{\partial y}\cos\gamma.$$

$$\therefore \quad \iint_S [\nabla \times (F_1 i).n\,dS$$

$$= \iint_S \left(\frac{\partial F_1}{\partial z}\cos\beta - \frac{\partial F_1}{\partial y}\cos\gamma\right) dS$$

We shall prove that

$$\iint_S \left(\frac{\partial F_1}{\partial z}\cos\beta - \frac{\partial F_1}{\partial y}\cos\gamma\right) dS$$

$$= \oint_C F_1 dx.$$

Let R be the orthogonal projection of S on the xy-plane and let Γ be its boundary which is oriented as shown in the figure. Using the representation z = f(x, y) of S, we may write the line integral over C as a line integral over Γ. Thus

$$\oint_C F_1(x, y, z)dx = \oint_\Gamma F_1[x, y, f(x, y)]dx$$

$$= \oint_{\Gamma} \{F_1[x, y, f(x, y)]dx + 0dy\}$$

$$= -\iint_R \frac{\partial F_1}{\partial y} dx\, dy, \text{ by Green's theorem in plane for the region R.}$$

But $\dfrac{\partial F_1[x, y, f(x, y)]}{\partial y}$

$$= \frac{\partial F_1(x, y, z)}{\partial y} + \frac{\partial F_1(x, y, z)}{\partial z} \frac{\partial f}{\partial y}. \qquad [\because z = f(x, y)]$$

$$\therefore \quad \oint_C F_1(x, y, z)dx = -\iint_R \left(\frac{\partial F_1}{\partial y} + \frac{\partial F_1}{\partial z} \frac{\partial f}{\partial y} \right) dx\, dy \qquad ...(1)$$

Now the equation $z = f(x, y)$ of the surface S can be written as $\phi(x, y, z) = z - f(x, y) = 0$.

We have grad $f = -\dfrac{\partial f}{\partial x} i - \dfrac{\partial f}{\partial y} j + k$.

Let [grad ϕ] = a.

Since grad ϕ is normal to S. Therefore, we get

$$n = \pm \frac{\text{grad}\phi}{a}.$$

But the components of both n and grad ϕ in positive direction of z-axis are positive. Therefore,

$$n = + \frac{\text{grad}\phi}{a}$$

or $\cos\alpha\, i + \cos\beta\, j + \cos\gamma\, k = -\dfrac{1}{a}\dfrac{\partial f}{\partial x} i - \dfrac{1}{a}\dfrac{\partial f}{\partial y} j + \dfrac{1}{a} k.$

$$\therefore \quad \cos\alpha = -\frac{1}{a}\frac{\partial f}{\partial x},$$

$$\cos\beta = -\frac{1}{a}\frac{\partial f}{\partial y}, \quad \cos\gamma = \frac{1}{a}.$$

Now $dS = \dfrac{dx\, dy}{\cos\gamma} = a\, dx\, dy.$

$$\therefore \quad \iint_S \left(\frac{\partial F_1}{\partial z} \cos\beta - \frac{\partial F_1}{\partial y} \cos\gamma \right) dS$$

$$= \iint_R \left[\frac{\partial F_1}{\partial z}\left(-\frac{1}{a}\frac{\partial f}{\partial y}\right) - \frac{\partial F_1}{\partial y}\frac{1}{a}\right] a\,dx\,dy$$

$$= -\iint_R \left[\frac{\partial F_1}{\partial z}\left(-\frac{1}{a}\frac{\partial f}{\partial y}\right) - \frac{\partial F_1}{\partial y}\frac{1}{a}\right] a\,dx\,dy. \qquad ...(2)$$

From (1) and (2), we get

$$\oint_C F_1 dx = \iint_S \left(\frac{\partial F_1}{\partial z}\cos\beta - \frac{\partial F_1}{\partial y}\cos\gamma\right) dS$$

$$= \iint_S [\nabla \times (F_1 i)].n\,dS \qquad ...(3)$$

Similarly, by projections on the other coordinate planes, we get

$$\oint_C F_2 dy = \iint_S [\nabla \times (F_2 j)].n\,dS \qquad ...(4)$$

$$\oint_C F_3 dz = \iint_S [\nabla \times (F_3 k)].n\,dS \qquad ...(5)$$

Adding (3), (4), (5), we get

$$\oint_C (F_1 dx + F_2 dy + F_3 dz) = \iint_S [\nabla \times (F_1 i + F_2 j + F_3 k)].n\,dS$$

or $$\oint_C F.dr = \iint_S (\nabla \times F).n\,dS$$

If the surface S does not satisfy the restrictions imposed above, even then Stok's theorem will be true provided S can be subdivided into surfaces $S_1, S_2, \ldots S_k$ with boundaries $C_1, C_2, \ldots C_k$ which do satisfy the restrictions, Stok's theorem holds for each such surface. The sum of surface integrals over $S_1, S_2, \ldots S_k$ will give us line integral over C.

Note: *Green's theorem in plane is special case of Stok's theorem.* If R is a region in the xy-plane bounded by a closed curve C, then in vector form Green's theorem in plane can be written as

$$\iint_R (\nabla \times F).k\,dR = \oint_C F.dr.$$

This is nothing but a special case of Stok's theorem because here k = n outward drawn unit normal to the surface of region R.

CONCEPTS OF CURVES

Oriented Curve: Supposes C is a curve in space. Let us orient C by taking one of the two directions along C as the *positive direction*; the opposite direction along C is then called the negative direction. Suppose A is the initial point and B the terminal point of C under the chosen orientation. In case these two points coincide, the curve C is called a *closed curve.*

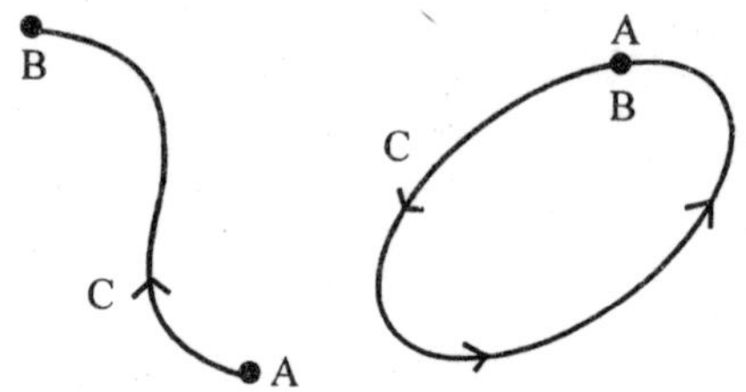

Fig. 5 : Oriented closed curve.

Smooth Curve: Let r (t) = r (t) i + y (t) j + z (t) k, where r (t) is the position vector of (r, y, z), be the parametric representation of a curve C joining the points A and B, where $t = t_1$ and $t = t_2$ respectively. We know that dr/dt is a tangent vector to this curve at the point r. Suppose the function r (t) is continuous and has a continuous first derivative not equal to zero vector for all values of t under consideration. Then the curve C possesses a unique tangent at each of its points. A curve satisfying these assumptions is called a smooth curve.

A curve C is said to be *piecewise smooth* if it is composed of a finite number of smooth curves.

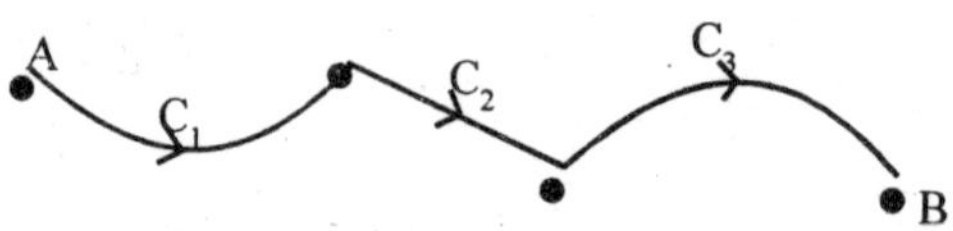

Fig. 6 : Piecewise smooth curve.

The curve C in the adjoining figure is piecewise smooth as it is composed of three smooth curves C_1, C_2 and C_3. The circle is a smooth closed curve while the curve consisting of the four sides of a rectangle is a peccaries smooth closed curve.

Smooth Surface: Suppose S is a surface which has a unique normal at each of its points and the direction of this normal depends continuously on the points of S. Then S is called a smooth surface.

If a surface S is not smooth but can be subdivided into a finite number of smooth surfaces, then it is called a piecewise smooth surface. The surface of a there is smooth while the surface of a cube is piecewise smooth.

Classification of Regions: A region R in which every closed curve can be contracted to a point without passing out of the region is called a simply connected region. Otherwise the region R is multiply-connected. The region interior to a circle is a simply-connected plane region. The region interior to a sphere is a simply-connected region in space. The region between two concentric circles lying in the same plane is a multiply connected plane region.

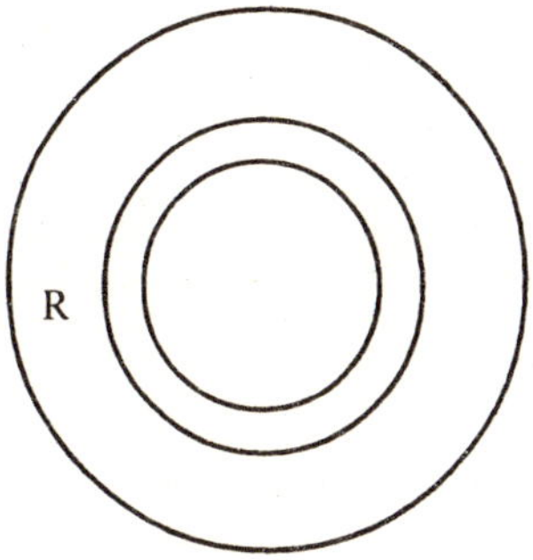

Fig. 7

If we take a closed curve in this region surrounding the inner circle, then it cannot be contracted to a point without passing out of the region. Therefore the region is not simply-connected. However the region between two concentric spheres is a simply-connected region in space. The region between two infinitely long coaxial cylinders is a multiply-connected region in space.

SOLVED PROBLEMS

1. *If* $F = \left(y\frac{\partial f}{\partial z} - z\frac{\partial f}{\partial y}\right)i + \left(z\frac{\partial f}{\partial x} - x\frac{\partial f}{\partial z}\right)j + \left(x\frac{\partial f}{\partial y} - y\frac{\partial f}{\partial x}\right)k$, *prove that*

(i) $F = r \times \nabla f$,

(ii) $F \cdot r = 0$,

(iii) $F \cdot \nabla f = 0$.

Solution: We have

$$\nabla f = \frac{\partial f}{\partial x}i + \frac{\partial f}{\partial y}j + \frac{\partial f}{\partial z}k$$

and $\quad r = xi + yj + zk.$

$$\text{(i) } r \times \nabla f = \begin{vmatrix} i & j & k \\ x & y & z \\ \partial f/\partial x & \partial f/\partial y & \partial f/\partial z \end{vmatrix}$$

$$= \left(y\frac{\partial f}{\partial z} - z\frac{\partial f}{\partial y}\right)i + \left(z\frac{\partial f}{\partial x} - x\frac{\partial f}{\partial z}\right)j + \left(x\frac{\partial f}{\partial y} - y\frac{\partial f}{\partial x}\right)k = F.$$

(ii) $F . r = (r \times \nabla f) . r$ $\quad [\because F = r \times \nabla f]$

$= 0,$

because the value of a scalar triple product having two vectors equal is zero.

(iii) $F . \nabla f = (r \times \nabla f) . \nabla f = [r, \nabla f, \nabla f] = 0,$

because the value of a scalar triple product having two vectors equal is zero.

2. *Find* $\int_C t.dr$ *where t is the unit tangent vector and C is the unit circle, in xy-plane, with centre at the origin.*

Solution:

For any curve, $\frac{dr}{ds}$ = unit tangent vector = t.

$$\therefore \int_C t.dr = \int_C \left[t.\left(\frac{dr}{ds}\right)\right]ds = \int_C (t.t)ds$$

$$= \int_C ds \qquad [\because t . t = 1, \text{ t being unit vector}]$$

$= \int_{x=0}^{2x} ds$, since along the unit circle C, s goes from 0 to 2π

$= [s]_0^{2x} = 2\pi.$

3. *Evaluate* $\int (x\,dy - y\,dx)$ *around the circle* $x^2 + y^2 = 1$.

Solution:

Let C denote the circle $x^2 + y^2 = 1$. The parametric equations of this circle are $x = \cos t$, $y = \sin t$.

To integrate around the circle C we should vary t from 0 to 2π.

$$\therefore \oint_C (x\,dy - y\,dx) = \int_0^{2\pi}\left(x\frac{dy}{dt} - y\frac{dx}{dt}\right)dt$$

$$= \int_0^{2\pi}(\cos^2 t + \sin^2 t)dt = \int_0^{2\pi} dt = 2\pi.$$

4. *Evaluate* $\int_C F.dr$, *where* $F = i \cos y - j\, x \sin y$ *and C is the curve* $y = \sqrt{(1 - x^2)}$ *in the xy-plane from (1, 0) to (0, 1).*

Solution:

We have $\int_C F.dr = \int_C (i\cos y - j x \sin y).(i\,dx + j\,dy)$

$$= \int_C (\cos y\,dx - x\sin y\,dy) = \int_C d(x\cos y)$$

$$= [x \cos y]_{(1,\,0)}^{(0,\,1)} = 0 - 1 = -1.$$

4(a). *Evaluate* $\int_C F.dr$, *where* $F = xy\, i + (x^2 + y^2)\, j$ *and curve C is the are of* $y = x^2 - 4$ *from (2, 0) to (4, 12).*

Solution:

We have $\int_C F.dr = \int_C [xy\,i + (x^2 + y^2)\,j].(dx\,i + dy\,j)$

$$= \int_C [xy\,dx + (x^2 + y^2)dy] = \int_C xy\,dx + \int_C (x^2 + y^2)dy.$$

Along C, $y = x^2 - 4$ and $x^2 = y + 4$.

$$\therefore \int_C F.dr = \int_{x=2}^{4} x(x^2 4)dx + \int_{y=0}^{12}(y + 4 + y^2)dy$$

$$= \left[\frac{x^4}{4} - 2x^2\right]_2^4 + \left[\frac{y^2}{2} + 4y + \frac{y^3}{3}\right]_0^{12} = 732.$$

5. *Verify Stoke's theorem for* $F = (x^2 + y^2)\, i - 2x\, y\, j$ *taken round the rectangle bounded by* $x = \pm a$, $y = 0$, $y = b$.

Solution : Here $F = (x^2 + y^2)\, i = 2xyj$ and $dr = (x)\, i + (dy)\, j$

$\therefore F\, .\, dr = [(x^2 + y^2)\, i - 2xyj]\, .\, [(dx)\, i + (dy)\, j]$

$$= (x^2 + y^2)\, dx - 2xy\, dy \qquad \text{. . . (i)}$$

Let C be the perimeter of the given rectangle and C_1, C_2, C_3, and C_4 be the lines from A to D, D to C, C to B and B to A respectively.

Then $\int_C F \cdot dr$

$$= \int_{C1} F \cdot dr + \int_{C2} F \cdot dr + \int_{C3} F \cdot dr + \int_{C4} F \cdot dr \qquad \text{. . . (ii)}$$

Now for C_1 we have y = b, dy = 0 and x varies from a to – a, so from

(i) we get $\int_{C1} F \cdot dr = \int_{x=0}^{-a} (x^2 + b^2)\, dx = \left[\frac{1}{3} x^3 + b^2 x\right]_0^{-a}$

$$= -\left(\frac{2}{3} a^3 + 2b^2 a\right) \qquad \text{. . . (iii)}$$

For C_2 we have x = –a, dx = 0 and y varies form b to 0, so from (i)

we get $\int_{C2} F \cdot dr = \int_{y=b}^{a} (2ay)\, dy = -2a\left(\frac{1}{2} y^3\right)_0^b = -ab^2$ ... (iv)

For C_3 we have y = 0, dy = 0 and x varies form –a to a, so form (i)

$$\int_{C3} F \cdot dr = \int_{x=-a}^{a} x^2\, dx = \left(\frac{1}{3} x^2\right)_{-a}^{a} = \frac{2}{3} a^3 \qquad \text{. . . (v)}$$

For C_4 we have x = a, dx = 0 and y varies forom a to 0 so, from (i)

we get $\int_{C4} F \cdot dr = \int_{y=0}^{b} 2ay\, dy = -\left(ay^2\right)_0^b = -ab^2$ . . . (vi)

∴ From (ii), (iii), (iv), (v) and (vi) we have

$$\int_C F \cdot dr = -\left(\frac{2}{3} a^3 + 2b^2 a\right) - ab^2 + \frac{2}{3} a^3 - ab^2 = -4ab^2 \qquad \text{... (vii)}$$

Again as is evident from the figure the unit outward normal vector n = k.

$$\text{Also curl } F = \begin{vmatrix} i & j & k \\ \frac{\partial}{\partial x} & \frac{\partial}{\partial y} & \frac{\partial}{\partial z} \\ x^2 + y^2 & -2xy & 0 \end{vmatrix}$$

$$= i[0] - j\,[0] + k\left[\frac{\partial}{\partial x}(-2xy) - \frac{\partial}{\partial y}(x^2 + y^2)\right]$$

$$= -4yk$$

∴ $n \cdot \text{curl } F = k \cdot (-4yk) = -4y$

Also dS = dx dy

$\therefore \int_S n \cdot \text{curl } F \, dS = \int_{x=-a}^{a} \int_{y=0}^{a} -4y \, dx \, dy$

$$= -\int_{-a}^{a} \left(\frac{1}{2}y^2\right)_0^b dx = -2b^2 (x)_{-a}^{a}$$

$$= -4 \, ab^2$$

From (vii) and (viii) we get

$$\int_C F \cdot dr = \int_S n \cdot \text{curl } F \, dS.$$ Hence proved.

6. *Evaluate $\int_c F \cdot dr$ by Stoke's theorem, where C is the boundary of the triangle whose vartices are O(0, 0, 0), A (i, 0, 0), B, (1, 1, 0) and $F = y^2 i + x^2 j = (x + z) k$.*

Solution : Here curl F = $\begin{vmatrix} i & j & k \\ \frac{\partial}{\partial x} & \frac{\partial}{\partial y} & \frac{\partial}{\partial z} \\ y^2 & x^2 & -(x+z) \end{vmatrix}$

$$= 0i = j(-1) + k(2x - 2y)$$

$$= j + 2(x - y) k$$

Again from the figure it is evident that here n = k and dS = dx dy,

$\therefore n \cdot \text{curl } F = 2(x - y)$

Also the equation of the line OB is y = x and for the trangle OAB we find that x varies from 0 to 1 and y varies from 0 to x.

$\therefore$ By Stoke's theorem, the given integral

$$= \iint_S n \cdot \text{curl} F \, dS = \int_{x=0}^{1} \int_{x=0}^{x} 2(x - y) \, dx \, dy$$

$$= 2 \int_{x=0}^{1} \left(xy - \frac{1}{2}y^2\right)_{y=0}^{x} dx = 2 \int_0^1 \left(x^2 - \frac{1}{2}x^2\right) dx$$

$$= \int_0^1 dx = \left(\frac{1}{3}x^2\right)_0^1 = \frac{1}{3} \quad \textbf{Ans.}$$

7. *Evaluate by Stoke's theorm $\int_c (y z \, dx + z x \, dy + x y \, dz)$, where C is the curve $x^2 + y^2 = 1$, $z = y^2$.*

Solution : Here $\int_C (yz \, dx + zx \, dy + xy \, dz)$

$$= \int_C (yzi + zxj + xyk) \cdot (idx + jdy + kdz)$$

$$= \int_C (yzi + zxj + xyk).dr$$

∴ $\text{"F"} = yzi + zxj + xyk$

∴ $$\text{curl } \vec{F} = \begin{vmatrix} i & j & k \\ \frac{\partial}{\partial x} & \frac{\partial}{\partial y} & \frac{\partial}{\partial z} \\ yz & zx & xy \end{vmatrix}$$

$$= (x - x)\, i\, (y - y)\, j + (z - z)\, k = 0$$

∴ By stoke's theorem, the given integral

$$= \int_C n \,.\, \text{curl } F \, dS$$

$$= 0, \text{ since curl } F = 0.$$

8. *Apply Stoke's theorm to prove that*

$\int_c (y\, dx + z\, dy + x\, dz) = -2\sqrt{2}\, \pi\, a^2$,

where C is te curve given by $x^2 + y^2 + z^2 - 2a\,x = 2a\,y = 4,\ x + y = 2a$ and begins at the point $(2a, 0, 0)$ and goes at first below the xy-plane.

Solution : The curve C is given by

$$x^2 + y^2 + z^2 - 2ax + 2ay = 0,\ x + y = 2a$$

or $$x^2 + y^2 + z^2 - 2ax + 2ay = 0,\ x + y = 2a$$

which is the curve of intersection of the sphere

$(x - a)^2 + (y - a)^2 + z^2 = 2a^2$ and the plane $x + y = 2a$

The centre D of the sphere is $(a, a, 0)$ and radius $\sqrt{(2a)^2} = a\sqrt{2}$ and the sphere passes through the orgin.

Also the plane $x + y = 2a$ passes through the centre D $(a, a, 0)$ and cuts the x and y axes at B $(2a, 0, 0)$ and A $(0, 2a, 0)$ respectively.

Hence C is the circle with D as centre and passing through A $(0, 2a, 0)$ and B $(2a, 0, 0)$.

Also $\int_C (ydx + zdy + xdz)$

$$= \int_C (yi + zj + xk) \,.\, (idx + jdy + kdz)$$

$$= \int_C (yi + zj + xk) \,.\, dr \qquad \text{. . . (i)}$$

Hence here $F = yi + zi + xk$ . . . (ii)

∴ By stoke's theorem we have the given integral

$\int_C (yi + zj + xk) \,.\, dr = \int_C n \,.\, \text{curl } F\, dS,$ . . . (iii)

where S is the surface of the hemisphere passing through A, B and P.

Now let n be the outward drawn unit normal vector at any point P (x, y, z) in the usual way. Then the direction cosines of the line DP

are $\frac{x-a}{DP}, \frac{y-a}{DP}, \frac{z-0}{DP}$, where DP = radius of the sphere = $a\sqrt{2}$

or $$\frac{x-a}{a\sqrt{2}}, \frac{y-a}{a\sqrt{2}}, \frac{z}{a\sqrt{2}}.$$

$\therefore$ $n = [(x-a)\,i + (y-a)\,j + z\,k]/(a\sqrt{2})$

Also $\text{curl } F = \begin{vmatrix} i & j & k \\ \frac{\partial}{\partial x} & \frac{\partial}{\partial y} & \frac{\partial}{\partial z} \\ y & z & x \end{vmatrix} = -i - j - k$ on simplifying

$\therefore$ $n \,.\, \text{curl } F = [(x-a)\,i + (y-a)\,j + z\,k\}/(a\sqrt{2})] \,.\, (-i - j - k)$

$$= \frac{-(x-a)-(y-a)-z}{\left(a\sqrt{2}\right)} = \frac{2a-x-y-z}{\left(a\sqrt{2}\right)}$$

$\therefore$ $\int_S n \,.\, \text{curl } F \, dS = \int_S \frac{(2a-x-y-z)}{\left(a\sqrt{2}\right)} \, dS$

$$= \frac{1}{a\sqrt{2}} \int_S (2a-x-y-z)\, dS$$

$$= \frac{1}{a\sqrt{2}} \left[\int_S 2a\, dS - \int_S y\, dS - \int_S z\, dS\right] \qquad \text{. . . (iv)}$$

Now $\int_S 2a\, dS = 2a \int_S dS = 2a\,S$, where S is the area of the surface of the hemisphere.

Also if G $(\bar{x}, \bar{y}, \bar{z})$ be the centroid of the hemisphere whose surface is S, then $\bar{x} = \frac{3a}{2}, \bar{y} = \frac{3a}{2}, \bar{z} = 0$

Also $\bar{x} = \frac{\int_S x\, dS}{S}$

or $\int_S x\, dS = \bar{x}S = \frac{3a}{2} S$

Similarly $\int_S y\, dS = \frac{3}{2} aS$ and $\int_S z\, dS = 0$

$\therefore$ from (iv) we have

$$\int_S n \,.\, \text{curl } F\, dS = \frac{1}{a\sqrt{2}} \left[2aS - \frac{3}{2}aS - \frac{3}{2}aS - 0\right]$$

$$= -\frac{1}{\sqrt{2}} S, \text{ where } S = 2\pi\ (a\sqrt{2})^2 = 4\ \pi\ a^2$$

$$= -\frac{1}{\sqrt{2}}\ (4\pi a^2) = -\ 2\sqrt{2}\ \pi a^2 \qquad \text{. . . (v)}$$

Hence from (iii) and (v) we have

$\int_C (ydx + zdy + xdz0 = 2\ \sqrt{2}\ \pi a^2$ Hence proved.

9. *Evaluate by Stoke's theorem*

$\int_c (e^x\ dx + 2\ y\ dy - dz)$, where C is the curve

$$x^2 + y^2 = 4,\ z = 2.$$

Solution : Here $\int_S (e^x\ dx + 2y\ dy = dz)$

$$= \int_C (e^x\ i + 2yj - k)\ .\ (idx + jdy + kdz)$$

$$= \int_C (e^x\ i + 2yj - k)\ .\ dr$$

Hence here 'F' $= e^x\ i + 2yj = k$

$\therefore$ By stoke's theorem we have the given integral

$\int_C (e^x\ i + 2y)\ .\ dr\ dr = \int_{S1} n\ .\ \text{curl F dS},$ . . . (i)

where S_1 is the surface whose boundary

C_1 is given by $x^2 + y^2 = 4,\ z = 2$

Here C is circle $x^2 + y^2 = 4$, $z = 2$ whose centre is D(0, 0, 2) and radius 2. Also C is the boundary for the plane region S.

$\therefore$ Here n = k (Note) and so we get n . curl

$$F = k.\ \text{curl}\ (e^x\ i + 2yj - k) \qquad \text{. . . (i)}$$

$$\text{Here curl}\ (e^x i + 2yj - k) = \begin{vmatrix} i & j & k \\ \frac{\partial}{\partial x} & \frac{\partial}{\partial y} & \frac{\partial}{\partial z} \\ e^x & 2y & -1 \end{vmatrix}$$

$= i\ (0) + j\ (0) + k\ (0) = 0$

$\therefore$ From (ii) we get n . curl F = 0

Hence from (i) the value of the given integral

$= \int_{S1} n\ .\ \text{curl F dS} = 0$ **Ans.**

10. *If $F = (y^2 + z^2 + x^2)\ i + (z^2 + x^2 - y^2)\ j + (x^2 + y^2 - z^2)\ k$, evaluate $\int$ curl F. n dS, integrated over the portion of the surface $x^2 + y^2 - 2ax + az = 0$ above the plane $z = 0$ and verify Stoke's theorem.*

Solution : Proceed exactly as in here the given integral $\int_C F \cdot dr$

$= \int_C [(y^2 + z^2 - x^2)i + (z^2 + x^2 - y^2)j + (x^2 + y^2 - z^2)k].[idx + jdy + kdz]$

$= \int_C [(y^2 + z^2 - x^2)\, dx + (z^2 + x^2 - y^2)\, dy + (x^2 + y^2 - z^2)\, dz]$

$= \int_C (y^2 - x^2)\, dx + (x^2 - y^2)\, dy;$ $\qquad [\because z = 0, dz = 0 \text{ on C.}]$

$= \int_C (y^2 - x^2)\,(dx = dy)$

$= \int_{\phi=0}^{2\pi} [a^2 \sin^2\theta - (a + a\cos\theta)^2]\,(-a\sin\theta - a\cos\theta)\, d\theta,$

putting $x = a + a\cos\theta$, $y = a\sin\theta$

$= -a^3 \int_0^{2\pi} (\sin^2\theta - 1 - \cos^2\theta - 2\cos\theta)(\sin\theta + \cos\theta)\, d\theta$

$= -a^3 \int_0^{2\pi} 2\cos\theta\,(1 + \cos\theta)(\sin\theta + \cos\theta)\, d\theta$

$= 2a^3 \int_0^{2\pi} \cos\theta\,(1 + \cos\theta)\sin\theta\, d\theta + 2a^3 \int_0^{2\pi} \cos\theta\,(1 + \cos\theta)\, d\theta$

$= 2a^3 \int_0^{2\pi} \cos^2\theta\, d\theta + 2a^3 \int_0^{2\pi} \cos^3\theta \cos^3\theta\, d\theta$

the value of 1st integral being zero.

$= a^3 \int_0^{2\pi} (1 + \cos 2\theta)\, d\theta + \frac{1}{2} a^3 \int_0^{2\pi} (\cos 3\theta + 3\cos\theta)\, d\theta$

$= a^3 \left[\theta + \frac{1}{2}\sin 2\theta\right]_0^{2\pi} + \frac{1}{2}a^3 \left[\frac{1}{3}\sin 3\theta + 3\sin\theta\right]_0^{2\pi} = 2a^3\pi$ **Ans.**

$$\text{Again curl } F = \begin{vmatrix} i & j & k \\ \dfrac{\partial}{\partial x} & \dfrac{\partial}{\partial y} & \dfrac{\partial}{\partial z} \\ y^2 + z^2 - x^2 & z^2 + x^2 - y^2 & x^2 + y^2 - z^2 \end{vmatrix} \quad \ldots \text{(i)}$$

$= i\,(2y - 2z) - j\,(2x - 2z) + k\,(2x - 2y)$

$= 2\,[(y - z)\, i + (z - x)\, j + (x - y)\, k]$

Also C being the boundary for the region S as well as the plane region S_1 we have "n" = k

$\therefore$ n . curl F = k . [2 {y – z) i + (z – x) j + (x – y) k})]

$= 2\,(x - y)$

$\therefore \int_S n \,.\, \text{curl } F = dS = \int_{S1} n \,.\, \text{curl } F\, dS$, where $dS = r\, d\theta\, dr$

$= 2 \int_{S1} (x - y)\, r\, d\theta\, dr$, where $x = a\,(1 + \cos\theta)$, $y = a\sin\theta$

$= 2 \int_{\theta=0}^{2\pi} \int_{r=0}^{a} a(1 + \cos\theta - \sin\theta)\, r\, d\theta\, dr$

$$= 2\,a\,[\theta + \sin\theta + \cos\theta]_0^{2\pi} \left(\frac{1}{2}\,r^2\right)_0^{\pi}$$

$$= 2\,a\,[2\pi]\left[\frac{1}{2}\,a^2\right] = 2\pi a^3 \qquad \ldots \text{(ii)}$$

Hence from (i) and (ii) stoke's theorem is verified.

11. *Show that the vector function F defined by $F = (2x\,y + z^3)\,i + x^2\,j + 3\,x\,z^2\,k$ is irrotational and find a function ϕ such that $F = \nabla\phi$.*

Solution :
$$\text{curl } F = \begin{vmatrix} i & j & k \\ \dfrac{\partial}{\partial x} & \dfrac{\partial}{\partial y} & \dfrac{\partial}{\partial z} \\ 2xy+z^3 & x^2 & 3xz^2 \end{vmatrix}$$

$$= i\left[\frac{\partial}{\partial y}(3xz^2) - \frac{\partial}{\partial z}(x^2)\right] + j\left[\frac{\partial}{\partial z}(2xy + z^3) - \frac{\partial}{\partial x}(3xz^2)\right]$$

$$+ k\left[\frac{\partial}{\partial x}(x^3) - \frac{\partial}{\partial y}(2x + z^3)\right]$$

$= i\,[0] + j\,[3z^2 - 2x^2] + k\,[2x = 2x] = 0$

Hence F is irrotational.

Now $F = \nabla\phi$ (given)

or $$F_1\,i + F_2\,j + F_3\,k = i\,\frac{\partial\phi}{\partial x} + j\,\frac{\partial\phi}{\partial y} + k\,\frac{\partial\phi}{\partial z}$$

$\therefore$ $$\frac{\partial\phi}{\partial x} = F_1,\ \frac{\partial\phi}{\partial y} = F_2,\ \frac{\partial\phi}{\partial z} = F_3$$

Also $$d\phi = \frac{\partial\phi}{\partial x}\,dx + \frac{\partial\phi}{\partial y}\,dy + \frac{\partial\phi}{\partial z}\,dz$$

$$= F_1\,dx + F_2\,dy + F_3\,dz$$

or $$d\phi = (2xy + z^3)\,dx + x^2\,dy + 3xz^2\,dz$$

$$= (2xy\,dx + x^2\,dy) + (z^3\,dx + 3xz^2\,dz)$$

or $$d\phi = d\,(x^2y) + d\,(xz^3) = d\,(x^2y + xz^3)$$

or $$\phi = x^2y + xz^3 + \text{constant}$$ **Ans.**

12. *If $F = (y\sin z = \sin x)\,i + (x\sin z + 2y\,z)\,j + (xy\cos z - y^2)\,k$, then prove that F is irrotational and find a function ϕ such that $F = \nabla\phi$.*

Solution : $\text{curl } F = \begin{vmatrix} i & j & k \\ \dfrac{\partial}{\partial x} & \dfrac{\partial}{\partial y} & \dfrac{\partial}{\partial z} \\ y \sin z - \sin x & x \sin z + 2yz & xy \cos z + y^2 \end{vmatrix}$

$$= i\left[\frac{\partial}{\partial y}(xy \cos z + y^2) - \frac{\partial}{\partial z}(x \sin z + 2yz)\right]$$

$$+ j\left[\frac{\partial}{\partial z}(y \sin z + \sin x) - \frac{\partial}{\partial x}(xy \cos z + y^2)\right]$$

$$+ k\left[\frac{\partial}{\partial x}(x \sin z + 2yz) - \frac{\partial}{\partial y}(y \sin z - \sin x)\right]$$

$$= i\,[x \cos z + 2y - x \cos z - 2y] + j\,[0] + k\,[0]$$

or $\text{curl } F = 0$. Hence F is irrotational.

Again $d\phi = F_1 dx + F_2\, dy + F_3\, dz$, if $F = \nabla\phi$

and $F = F_1\, i + F_2\, j + F_3\, k$.

$= (y \sin z - \sin x)\, dx = (x \sin z + 2y\, z)\, dy + (xy \cos z + y^2)\, dz$

or $d\phi = -\sin x\, dx + (y^2\, dz + 2y\, z dy) + (y \sin z\, dx$

$+ x \sin z\, dy + xy \cos z\, dz)$

$= d(\cos x) + d(y^2 z) + d(xy \sin z)$

$= d(\cos x + y^2 z + xy \sin z)$

or $\phi = \cos x + y^2 z + xy \sin z + \text{constant}$ **Ans.**

13. *Evaluate* $\int_C (e^{-x} \sin y\, dx + e^{-x} \cos y\, dy)$ *by Green's theorem when C is the rectangle whose vertices are*

$$(0, 0), (\pi, 0)\ \left(\pi, \frac{1}{2}\pi\right) \text{ and } \left(0, \frac{1}{2}\pi\right).$$

Solution : Comparing the given line integral with the integral on the left of we have

$$F_1 = e^{-x} \sin y,\ F_2 = e^{-x} \cos y.$$

$$\therefore \quad \frac{\partial F_1}{\partial y} = e^{-x} \cos y \text{ and } \frac{\partial F_2}{\partial x} = -e^{-x} \cos y$$

Hence, by Green's theorem (cartesian form) we have the given line integral

$$= \int_S (-e^{-x}\cos y - e^{-x}\cos y)\,dx\,dy$$

$$= -2\int_{x=0}^{\pi}\int_{y=0}^{\pi/2} e^{-x}\cos y\,dx\,dy$$

$$= -2\left[-e^{-x}\right]_0^{\pi}(\sin y)_0^{\pi/2}$$

$$= 2(e^{-\pi} - 1)(1) = 2(e^{-\pi} - 1).$$ **Ans.**

14. *Evaluate* $\int_c [(y - \sin x)\,dx + \cos x\,dy]$, *where C is the triangle formed by teh lines* $y = 0$, $x = \frac{1}{2}\pi$ and $y = \frac{2}{\pi}x$.

Solution : We know

$$\int_C (F_1\,dx + F_2\,dy) = \iint_S \left(\frac{\partial F_2}{\partial x} - \frac{\partial F_1}{\partial y}\right)dx\,dy$$

Here $F_1 = y - \sin x;\ F_2 = \cos x$

$\therefore$ $\frac{\partial F_1}{\partial y} = 1$ and $\frac{\partial F_2}{\partial x} = -\sin x$

$\therefore$ $\int_C [(y - \sin x)\,dx + \cos x\,dy]$

$$= \iint_S (-\sin x - 1)\,dx\,dy$$

where S is the area of the given triangle

$$= -\int_{x=0}^{\pi/2}\int_{y=0}^{2x/\pi} \cdot \frac{2x}{\pi}\,dx = -\frac{2}{\pi}\int_0^{\pi/2}(x + x\sin x)\,dx$$

$$= -\frac{2}{\pi}\left[\left(\frac{1}{2}x^2\right)_0^{\pi/2} + \{x(-\cos x\}_0^{\pi/2} - \int_0^{\pi/2} 1.(-\cos x)\,dx\right]$$

$$= -\frac{2}{\pi}\left[\frac{1}{2}\cdot\frac{\pi^2}{4} + (\sin x)_0^{\pi/2}\right] = -\frac{2}{\pi}\left[\frac{1}{8}\pi + 1\right] = -\left(\frac{\pi}{4} + \frac{2}{\Pi}\right)$$ **Ans.**

15. *State Gauss's divergence theorem and deduce that if g is harmonic in a domain containing a closed bounded region T and if S be the boundary of T, then*

$$\iint_S \frac{\partial g}{\partial n}\,dA = \iiint_T |\text{grad } g|^2\,dV.$$

Solution : We know Gauss' divergence theorem is

$$\iint_S F \cdot n\,dA = \iiint_T (\text{div } F)\,dV$$

Let $F = g\,\nabla g$

Then $\iiint_S (g\nabla g) . n\, dA = \iiint_T \text{div}\,(g\nabla g)\, dV$

or $\iint_S g \dfrac{\partial g}{\partial n}\, dA = \iiint_T [g\, \text{div}\, \nabla g + (\nabla g) . \nabla g]\, dA$

$\partial g/\partial n$ denotes the derivative of g in the direction of n, and div (ϕA) = ϕ div A + A $\nabla\phi$.

$\iiint_T [g\,(g\nabla . (\nabla g) + (\nabla g)^2]\, dV$

$\iiint_T [g\nabla^2 g + (\nabla g)^2]\, dV$

$\iiint_T [\nabla g]^2\, dV,$ $\qquad$ [$\because$ $\nabla^2 g = 0$, if g is harmonic]

$\iiint_T |\text{grad}\, g|^2\, dV,$

$\therefore$ square of a vector = square of its module Hence proved.

16. *$F = ax^2 i + zy^2 j + b y^2 k$ and let C_1 be the straight line joining O (0, 0, 0) to the point D (c, c, c) and let C_2 be the curve composed of three segments of straight lines*

$x = 0,\ y = 0,\ 0 < z < c$

$x = 0,\ 0 < y < c,\ z < c$

$0 < x < c,\ y = c,\ z = c$

Find $\int F.\, dr$ over both the curvesx and explain why despite the fact that the beginning and end points of curves are the same, the two integrals are not equal.

Solution : We find that C_1 is the straight line OD and C_2 is composed of three lines OA, AB and BD.

Let C_3, C_4, C_5 represent the lines OA, AB and BD respectively.

Now $F = ax^2 i + zy^2 j + by^2 k$

$\therefore$ $F . dr = (ax^2 i + zy^2 j + by^2 k) . (i\, dx + j\, dy + k\, dz)$

or $F . dr = ax^2\, dx + zy^2 + zy^2\, dy + by^2\, dz$ $\qquad$ . . . (i)

Also for the line OD we have x = y = z

$\therefore$ For C_1 we have

$\int_{C1} F . dr = \int_{C1} (ax^2\, dx + zy^2\, dy + by^2\, dz)$, where x = y = z

$= \int_{x=0}^{C} (ax^2 + x^2 + bx^2)$, putting y = z = x

$= \left(\frac{1}{3} zx^2 + \frac{1}{4} x^4 + \frac{1}{3} bx^3\right)_0^c = \frac{1}{2} c^3\, (4a + 3c\ 4b)$ $\qquad$ . . .(ii)

And $\int_{C2} F.dr = \int_{C3} F.dr + \int_{C4} F.dr + \int_{C5} F.dr$ . . . (iii)

Now for C_3 we find x = 0, y = 0

$$\therefore \qquad \int_{C3} F \; dr = 0 \qquad \ldots (iv)$$

For C_4 we have x = 0, z = c [∵ dz = 0 and]

$$\int_{C4} F.dr = \int_{y=0}^{c} cy^2 \, dy = c\left(\frac{1}{3}y^3\right)_0^c = \frac{1}{3}c^4 \qquad \ldots (v)$$

For C_5 we have y c, z = c [∵ dy = 0 = dz and]

$$\therefore \int_{C5} F . dr = \int_{x=0}^{c} dx = a\left(\frac{1}{3}x^3\right)_0^c = \frac{1}{3}ac^3 \qquad \ldots (vi)$$

∴ From (iii), (iv), (v) and (vi) we have

$$\int_{C5} F . dr = 0 + \frac{1}{3}c^4 + \frac{1}{3}ac^3 = \frac{1}{3}c^3(c + a) \qquad \ldots (vii)$$

From (ii) and (vii) we find that the two integrals are not equal.

Now let $d\phi = F . dr$

$$= ax^2 \, dx + zy^2 \, dy + by^2 \, dz,$$

which is not an exact differential (It could have been shown otherwise by curl $F \neq 0$).

Hence F is not conservative and consequently the values of the two integrals do not depend on the end points.

17. *Find the area of the loop of the folium $x^3 + y^3 = 3\,a\,x\,y,\ a > 0$.*

Solution : Required area

$$= \frac{1}{2}\int_C (x\,dy - y\,dx),$$

$$= \frac{1}{2}\int_C x^2\left[\frac{x\,dy - y\,dx}{x^2}\right] = \frac{1}{2}\int_C x^2 \, d\left(\frac{y}{x}\right)$$

$$= \frac{1}{2}\int_C x^2 \; dt,$$

Putting y = tx

Putting y = tx in the given equation of the folium we have

$$x^3 + t^3 x^3 = 3a \; x \; tx$$

or $$x^3 (1 + t^3) = 3atx^2$$

or $$x = \frac{3at}{1 + t^3}$$

Also for half the loop t varies from 0 to 1

∴ From (i) the required area

$$= \frac{1}{2} \int_C \left(\frac{3at}{1 + r^3} \right)^2 dt$$

$$= \frac{1}{2} . 2 \int_{t=0}^{1} \frac{9a^2t^2}{(1 + t^3)^2} dt,$$ [∵ loop is symmetrical about the line y = x]

$$= 3a^2 \int_0^1 \frac{3t^2}{(1 + t^3)^2} dt$$

$$= 3a^2 \left[-(1 + t^3)^{-1} \right]_0^1 = 3a^2 \left[-\frac{1}{2} + 1 \right]$$

$= (3/2)\ a^2$ **Ans.**

18. *V is the volume enclosed by a closed surface S and F = x i + 2 y j + 3 z k, show that* $\int_s F . n\ dS = 6\ V$.

Solution : Given F = xi + 2yj + 3zk

$$\therefore \text{div } F = \nabla . F = \left(i \frac{\partial}{\partial x} + j \frac{\partial}{\partial y} + k \frac{\partial}{\partial z} \right) . (xi + 2yj + 3zk)$$

$$= \frac{\partial}{\partial x}(x) + \frac{\partial}{\partial y}(2y) + \frac{\partial}{\partial z}(3z) = 1 + 2 + 3 = 6$$

∴ By Gauss' Theorem we have

$$\int_s F. n\ d\ S = \int_1 \text{div } F\ d\ V = \sqrt_v 6\ d\ V = 6\ V$$

Hence proved.

19. *If F = 4x z i = y² j + y z k, then evaluate* $\int_s F . n\ dS$ *where S is the surface of the cube enclosed by x = 0, x = 1, y = 0, y = 1 z = 0 and z = 1.*

Solution : Given $F = 4xzi - y^2 j + yzk$.

$$\therefore \text{div } F = \left(i \frac{\partial}{\partial x} + j \frac{\partial}{\partial y} + k \frac{\partial}{\partial z} \right) . (4xzi - y^2j + yzk)$$

$$= \frac{\partial}{\partial x}(4xz) + \frac{\partial}{\partial y}(-y^2) + \frac{\partial}{\partial z}(yz)$$

$= 4z - 2y + y = 4z - y.$

∴ By Gauss' Theorem, we have

$$\int_s F . n\ dS = \int_t \text{div } F\ dV$$

$$=\int_{x=0}^{1}\int_{y=0}^{1}\int_{x=0}^{1}(4x-y)dx\ dy\ dz$$

$$=\int_{x=0}^{1}\int_{y=0}^{1}\left(2z^2-yz\right)_{x=0}^{1}dx\ dy$$

$$=\int_{x=0}^{1}\int_{y=0}^{1}(2-y)\ dx\ dy=\int_{y=0}^{1}\left(2y-\frac{1}{2}y^2\right)_{x=0}^{1}dx$$

$$=\int_0^1\left(2-\frac{1}{2}\right)dx=\frac{3}{2}\left[x\right]_0^1=\frac{3}{2}$$

20. *Evaluate $\int_s (y^2z^2\ i + z^2x^2\ j + z^2y^2\ k)\ .\ \hat{n}\ dS$, where S is the part of the sphere $x^2 + y^2 + z^2 = 1$ above the xy-plane and bounded by this plane.*

Solution : Let $F = y^2z^2i + z^2x^2j + z^2y^2k$

$\therefore$ div $F = \nabla\ .\ F$

$$=\left(i\frac{\partial}{\partial x}+j\frac{\partial}{\partial y}+k\frac{\partial}{\partial z}\right).(y^2z^2i+z^2x^2j+z^2y^2k)$$

$$=\frac{\partial}{\partial x}(y^2z^2)\frac{\partial}{\partial y}(z^2x^2)\frac{\partial}{\partial z}(z^2y^2)=2y^2$$

$\therefore$ By Gauss' divergence theorem, we have

$$\int_S F\ .\ n\ dS=\int_V \text{div}\ F\ dV=\int_V 2zy^2\ dx\ dy\ dz.$$

where V is the volume enclosed by the surface S *i.e.* V is the volume of the hemisphere $x^2 + y^2 + z^2 = 1$ above the xy-plane.

Here as is evident from figure 31 A and 31 B, the limits for integration are for z from 0 to $\sqrt{(1 - x^2 - y^2)}$; for y from $-\sqrt{(1 - x^2)}$ to $+\sqrt{(1 - x^2)}$ and for x from -1 to $+1$.

$\therefore$ From (i) we have

$$\int_S F\ .\ \hat{n}\ dS=\int_V 2zy^2\ dx\ dy\ dz$$

$$=\int_{x=-1}^{1}\int_{y=-\sqrt{(1-x^2)}}^{\sqrt{(1-x^2)}}\int_{z=0}^{\sqrt{(1-x^2-y^2)}}2zy^2\ dx\ dy\ dz$$

$$=\int_{x=-1}^{1}\int_{y=-\sqrt{(1-x^2)}}^{\sqrt{(1-x^2)}}y^2\left(z^2\right)_0^{\sqrt{(1-x^2-y^2)}}dx\ dy$$

$$=\int_{x=-1}^{1}\int_{y=-\sqrt{(1-x^2)}}^{\sqrt{(1-x^2)}}y^2\ (1-x^2-y)\ dx\ dy$$

$$= \int_{x=-1}^{1}\left[\frac{1}{3}(1-x^2)\,y^3 - \frac{1}{5}y^5\right]_{y=-\sqrt{(1-x^2)}}^{\sqrt{(1-x^2)}} dx$$

$$= 2\int_{-1}^{1}\left[\frac{1}{3}(1-x^2)^{5/2} - \frac{1}{5}(1-x^2)^{5/2}\right] dx$$

$= \frac{4}{15}\int_{-1}^{1}(1-x^2)^{5/2}\,dx\ \frac{8}{15}\int_0^1 (1-x^2)^{5/2}\,dx$, by property of definite integral.

$= \frac{8}{15}\int_{\phi=0}^{\pi/2}(1-\sin^2\phi)^{5/2}\cos\phi\,d\phi$, putting $x = \sin\phi$

$= \frac{8}{15}\int_0^{\pi/2}\cos^2\phi\,d\phi = \frac{8}{15}\cdot\frac{5}{6}\cdot\frac{3}{4}\cdot\frac{1}{2}\cdot\frac{\pi}{2} = \frac{\pi}{12}$. **Ans.**

21. *Evaluate $\int_s F \cdot n\,dS$, where $F = xi + yi + z^2 k$ and is the closed surface bounded by the cone $x^2 + y^2 = z^2$ and the plane $z = 4$.*

Solution : Here div F = ∇ . F

$$= \left(i\frac{\partial}{\partial x} + j\frac{\partial}{\partial y} + k\frac{\partial}{\partial z}\right) \cdot (xi + yj + z^2k)$$

$$= \frac{\partial}{\partial x}(x)\ \frac{\partial}{\partial y}(y)\ \frac{\partial}{\partial z}(z^2) = 1 + 1 + 2z = 2 + 2z$$

∴ By Gauss' divergence theorem, we have

$\int_S F \cdot n\,dS = \int_V \text{div } F\,dV = \int_t (2 + 2z)\,dV$,

where V is the volume enclosed by the surface S.

or $\int_S F \cdot n\,dS = 2\int_V dV + 2\int_t z\,dV = 2V + 2z\,V$ . . . (A)

where $\bar{z} = \frac{\int_V z\,dV}{\int_t dV}$ gives the distance of the centroid of the cone from the origin O

i.e. $\bar{z}\,V = \int_V dV$, $[\because \int_V dV = v]$

or $\int_S F \cdot n\,dS = 2V(1 + \bar{z})$, from (A) . . . (i)

Now from the adjoining figure it is evident that if G be the centroid of the cone, then $\bar{z}$ = OG = 3/4 (OA) = 3/4 (4) = 3, where A is the centre of the circular section of the cone by the plane z = 4 and as such OA = 4.

Also V = volume of the cone with O as veretex and A as the centre of its base.

$$= 1/3\ \pi\ (AB)^2.\ OA$$

Now the equation of the circle with centre A is $x^2 + y^2 = 4$, $z = 4$.

∴ The radius of this circle = AB = 2

Hence from (ii) we have $V = 1/3\ \pi\ (2)^2\ .\ 4 = 16/3\ \pi$

∴ From (i) we have $\int_S F\ .\ n\ dS = 2V\ (1 + \bar{z})$

$$= 2\ .\ \frac{16}{3}\pi\ (1 + 3) = \frac{128}{3}\pi \quad \textbf{Ans.}$$

22. *Evaluate* $\int_s F\ .\ n\ dS$, *where* $F = x^2 + y^2 + z^2\ k$ *and S is the surface of the solid cut off by the plane* $x + y + z = a$ *from the first octant.*

Solution : Here $F = x^2i + y^2j + z^2k$

$$\therefore\ \text{div } F = \left(i\frac{\partial}{\partial x} + j\frac{\partial}{\partial y} + k\frac{\partial}{\partial z}\right).\ (x^2i + y^2j + z^2k)$$

$$= \frac{\partial}{\partial x}(x^2)\ \frac{\partial}{\partial y}(y^2)\ \frac{\partial}{\partial z}(z^2) = 2\ (x + y + z)$$

∴ By Gauss' divergence theorem, we have

$$\int_S F\ .\ n\ dS = \int_V \text{div } F\ dV = 2\int_V (x + y + z)\ dx\ dy\ dz \qquad \ldots \text{(i)}$$

where V is the volume cut off by the plane $x + y + z = a$ from the first octant.

The equation of the line AB from B is

∴ A and B it is evident that the limits for integration are for z from 0 to $a - x - y$; for y from 0 to $a = x$ and for x from 0 to a.

∴ From (i) we have $\int_S F\ .\ n\ dS$

$$= 2\int_{x=0}^{a}\int_{y=0}^{a-x}\int_{z=0}^{a-x-y} (x + y + z)\ dx\ dy\,dz$$

$$= 2\int_{x=0}^{a}\int_{y=0}^{a-x}\left[xz + yz + \frac{1}{2}z^2\right]_{z=0}^{a-x-y} dx\ dy$$

$$= 2\int_{x=0}^{a}\int_{y=0}^{a-x} [x(a - x - y) + (a - x - y) + \frac{1}{2}(a - x - y)^2]\ dx\ dy$$

$$= 2\int_{x=0}^{a}\int_{y=0}^{a-x} \frac{1}{2}(a^2 - x^2 - y^2 - 2xy)\ dx\ dy$$

$$= \int_{x=0}^{a}\left[a^2y - x^2y - \frac{1}{3}y^3 - xy^2\right]_{y=0}^{a=x} dx$$

$$= \int_0^a [a^2 (a-x) - x^2 (a-x) - \frac{1}{3}(a-x)^3 - x(a-x)^2]\, dx$$

$$= \frac{1}{3}\int_0^a (2a^3 - 3a^2x + x^3)\, dx = \frac{1}{3}\left[2a^3x - \frac{3}{2}a^2 + \frac{1}{4}x^4\right]_0^a$$

$$= \frac{1}{3}\left[2a^4 - \frac{3}{2}a^4 + \frac{1}{4}a^4\right] = \frac{1}{4}a^4.$$ **Ans.**

23. *Prove that $\int_s r \times n\, n\, dS = 0$ for any closed surface S.*

Solution : Ifd A is any arbitrary constant vector, then

$A \,.\, \int_S r \times n\, dS = \int_S A \,.\, (r \times n)\, dS,$ [$\because$ A is constant]

$= \int_S (A \times r) \,.\, n\, dS.$ interchanging dot and cross

$= \int_S \text{div}\,(A \times r)\, dV$ by Gauss' Div. Theorem

Now $\text{div}\,(a \times r) = r \,.\, \text{curl } A - A \,.\, \text{curl } r$ . . . (i)

$= 0$, since curl A = 0, A being constant

and curl r = 0.

$\therefore$ From (i) we have

$A \,.\, \int_S r \times n\, dS = 0$, where A is an arbitrary constant vector

or $\int_S r \times n\, dS = 0$ Hence proved.

24. *Show that*

$$V = \iint_S x\, dy\, dz = \iint_S z\, dx\, dy$$

$$= \frac{1}{3}\iint_S (x\, dy\, dz + y\, dz\, dx + z\, dx\, dy),$$

where V is the volume of a region T bounded by a surface S.

Solution : By Gauss' divergence theorem, we have

$$\iint_S x\, dy\, dz = \iiint_V \frac{\partial}{\partial x}(x)\, dx\, dy\, dz$$

$\iiint_V dV,$ [$\because$ dV = dx dy dz]

$= V$

Similarly we can prove that

$$\iint_S y\, dy\, dz = \text{and} \iint_S z\, dx\, dy = V$$

Adding these results, we get

$$3\,V = \iint_S x\, dy\, dz + \iint_S y\, dz\, dx + \iint_S z\, dx\, dy$$

or $$V = \frac{1}{3} \iint_S (x\, dy\, dz + y\, dz\, dx \,|+ z\, dx\, dy)$$

25. *Evaluate* $\iint_S (x\, dy\, dz + y\, dz\, dx + z\, dx\, dy)$, *where S is the surface of the sphere* $x^2 + y^2 + z^2 = a^2$.

Solution : We know

$$\iint_S (F_1\, dy\, dz + F_2\, dz\, dx + F_3\, dx\, dy)$$

$$= \iiint_V \left(\frac{\partial F_1}{\partial x} + \frac{\partial F_2}{\partial y} + \frac{\partial F_3}{\partial z}\right) dx\, dy\, dz \qquad \ldots \text{(i)}$$

Here $F_1 = x$, $F_2 = y$ and $F_3 = z$

$$\therefore \left(\frac{\partial F_1}{\partial x} = 1, \frac{\partial F_2}{\partial y} = 1, \frac{\partial F_3}{\partial z}\right) = 1$$

$\therefore$ From (i) we have

$$\iint_S (x\, dy\, dz + y\, dz\, dx + z\, dx\, dy)$$

$= \iint_V (1 + 1 + 1)\, dx\, dy\, dz$, where V is the volume enclosed by the surface S.

$$= 3 \iiint_V dx\, dy\, dz,$$

$$= 3 \text{ (volume of the given sphere)}$$

$$= 3\left(\frac{4}{3} \pi a^3\right) = 4 \pi a^3 \quad \textbf{Ans.}$$

26. *By transforming into a triple integral, evaluate* $\iint_s (x^3\, dy\, dz + x^2 y\, dz\, dx + x^2 z\, dx\, dy)$, *where S is the closed surface consisting of the cylinder* $x^2 + y^2 = a^2$ $(0 \leq z \leq b)$ *and the circular dises* $z = 0$ *and* $z = b$ $(x^2 + y^2 \leq a)$.

Solution : Here $F_1 = x^3$, $F_2 = x^2 y$, $F_3 = x^2 z$

$$\therefore \frac{\partial F_1}{\partial x} = 3x^2, \frac{\partial F_2}{\partial y} = x^2, \frac{\partial F_3}{\partial z} = x^2$$

$\therefore$ From cartesian form of Gauss' divergence theorem viz.

$$\int_S (F_1\, dy\, dz + F_2\, dz\, dx + F_3\, dx\, dy)$$

$$= \iiint_V \left(\frac{\partial F_1}{\partial x} + \frac{\partial F_2}{\partial y} + \frac{\partial F_3}{\partial z}\right) dx\, dy\, dz, \text{ we have}$$

$\iint_s (x^3\,dy\,dz + x^2 y\,dz\,dx + x^2 z\,dx\,dy)$

$= \iiint_V (3x^2 + x^2 + x^2)\,dx\,dy\,dz,$

where V is the volume enclosed by S.

$= 5 \int_{x=-a}^{a} \int_{y=-\sqrt{(a^2-x^2)}}^{\sqrt{(a^2-x^2)}} \int_{z=0}^{b} x^2\,dx\,dy\,dz$

$= 5 \times 2 \times 2 \int_{x=0}^{a} x^2 \int_0^{\sqrt{(a^2-x^2)}} (z)_0^b\,dx\,dy$

$= 20\,b \int_{x=0}^{a} x^2 (y)_0^{\sqrt{(a^2-x^2)}}\,dx = 20\,b \int_0^a x^2 \sqrt{(a^2-x^2)}\,dx$

$= 20\,b \int_{\phi=0}^{\pi/2} a^2 \sin^2\phi \sqrt{(a^2 - a^2\sin^2\phi)}\,a\cos\phi\,d\phi,$

putting $x = a \sin\phi$

$= 20\,a^1\,b \int_0^{\pi/2} \sin^2\phi \cos\phi\,d\phi = 20\,a^2\,b\,.\,\frac{\Gamma\left(\frac{3}{2}\right)\Gamma\left(\frac{3}{2}\right)}{2\,\Gamma\,(3)}$

$= 20\,a^4\,v\,\frac{\frac{1}{2}\sqrt{\pi}\,.\,\frac{1}{2}\sqrt{\pi}}{2\,.\,2\,.\,1} = \frac{5}{4}\,a^4\,b.$ **Ans.**

27. *Verify Stoke's theorm when $F = (2x - y)\,i = y\,z^2\,j - y^2\,z\,k$ where S is the upper half of the sphere $x^2 + y^2 + z^2 = i$ and C is the boundary.*

Solution : Given

$F = (2x - y)\,i - yz^2 j - y^2 zk$

Also we know

$r = xi + yj + zk$

$\therefore$ $F\,.\,dr = [(2x - y)\,i - yz^2 j - y^2 j - y^2 zk]\,.\,[i\,dx + j\,dy + k\,dz]$

or $F\,.\,dr = (2r = y)\,dx - yz2\,dy - y^2 - y^2 zdz$ . . . (i)

Also here C is the circle $x^2 + y^2 = 1$; $z = 0$ *i.e* a circle of unit radius on x y-plane.

$\therefore$ For C we have $z = 0$, $dz = 0$. consequently from (i) we have

$F\,.\,dr = (2x - y)\,dx$ . . . (ii)

Also on C we have $x = \cos\phi$ and $y = \sin\phi$ and ϕ varies from 0 to 2π.

$\therefore$ From (ii) we have

$$\int_C F \cdot dr = \int_C (2x - y)\, dx$$

$$= \int_{\phi=0}^{2\pi} (2\cos\phi - \sin\phi)(-\sin\phi)\, d\phi$$

$$= \int_{\phi=0}^{2\pi} (\sin^2\phi - \sin 2\phi)\, d\phi$$

$$= 4\int_0^{\pi/2} \sin^2\phi\, d\phi + \left[\frac{1}{2}\cos 2\phi\right]_0^{2\pi}$$

$$= 4 \cdot \frac{1}{2} \cdot \frac{1}{2}\pi + \frac{1}{2}[1 - 1] = \pi$$

Hence $\int_C F \cdot dr = \pi$ . . . (iii)

Also the direction cosines of the line OP are

$$\sin\theta\cos\phi,\ \sin\theta\sin\phi,\ \cos\theta$$

$\therefore$ $n = (\sin\theta\cos\phi)\, i + (\sin\theta\sin\phi)\, j + (\cos\theta)\, k$, where n is the unit outward drawn normal vector at P.

$$\text{Again curl } F = \begin{vmatrix} i & j & k \\ \dfrac{\partial}{\partial x} & \dfrac{\partial}{\partial y} & \dfrac{\partial}{\partial z} \\ 2x - y & -yz^2 & -y^2 z \end{vmatrix}$$

$$= i\left[\frac{\partial}{\partial y}(-y^2x) - \frac{\partial}{\partial z}(-yz^2)\right] + j\left[\frac{\partial}{\partial z}(2x - y) - \frac{\partial}{\partial x}(-y^2z)\right]$$

$$+ k\left[\frac{\partial}{\partial x}(-yz^2) - \frac{\partial}{\partial y}(2x - y)\right]$$

$$= 0i + 0j + k = k$$

$$\therefore\ n \cdot \text{curl } F = [(\sin\theta\cos\phi)\, i + (\sin\theta\sin\phi)\, j + (\cos\theta)\, k] \cdot k$$

$$= \cos\theta$$

$$\therefore \int_S n \cdot \text{curl } F\, dS$$

$$= \int_S \cos\theta\, dS, \text{ where } dS = \sin\theta\, d\theta\, d\phi$$

$$= \int_{\theta=0}^{\pi/2}\int_{\phi=0}^{2\pi} \cos\theta\sin\theta\, d\theta\, d\phi$$

$$= \int_{\theta=0}^{\pi/2}\int_{\phi=0}^{2\pi} \frac{1}{2}\sin 2\theta\, d\theta\, d\phi = \frac{1}{2}\left[-\frac{1}{2}\cos 2\theta\right]_0^{\pi/2} [\phi]_0^{2\pi}$$

$$= \frac{1}{2}\left[\frac{1}{2}\cos\theta + \frac{1}{2}\cos\theta\right] 2\pi = \frac{1}{2}\left(\frac{1}{2} + \frac{1}{2}\right) \cdot 2\pi = \pi \quad \text{. . . (iv)}$$

Hence from (iii) and (iv) we have

$$\int_C F \cdot dr = \pi = \int_S n \cdot \text{curl } F \, dS$$

which verifies Stoke's Theorem.

28. *Verify Stocke's theorem when $F = x^2 i + xy j$, where C is the perimeter of the square in xy-plane whose sides are along the lines $x = 0$, $y = 0$, $x = a$, $y = a$.*

Solution : Here $F = x^2 i + xyj$ and $dr = (dx)\, i + (dy)\, j$

$\therefore F \cdot dr = (x^2 i + xyj) \cdot (idx + jdy) = x^2 dx + xy^2 dy$ . . . (A)

$\therefore \int_C F \cdot dr$

$= \int_{C1} F \cdot dr + \int_{C2} F \cdot dr + \int_{C3} F \cdot dr + \int_{C4} F \cdot dr$ . . . (i)

where C_1, C_2, C_3, and C_4 are the lines from a to O, O to D, D to B and B to A respectively

Now for C_1 we have $x = 0$, $dx = 0$ and y varies from a to 0, so from (A)

we get $\int_{C1} F \cdot dr = 0$ . . . (ii)

For C_2 we have $y = 0$, $dy = 0$ and x varies form 0 to a, so from (A) we get

$$\int_{C2} F \cdot dr = \int_{x=0}^{a} x^2 dx = \left(\frac{1}{3}x^3\right)_0^a = \frac{1}{3}a^3 \quad \text{. . . (iii)}$$

For C_3 we have $x = a$, $dx = 0$ and x varies form 0 to a, so form (A)

$$\text{we get } \int_{C3} F \cdot dr = \int_{y=0}^{a} ay\, dy = \frac{1}{2}a\left(y^2\right)_0^1 = \frac{1}{2}a^3 \quad \text{. . . (iv)}$$

For C_4 we have $y = a$, $dy = 0$ and x varies forom a to 0 so, from (A)

$$\text{we get } \int_{C4} F \cdot dr = \int_{x=0}^{a} x^2\, dx = \frac{1}{2}\left(x^3\right)_a^0 = -\frac{1}{3}a^3 \quad \text{. . . (v)}$$

$\therefore$ From (i), (ii), (iii), (iv) and (v) we have

$$\int_C F \cdot dr = 0 + \frac{1}{3}a^3 + \frac{1}{2}a^3 - \frac{1}{3}a^3 = \frac{1}{2}a^3 \quad \text{. . . (vi)}$$

Again for the square ABDO, the unit outward drawn normal vector $n = k$.

$$\text{Also curl } F = \begin{vmatrix} i & j & k \\ \frac{\partial}{\partial x} & \frac{\partial}{\partial y} & \frac{\partial}{\partial z} \\ x^2 & xy & 0 \end{vmatrix} = i\,(0) + j(0) + k(y)$$

or $\quad \text{Curl } F = yk$

$\therefore \quad n \cdot \text{curl } F = k \cdot yk = y$

Also $\quad dS = dx\, dy$

$\therefore \int_S n \cdot \text{curl } F\, dS = \int_{x=0}^{a} \int_{y=0}^{a} y\, dx\, dy$

$$=(x)_0^1 \left(\frac{1}{2} y^2\right)_0^1 = \frac{1}{2} a^3 \qquad \ldots \text{(vii)}$$

From (vi) and (vii) we get $\int_C F \cdot dr = \int_S n \cdot \text{curl } F\, dS$

29. *Evaluate $\iint_s (x^3\, dy\, dz + y^2\, dz\, dx + z^2\, dx\, dy)$, over the surface S of a cube bounded by the coordinate planes and the planes $x = y = z = a$.*

Solution : Here $F_1 = x^3$, $F_2 = y^3$, $F_3 = z^3$.

$$\therefore \quad \frac{\partial F_1}{\partial x} = 3x^2, \frac{\partial F_2}{\partial y} = 3y^2 \text{ and } \frac{\partial F_3}{\partial z} = 3z^2$$

$\therefore$ By cartesian form of Gauss' divergence theorem, we have

$\iint_S (xh3\, dy\, dz + y^3\, dz\, dx + z^3\, dx\, dy)$

$= \iiint_V (3x^2 + 3y^2 + 3z^2)\, dx\, dy\, dz,$

where V is the volume enclosed by S,

$= 3 \int_{x=0}^{a} \int_{y=0}^{a} \int_{z=0}^{a} (x^2 + y^2)\, dx\, dy\, dz$

$= 3 \int_{x=0}^{a} \int_{y=0}^{a} \left(x^2 z + y^2 z + \frac{1}{3} z^3 \right)_{z=0}^{a} dx\, dy$

$= 3 \int_{x=0}^{a} \int_{y=0}^{a} \left(x^2 a + y^2 a + \frac{1}{3} a^3 \right) dx\, dy$

$= 3 \int_{x=0}^{a} \left(x^2 ay + \frac{1}{3} ay^3 + \frac{1}{3} a^3 y \right)_0^a dx = 3 \int_0^a \left(a^2 x^2 + \frac{2}{3} a^4 \right) dx$

30. *Evaluate $\iint_s (x^3\, dy\, dz + y^2\, dz\, dx)$, where S is the surface of the sphere $x^2 + y^2 + z^2 = a^2$.*

Solution : Here $F_1 = x^3$, $F_2 = y^3$ and $F_3 = 0$

$$\therefore \quad \frac{\partial F_1}{\partial x} = 3x^2, \frac{\partial F_2}{\partial y} = 3y^2 \text{ and } \frac{\partial F_3}{\partial z} = 0$$

$\therefore$ By cartesian form of Gauss' theorem, we have

$$\iint_S (x^3\, dy\,dz + y^3\, dz\,dx)$$

$= \iiint_V (3x^2 + 3y^2)dx\, dy\, dz,$ where V is the volume enclosed by the sphere $x^2 + y^2 + z^2 = a^2$.

$= 3 \iiint_V (x^2 + y^2)dx\, dy\, dz,$

$= \frac{3}{\rho} \iiint_V (x^2 + y^2)\, \rho\, dx\, dy\, dz,$ where ρ is the density of the material of the sphere assmed constant

$= \frac{3}{\rho} \iiint_V (x^2 + y^2)\, dm,$ where $dm = \rho\, dx\, dy\, dz$

$= \frac{3}{\rho}$ [Moment of inertia of the given sphere about z-axis]

$= \frac{3}{\rho}\, [(2/5)\, Ma^2]$, where M is the mass of the sphere

$= \frac{3}{\rho}\, [(2/5)\, .\, (4/3)\, \pi\, a^3\, \rho\, .\, a^2],$ $\quad [\because M = (4/3)\, \pi\, a^3\, \rho.]$

$= (8/5)\, \pi\, a^5.$

31. *Prove that*

$$\iint_S [lu + mv + nw]\, dS = \iiint_V \left(\frac{\partial u}{\partial x} + \frac{\partial v}{\partial y} + \frac{\partial w}{\partial z}\right) dx\, dy\, dz$$

where i, m, n, are the direction cosines of the outward drawn normal at a point P (x, y, z) of the closed surface S enclosing hte volume V.

Solution : ∵ (l, m, n) are the direction cosines of the outward drawn normal vector at any point P (x, y, z) on the surface S, so we have

$$n = l\, i + mj + nk \quad \text{. . . (i)}$$

Also let

$$F = u\, i + vj + wk \quad \text{. . . (ii)}$$

∴ From (i) and (ii) we have

$$F\, .\, n = (ui + yj + wk)\, .\, (li + mj + nk)$$

or

$$F\, .\, n = lu + mv + nw \quad \text{. . . (iii)}$$

And $\text{div } F = \nabla\, .\, F = \left(i\frac{\partial}{\partial x} + j\frac{\partial}{\partial y} + k\frac{\partial}{\partial z}\right).\, (ui + vj + wk)$

$$\text{div } F = i\frac{\partial u}{\partial x} + \frac{\partial v}{\partial y} + \frac{\partial w}{\partial z} \quad \text{. . . (iv)}$$

$\therefore$ Substituting these values of F . n and div F in Gauss' theorem $\int_S$ F . n dS = $\int_V$ div F dV, where dV = dx dy dz

we have $\int_S (lu + mv + nw)\, dS = \left(\frac{\partial u}{\partial x} + \frac{\partial v}{\partial y} + \frac{\partial w}{\partial z}\right) dx\, dy\, dz$

Hence proved.

32. *Prove that* $\int_C \phi \nabla \psi \, . \, dr = \int_S (\nabla \phi \times \nabla \psi) \, . \, n\, dS$

Solution : By Stoke's theroem, we have

$$\int_S F \, . \, dr = \int_S \text{curl } F \, . \, n\, dS \qquad \ldots \text{(i)}$$

Let $F = \phi \nabla \psi$. then

curl $F = \text{curl } (\phi \nabla \psi)$

$= \nabla\phi \times \nabla\psi + \phi \text{ curl } (\nabla\psi)$.

$\therefore \nabla \times (\psi\nabla) = \nabla\phi \times V + \phi (\nabla \times V)$.

or curl $F = \nabla\phi \times \nabla\psi$, [$\because$ curl (grad ψ) = 0]

Subsituting these values in (i) we get

$$\int_S \phi \nabla \psi \, . \, dr = \int_S (\nabla \phi \times \nabla \psi) \, . \, n\, dS$$

Hence proved.

33. *Evaluate* $\int_s F \, . \, n\, dS$, *where* $F = a x i + b y j + c z k$ *and S is the surface of sphere* $x^2 + y^2 + z^2 = 1$.

Solution : Here div F = ∇ . F

$$= \left(i \frac{\partial}{\partial x} + j \frac{\partial}{\partial y} + \frac{\partial}{\partial z}\right) . (axi + byj + czk)$$

$$= \frac{\partial}{\partial x}(ax) + \frac{\partial}{\partial y}(by) + \frac{\partial}{\partial z}(cz) = a + b + c$$

$\therefore$ By Gauss' theorem we have

$$\int_S F \, . \, n\, dS = \int_V \text{div } F\, dV,$$

where V is the volume enclosed by the surface S

$= \int_V (a + b + c)\, dV = (a + b + c) \int_V dV$

$= (a + b + c) V$, where V = volume enclosed by S = $4/3\, \pi\, (1)^3$

$= 4/3\, \pi\, (a + b + c)$.

34. *Verify the divergence theorem for* $F = 4x\, i - 2y^2\, j + z^2\, k$ *taken over the region S bounded by* $x^2 + y^2 = 4$, $z = 0$ *and* $z = 3$.

Solution : Here $\nabla F = \frac{\partial}{\partial x}(4x) + \frac{\partial}{\partial y}(-2y^2) + \frac{\partial}{\partial z}(z^2)$

$= 4 - 4y + 2z = 2(2 - 2y + z)$

$\therefore$ By Gauss' Divergence theorem we have

$\int_S F \cdot n \, dS = \int_V \nabla \cdot F \, dV$, where V is the volume enclosed by S.

Now

$$\int_t \nabla \cdot F dV = \int_{x=-2}^{2} \int_{y=-\sqrt{(4-x^2)}}^{\sqrt{(4-x^2)}} \int_{z=0}^{3} (2(2 - 2y + z) \, dx \, dy \, dz$$

$$= 2 \int_{x=-2}^{2} \int_{y=-\sqrt{(4-x^2)}}^{\sqrt{(4-x^2)}} \left[2z - 2yz + \frac{1}{2} z^2\right]_0^3 dx\, dy$$

$$= 2 \int_{x=-2}^{2} \int_{y=-\sqrt{(4-x^2)}}^{\sqrt{(4-x^2)}} [6 - 6y + (9/2)] \, dx\, dy$$

$$= 21 \int_{x=-2}^{2} \int_{y=-\sqrt{(4-x^2)}}^{\sqrt{(4-x^2)}} dx\, dy \; - 12 \int_{x=-2}^{2} \int_{y=-\sqrt{(4-x^2)}}^{\sqrt{(4-x^2)}} dx\, dy$$

$$= 21 \times 4 \int_{x=-2}^{2} \int_{y=0}^{\sqrt{(4-x^2)}} dx\, dy - 1\,[0],$$

as y is an odd function in the second integral

$$= 84 \int_{x=0}^{2} \int_{y=0}^{\sqrt{(4-x^2)}} dx = 84 \int_0^2 \sqrt{(2^2 - x^2)} \, dx$$

$$= 84 \left[\frac{1}{2} x \sqrt{(2^2 - x^2)} + \frac{1}{2} \cdot 2^2 \cdot \sin^{-1}\left(\frac{1}{2} x\right)\right]_0^2$$

$= 84\,[2 \sin^{-1} 1] = 84\,[2\,(\pi/2] = 84\,\pi$ . . . (i)

Again the surface S is the surface of the given cylinder consisting of the curved surface S_1 (say) and two equal circular ends S_2 and S_3 (say0 on the planes z = 0 and z = 3 respectively.

$\therefore \int_S F \cdot n \, dS = \int_{S_1} F \cdot n \, dS + \int_{S_2} F \cdot n \, dS + \int_{S_3} F \cdot n \, dS$ ...(ii)

Now for the surface S_2 *i.e.* the lower circular end of the cylinder on the plane z = 0, the outward drawn normal is downwards and parallel to z-axis so n = – k.

$\therefore \int_{S_2} F \cdot n \, dS = \int_{S_2} (4xi - 2y^2 j + z^2 k) \cdot (-k) \, dS$

$$= \int_{S_2} z^2 \, dS = 0, \qquad \ldots \text{(iii)}$$

as $z = 0$ for S_2.

Also for the surface S_3 *i.e.* the upper curcular end of the cylinde on the plane $z = 3$, the outward drawn normal is upwards and parallel to z-axis, so $n = k$

$\therefore \int_{S_3} F \, . \, n \, dS = \int_{S_3} (4xi - 2y^2 j + z^2 k) \, . \, k \, dS$

$= 9 \int_{S_3} dS,$ $\qquad [\because z = 3 \text{ for } S_3]$

$= 9 \, S_3$, where S_3 is the area of the upper end of the cylinder $x^2 + y^2 = 4$ *i.e.* a circle of radius 2

$$= 9 \, [\pi \, (2)^2] = 36\pi \qquad \ldots \text{(iv)}$$

And for the curved surface S_1 of the cylinder, we have

$$n = \frac{\nabla \phi}{|\nabla \phi|}, \text{ where } \phi \equiv x^2 + y^2 = 4$$

$$= \frac{2(xi + yj)}{|2(xi + yj)|},$$

$$\left[\because \nabla \phi = i \frac{\partial \phi}{\partial x} + j \frac{\partial \phi}{\partial y} + k \frac{\partial \phi}{\partial z} = 2xi + 2yj\right]$$

$$= \frac{si + yj}{\sqrt{(x^2 + y^2)}} = \frac{xi + yj}{2}, \qquad [\because x^2 + y^2 = 4]$$

Also for S_1, z varies from $z = 0$ to $z = 3$ and take $x = 2 \cos \theta$ $y = 2 \sin \theta$ so that θ varies from 0 to 2π.

$\therefore \int_{S_1} F \, . \, n \, dS = \int_{S_1} (4xi - 2y^2 j + z^2 k) \, . \, \frac{1}{2} (xi + j) \, dS$

$= \int_{S_1} (2x^2 - y^2) \, dS$

$= \int_{z=0}^{3} \int_{\theta=0}^{2\pi} [2(2 \cos\theta)^2 - (2 \sin \theta)^3] \, 2 \, d\theta \, dz,$

$\because$ dS = element of the area on the curved surface

$= (2 \, d \, \theta) \, dz$

$= \int_{z=0}^{3} \int_{\theta=0}^{2\pi} 8 \, (\cos^2 \theta - \sin^3 \theta) \, d\theta \, dz$

$= 16 \, [z]_0^3 \int_0^{2\pi} (\cos^2 \theta - \sin^3 \theta) \, d\theta$

$$= 16 \times 3 \times 2 \int_0^{\pi} (\cos^2 \theta \, d\theta, \qquad [\because \{\sin(2\pi - \theta)\}^3 = -\sin^3\theta,]$$

so by the property of definite integral value of the second integral is zero

$$= 96 \times 2 \int_0^{\pi/2} \cos^2 \theta \, d\theta = 192 \left[\frac{1}{2} \cdot \frac{\pi}{2}\right] = 48\pi \qquad \ldots (v)$$

$\therefore$ From (ii), (iii), (iv), and (v) we get

$$\int_S F \cdot n = 0 + (36\pi) + (48\pi) = 84\pi \qquad \ldots (iv)$$

$\therefore$ From (i) and (vi) we have

$$\int_S F \cdot n \, dS = \int_V \nabla \cdot F \, dV,$$

and thus divergence theorem is verifed.

35. *Verify the divergence theorem for $F = 2x^2 i - y^2 j + 4xz\, k$ taken over the region S in the first octant bounded by $y^2 + z^2 = 9$, $x = 0$ and $x = 2$.*

Solution : Here $\nabla \cdot F = \frac{\partial}{\partial x}(2x^2) + \frac{\partial}{\partial y}(-y^2) + k \frac{\partial}{\partial z}(4xz)$

$$= 4x - 2y + 4x = 8x - 2y$$

$$\therefore \int_V \nabla \cdot F \, dV = \int_V (8x - 2y) \, dV,$$

where V is the colume enclosed by S

$$= \int_{x=0}^{2} \int_{y=0}^{3} \int_{z=0}^{\sqrt{(9-y^2)}} (8x - 2y) \, dx \, dy \, dz,$$

as in last example.

$$= \int_{x=0}^{2} \int_{y=0}^{3} (8x - 2y) \sqrt{(9 - y^2)} \, dx \, dy$$

$$= 8 \int_{x=0}^{2} \int_{y=0}^{3} x \sqrt{(9 - y^2)} \, dx \, dy - 2 \int_{x=0}^{2} \int_{y=0}^{3} y \sqrt{(9 - y^2)} \, dx \, dy$$

$$= 8 \left[\frac{1}{2} x^2\right]_0^2 \left[\frac{1}{2} y \sqrt{(9 - y^2)} + \frac{1}{2} 9 \sin^{-1}\left(\frac{1}{3} y\right)\right]_0^3$$

$$- 2 [x]_0^2 \int_0^2 y \sqrt{(9 - y^2)} \, dy$$

$$= 8 \left[9\left(\frac{1}{2}\pi\right) + 4 \int_3^0 t^2 \, dt,\right] \text{ Putting } 9 - y^2 = t^2$$

$$= 36\pi - 4 \left[\frac{1}{3} t^3\right]_0^3 = 36\pi - 36 \qquad \ldots (i)$$

Again the surface S is the surface consisting of two equal quadrants of circular ends of the cylinder given by x = 0 and x = 2 and let us call them as S_2 and S_3 respectively; the curved surface S_1 (say of the cylinder lying between the planes y = 0 and y = 0 respectively

For the surface S_1 as defined above, we have

$\phi \equiv y^2 + z^2 = 9$, [$\because$ the equation of the cylinder is $y^2 + z^2 = 9$]

$$\therefore \quad \nabla \phi = \left(i \frac{\partial}{\partial x} + j \frac{\partial}{\partial y} + k \frac{\partial}{\partial z} \right) \phi$$

$$= i \frac{\partial}{\partial x} (y^2 + z^2 - 9) + j \frac{\partial}{\partial y} (y^2 + z^2 - 9) + k \frac{\partial}{\partial z} (y^2 + z^2 - 9)$$

or $$\nabla \phi = j\,(2y) + (2z) = 2\,(yj + zk)$$

$$\therefore \quad 'n' = \frac{\nabla \phi}{|\nabla \phi|} = \frac{2(yj + zk)}{\sqrt{\left[(2y)^2 + (2z)^2\right]}} = \frac{2\,(yj + zk)}{2\sqrt{\left(y^2 + z^2\right)}}$$

$$= \frac{1}{3} (yj + zk), \text{ since } y^2 + z^2 = 9$$

$$\therefore \quad F \cdot n = (2x^2\, i - y^2\, j + 4\, xzk) \cdot 1/3\, (yj + zk)$$

$$= \frac{1}{3} (-y^3 + 4xz^2)$$

Also as in take $y = r \cos \phi$, $z = r \sin \phi$ and on the cylinder $y^2 + z^2 = 9$ $r = 3$ ϕ varies from 0 to $1/2\ \pi$.

Also dS = elementary area on the surface of the cylinder
$= 3\, d\phi\, dx$ $\therefore$

$$\int_{S_1} F \cdot n\, dS = \int_{x=0}^{2} \int_{\phi=0}^{\pi/2} \frac{1}{3} (-y^3 + 4x z^2)\, 3 d\phi\, dx$$

$$= \int_{x=0}^{2} \int_{\phi=0}^{\pi/2} (-27 \cos^3 \phi + 36 x \sin^2 \phi)\, d\phi\, dx,$$

Putting $y = 3 \cos \phi$, $z = 3 \sin \phi$

$$= -27 (x)_0^2 \int_0^{\pi/2} \cos^3 \phi\, d\phi + 36 \left(\frac{1}{3} x^2 \right)_0^2 \int_0^{\pi/2} \sin^2 \phi\, d\phi$$

$$= -(2) \left(\frac{2}{3} \right) + 36\,(2) \left(\frac{1}{2} \cdot \frac{1}{2} \pi \right) = 18\pi - 36. \qquad \ldots \text{(ii)}$$

On the area S_2, which is a quadrant of the circular cross section of the cylinder $y^2 + z^2 = 9$ by the plane x = 0, we have n = outwad drawn unit normal vector to $S_2 = -i$, as the outward drawn normal to this surface will be in the negative side of x-axis.

$\therefore \int_{S_2} F \cdot n\, dS = \int_{S_2} (2x^2 i - y^2 j + 4xzk) \cdot (-i)\, dS$

$= -2 \int_{S_2} x^2\, dS = 0$, as $x = 0$, on S_2 . . . (iii)

Similarly, on the area S_3, which is a quadrant of the circular cross-section of the cylinder $y^2 + z^2 = 9$, by the plane $x = 2$, we have 'n' = i, as the outward drawn normal to the area S_3 will be in the positive direction of x-axis.

$\therefore \int_{S_3} F \cdot n\, dS = \int_{S_3} (2x^2 i + y^2 j + 4xzk) \cdot i\, dS$

$= 2 \int_{S_3} x^2\, dS$, on the plane $x = 2$

$= 8 \int_{S_3} dS = 8\, S_3$, where S_3 is as defined above

$$= 8\left[\frac{1}{4}\pi\,(3)^2\right] = 18\,\pi \qquad \text{. . . (iv)}$$

Again for S_4 which is a rectangle on the plane $z = 0$, we have n = k, as the outward drawn unit normal vector to S_1 will be in the negative direction of z-axis.

$\therefore \qquad \int_{S_4} F \cdot n\, dS = \int_{S_4} (2x^2 i - y^2 j + 4xzl) \cdot (-k)\, dS$

$= \int_{S_4} (-4xz)\, dS$, where S_4 lies on $z = 0$

$= 0$, as $z = 0$ . . . (v)

And for S_5, which is a rectangle on the plane $y = 0$, we have n = – j, as the outward drawn normal vector to S_5 will be in the negative direction of y-axis.

$\therefore \qquad \int_{S_5} F \cdot n\, dS = \int_{S_5} (2x^2 i - y^2 j + 4xzk) \cdot (-j)\, dS$

$= \int_{S_5} y^2\, dS$, on the plane $y = 0$

$= 0$ . . . (vi)

$\therefore$ Adding (ii), (iii), (iv), (v) and (vi) we have for the entire surface S of the quadrant of the cylinder

$\int_S F \cdot dS = (18\pi - 36) + 0 + 18\pi + 0 + 0 = 36\pi - 36$

$= \int_V \nabla \cdot F\, dV$, from (i).

Hence Gauss' Theorem is verified.

36. *Evaluate $\int_s F \cdot n\, dS$ where S denotes the surface of the ellipsoid $(x^2/a^2) + (y^2/b^2) + (z^2/c^2) = 1$ and $F = x^2 i + y^2 j + z^2 k$.*

Solution : Given $F = x^2i + y^2j + z^2k$

$$\therefore \text{ div } F = \nabla . F = \left(i\frac{\partial}{\partial x} + j\frac{\partial}{\partial y} + k\frac{\partial}{\partial z}\right) . (x^2i + y^2j + z^2k)$$

$$= \frac{\partial}{\partial x}(x^2) + \frac{\partial}{\partial y}(y^2) + \frac{\partial}{\partial z}(z^2) = 2(x + y + z) \quad \ldots \text{(i)}$$

Now by Gauss' divergence theorem, we have

$$\int_S F . n \, dS = \int_V \text{div } F \, dV,$$

where V is the volume enelosed by the given ellipsoid.

Also we know that if $(\bar{x}, \bar{y}, \bar{z})$ be the corrdinates of the centroid of a solid then.

$$\bar{x} = \frac{\int_V x dV}{\int_V dV}; \bar{y} = \frac{\int_V y dV}{\int_V dV}; \bar{z} \frac{\int_V z dV}{\int_V dV}$$

In the case of the given ellipsoid, its centroid is the origin

$\therefore \ \bar{x} = 0 = \bar{y} = \bar{z}$

Consequently, $\int_V dV = 0 = \int_V y \, d \, V = \int_V z \, d \, V$ . . . (ii)

Now $\int_S F . n \, d \, S = \int_V \text{div } F \, dV = \int_V 2(x + y + z) \, dV$, from (i)

$= 2 \int_V (x + y + z) \, dV = 2 \int_V x \, dV + 2 \int_V y \, d \, V + 2 \int_V z . d \, V$

$= 2(0) + 2(0) + 2(0)$, from (ii)

$= 0$ **Ans.**

37. *Verify divergence theorm for $F = (x^2 - yz) i + (y^2 - z x) j + (z^2 - xy) k$ taken over the rectangular parallelopiped $0 \leq x \leq a$, $0 \leq y \leq b$, $0 \leq z \leq c$.*

Solution : (consider it as rectangular parallelopiped whose edges are OB = a, OC = b and OF = c).

Now $F = (x^2 - yz) i + (y^2 - zx) j + (z^2 - xy) k$

$$\therefore \ F = \frac{\partial}{\partial x}(x^2 - yz) + \frac{\partial}{\partial y}(y^2 - zx) + \frac{\partial}{\partial z}(z^2 - xy)$$

$$= 2x + 2y + 2z = 2(x + y + z)$$

$\therefore \int_V \text{div } F \, d \, V$, where V is the volume of the rectangular parallelopiped.

$$= \int_{x=0}^{a} \int_{y=0}^{b} \int_{z=0}^{c} 2(x + y + z) \, dx \, dy \, dz$$

$$= 2 \int_{x=0}^{a} \int_{y=0}^{b} \left[xz + yz + \frac{1}{2}z^2\right]_{z=0}^{c} dx \, dy$$

$$= 2\int_{x=0}^{a}\int_{y=0}^{b}\left(xc + yc + \frac{1}{2}c^2\right)dx\,dy$$

$$= 2\int_0^a\left(cxy + \frac{1}{2}cy^2 + \frac{1}{2}c^2y\right)_{y=0}^{b}dx$$

$$= 2\int_0^a\left(cbx + \frac{1}{2}cb^2 + \frac{1}{2}c^2b\right)dx$$

$$= 2\left[\frac{1}{2}cbx^2 + \frac{1}{2}cb^2x + \frac{1}{2}c^2bx\right]_0^a$$

$$= cba^2 + cb^2a + c^2ba = abc\,[a + b + c] \quad \ldots \text{(i)}$$

Now, we shall calculate $\int F \,.\, n\, dS$ over the six faces of the rectangular parallelopiped separately and then add them to get the value of $\int_S F \,.\, n\, dS$, where S is the entire surface of the rectangular parallelopiped.

For the face ABED, we have n = 1 x = a, dS = dy dz

$\therefore \int F \,.\, n\, dS$ over the face ABED

$$= \int_{z=0}^{c}\int_{y=0}^{b}[(x^2 - yz)\,i + (y^2 - zx)\,j + (z^2 - xy)\,k]\,.\,i\,dy\,dz$$

$$= \int_{z=0}^{c}\int_{y=0}^{b}(x^2 - yz)\,dy\,dz, \text{ over the face ABED for which } x = a$$

$$= \int_{z=0}^{c}\int_{y=0}^{b}(a^2 - yz)\,dy\,dz = \int_{z=0}^{c}\left(a^2y - \frac{1}{2}y^2z\right)_{y=0}^{b}dz$$

$$= \int_0^c\left(a^2b - \frac{1}{2}b^2z\right)dz = \left(a^2bz - \frac{1}{4}b^2z^2\right)_0^a = a^2bc - \frac{1}{4}b^2c^2$$

$$= \frac{1}{4}bc\,(4a^2 - bc) \quad \ldots \text{(ii)}$$

For the face OCKF, we have n = – i, x = 0, dS = dy dz

$\therefore \int F \,.\, n\, dS$ over the face OCKF

$$= \int_{z=0}^{c}\int_{y=0}^{b}[(x^2 - yz)\,i + (y^2 - zx)\,j + (z^2 - xy)\,k]\,.\,(-i)\,dy\,dz$$

$$= \int_{z=0}^{c}\int_{y=0}^{b}(x^2 - yz)\,dy\,dz, \quad \text{for } x = 0$$

$$= \int_{z=0}^{c}\int_{y=0}^{b}yz\,dy\,dz = \left(\frac{1}{2}y^2\right)_0^b\left(\frac{1}{2}z^2\right)_0^a = \frac{1}{4}b^2c^2 \quad \ldots \text{(iii)}$$

For the face ABOC, we have n = – k, dS = dx dy and z = 0

$\therefore \int$ F . dS over the face ABOC

$= \int_{x=0}^{a} \int_{y=0}^{b} [(x^2 - yz)\, i + (y^2 - zx)\, j + (z^2 - xy)\, k] . (-k)\, dx\, dy$

$$= - \int_{x=0}^{a} \int_{y=0}^{b} (z^2 - xy)\, dx\, dy, \text{ for } z = 0$$

$$= \int_{x=0}^{a} \int_{y=0}^{b} xy\, dx\, dy = \left(\frac{1}{2}x^2\right)_0^a \left(\frac{1}{2}y^2\right)_0^b = \frac{1}{4} a^2 b^2 \qquad \text{. . . (iv)}$$

For the face DEFK, we have n = k, dS = dx dy and z = c

$\therefore \int$ F . n dS over the face DEFK

$$= \int_{x=0}^{a} \int_{y=0}^{b} (z^2 - xy)\, dx\, dy, \text{ for } z = c, \text{ as above}$$

$$= \int_{x=0}^{a} \int_{y=0}^{b} (c^2 - xy)\, dx\, dy = \frac{1}{4} ab\, (4c^2 - ab) \qquad \text{. . . (v)}$$

For the face OADK we have n = j, dS = dx dz and y = b

$\therefore \int$ F . n dS over the face OADK

$$= \int_{x=0}^{a} \int_{z=0}^{c} (b^2 - xy)\, dx\, dz, \text{ proceeding as for face ABED}$$

$$= \frac{1}{4} ac\, (4b^2 - ac) \qquad \text{. . . (vi)}$$

And for the face OBEF, n = = j dS = dx dz and y = 0

$\therefore \int$ F . n dS over the face OBEF

$$= - \int_{x=0}^{a} \int_{z=0}^{c} (y^2 - xz)\, dx\, dz, \text{ for } y = 0 \text{ as for face OCKF}$$

$$= \int_{x=0}^{a} \int_{z=0}^{c} xz\, dx\, dz = \frac{1}{4} a^2 c^2 \qquad \text{. . . (vii)}$$

Adding (ii) to (vii) we get for the whole surface S of the rectangular parallelopiped.

$$\int_S F . n\, dS = \frac{1}{4} bc\, (4a^2 + bc) + \frac{1}{4} b^2c^2 + \frac{1}{4} a^2b^2 + \frac{1}{4} ab\, (4c^2 - ab)$$

$$+ \frac{1}{4} ac\, (4b^2 - ac) + \frac{1}{4} a^2c^2$$

$$= a^2 bc + abc^2 + acb^2 = abc\, (a + b + c) \qquad \text{. . . (viii)}$$

Hence from (i) and (vii) we find that

$$\int_S F . n\, dS = \int_V \text{div } F\, dV$$

Hence divergence theorem is verified.

38. *Verify Stokes' theorem for the function F = x (ix + jy), integrated round the square, in the plane z = 0, whose sides are along the lines x = 0, y = 0, x = a, y = a.*

Solution: We clearly have

$$\oint_C F.dl + \int_{OA} F.dl + \int_{AB} F.dl + \int_{BC} F.dl + \int_{CO} F.dl,$$

where, $\int_{OA} F.dl = \int_o^a x(ix + jy).idx = \int_o^a x^2 dx = \frac{1}{3}a^3$,

$$\int_{AB} F.dl = \int_o^a x(ix + jy).jdy = \int_o^a aydy = \frac{1}{2}a^3,$$

$$\int_{BC} F.dl = \int_a^o x(ix + jy).idx = -\int_o^a x^2 dx = -\frac{1}{3}a^3 \text{ and}$$

$$\int_{CO} F.dl = \int_a^o x(ix + jy).jdx = 0$$

So that, $\oint_c F.dl = \frac{1}{3}a^3 + \frac{1}{2}a^3 - \frac{1}{3}a^3 + 0 = \frac{1}{2}a^3$.

Now, in accordance with Stokes' theorem, $\oint c\, F.d\, l = \iint S$ curl FdS.

Since curl x (ix + jy) = ky, we have

$$\iint_S \text{curl } FdS = \iint_S \text{curl}\, x(ix + jy)dS = \int_o^a \int_o^a ky.kdxdy$$

$$= \int_o^a \int_o^a ydxdy = \frac{1}{2}a^3.$$

Stokes' theorem thus stands verified.

39. *If r = a cos wt + b sin wt show that*

$$r\frac{dr}{dt} = w\, a \times b \text{ and } \frac{d^2r}{dt^2} = -w^2 r$$

Solution. Given r = a cos wt + b sin wt . . . (i)

$$\therefore \frac{dr}{dt} = -a\, w \sin wt + b\, w \cos wt \quad ...(2)$$

$$\therefore\ r \times \frac{dr}{dt} = (a \cos wt + b \sin wt) \times (-aw \sin wt + bw\, bw \cos wt)$$

$= a \times bw \cos^2 wt - b \times aw \sin^2 wt$, [$\because$ a × a = 0 etc.]

$= (w \cos^2 wt + w \sin^2 wt)\, a \times b$, [$\because$ –b × a = a × b]

= w a × b.

Also from (ii) we get

$$\frac{d^2r}{dt^2} = -aw^2 \cos wt - bw^2 \sin wt$$

$= - w^2$ (a cos wt + b sin wt) $= - w^2r$, from (i).

40. *Evaluate* $\int_C F.dr$, *where* $F = x^2i + y^3j$ *and curve C is the of the parabola* $y = x^2$ *in the x-y plane from (0, 0) to (1, 1).*

Solution:

We shall illustratic two methods for the solution of such a problem.

Method 1. The curve C is the parabola $y = x^2$ from (0, 0) to (1, 1).

Let x = t; then $y = t^2$. If r is the position vector of any point (x, y) on C, then

$$r(t) = xi + yj = ri + t^2j.$$

$$\therefore \frac{dr}{dt} = i + 2r\ j.$$

Also in terms of t. $F = t^2i + t^2j$.

At the point (0, 0), t = x = 0. At the point (1, 1), t = 1.

$$\therefore \int_C F.dr = \int_C \left(F.\frac{dr}{dt}\right)dt = \int_0^1 (t^2i + t^6j).(i + 2tj)dt$$

$$= \int_0^1 (t^2 + 2t^7)dt = \left[\frac{t^3}{3} + \frac{2t^8}{8}\right]_0^1 = \left[\frac{t^3}{3} + \frac{2t^8}{8}\right]_0^1 = \frac{4}{3} + \frac{1}{4} = \frac{7}{12}.$$

Method 2. In the x y-plane we have r = xi + yj.

$\therefore$ dr = dx i + dy j.

Therefore $F.dr = (x^2i + y^3j) . (dx\ i + dy\ j) = x^2dx + y^3dy$.

$$\therefore \int_C F.dr = \int_C (x^2dx + y^3dy).$$

Now along the curve C, $y = x^2$. Therefore dy = 2x dx.

$$\therefore \int_C F.dr = \int_{x=0}^1 [x^2dx + x^6(2x)dx]$$

$$= \int_0^1 (x^2 + 2x^7)dx = \left[\frac{x^3}{3} + \frac{2r^8}{8}\right]_0^1 = \frac{7}{12}.$$

41. *Evaluate* $\int_C F.dr$, *where* $F = (x^2 - y^2)\ i + xy\ j$ *and curve C is the are of the curve* $y = x^3$ *from (0, 0) to (2, 8).*

Solution:

The curve C is the curve $y = x^3$ from (0, 0) to (2, 8). Let x = t, then $y = j^3$. If r is the position vector of any point (x, y) on C, then

$r(t) = xi + yj = ti + t^3j.$

$$\therefore \int_C F.dr = i + 3t^2j.$$

All son in terms of t, $F = (t^2 - t^6)\, i + t^4j.$

At the point (0, 0), t = x = 0. At the point (2, 8), t = 2.

$$\therefore \int_C F.dr = \int_C \left(F.\frac{dr}{dt}\right)dt = \int_0^2 [(t^2 - t^6)i + t^4 j].(i + 3t^2 j)dt$$

$$= \int_0^2 [(t^2 - t^6) + 3t^6]dt = \int_0^2 [t^2 + 2t^6]dt$$

$$= \left[\frac{t^3}{3} + \frac{2t^7}{7}\right]_0^2 = \left[\frac{8}{3} + \frac{256}{7}\right] = \frac{824}{21}.$$

42. *If $F = 3xy\, i - y^2\, j$, evaluate $\int_C F.dr$, where C is the curve in the x y-plane, $y = x^2$, from (0, 0) to (1, 2)*

Solution: The parametric equations of the parabola $y = 2x^2$ can be taken as

$x = t,\ y = 2t^2.$

At the point (0, 0), x = 0 and so t = 0. Again at the point (1, 2), x = i and so t = 1.

Now $$\int_C F.dr = \int_C (3xy\, i - y^2 j).(dx\, i + dy\, j)$$

[Θ r = xi + yj, so that dr = dx i + dy j]

$$= \int_C (3xy\, dx - y^2 dy) = \int_{t=0}^1 \left(3xy\frac{dx}{dt} - y^2\frac{dy}{dt}\right)dt$$

$$= \int_0^1 (3.t.2t^2.1 - 4t^4.4t)dt$$

[Θ x = t, $y = 2t^2$ so that dx/dt = 1 and dy/dt = 4t]

$$= \int_0^1 (6t^3 - 16t^5)dt = \left[6.\frac{r^4}{4} - 16.\frac{r^6}{6}\right]_0^1 = \frac{6}{4} - \frac{16}{6} = \frac{3}{2} - \frac{8}{3} = -\frac{7}{6}.$$

43. *Evaluate* $\int_C F.dr$ *where F is* $x^2y^2\ i + y\ j$ *and C is* $y^2 = 4x$ *in the x y-plane from (0, 0) to (4, 4).*

Solution: In the x y-plane, we have r = x i + j so that dr = dx i + dy j.

$\therefore$ $F.dr = (x^2y^2\ i + j) . (dx\ i + dy\ j)$

$= x^2y^2\ dx + y\ dy.$ [Θ i . i = i, i . j = 0 etc.]

$$\therefore \int_C F.dr = \int_C (x^2y^2dx + y\,dy),$$

where C is the given curve $y^2 = 4x$ from (0, 0), to (4, 4)

$$= \int_C x^2y^2dx + \int_C y\,dy = \int_{x=0}^{4} x^2(4x)dx + \int_{y=0}^{4} y\,dy \qquad [\Theta\ y^2 = 4x]$$

$$= 4\left[\frac{1}{4}x^4\right]_0^4 + \left[\frac{1}{2}y^2\right]_0^4 = 256 + 8 = 264.$$

44. *If* $F = (3x^2 + 6y)\ i - 14yz\ j + 20xz^2\ k$, *evaluate* $\int_C F.dr$ *where C isthe consisting of the straight lines from (0, 0, 0) to (1, 0, 0) then to (1, 1, 0) and then to (1, 1, 1).*

Solution:

We have $F . dr = [(3x^2 + 6y)\ i - 14yz\ j + 20xz^2\ k] . (dx\ i + dy\ j + dzk)$

$= (3x^2 + 6y)\ dx - 14yz\ dy + 20xz^2\ dz.$

Let C_1 denote the straight line joining (0, 0, 0) to (1, 0, 0). Then alorg C_1, x = 1, z = 0 and x goes from 0 to 1. Obviously along C_1, dy = 0 and dz = 0.

Let C_2 denote the straight line joining (1, 0, 0) to (1, 1, 0). Then along C_2, x = 1, z = 0 and y varies from 0 to 1. Obviously along C_2, dx = 0 and dz = 0.

Again let C_3 denote the straight line joining (1, 1, 0) to (1, 1, 1). Along C_3, we have

x = 1, y = 1 so that dx = 0, dy= 0. Obviously along C_3, z varres from 0 to 1.

Now $$\int_C F.dr = \int_C F.dr + \int_{C_1} F.dr + \int_{C_2} F.dr + \int_{C_3} F.dr$$

We have $\int_{C_1} F.dr = \int_{C_1} [(3x^2 + 6y)dx - 14yz\,dy + 20xz^2dz]$

$= \int_{x=0}^{1} 3x^2dx$ [$\because$ along C_1, y = 0, z = 0, dy = 0 and x varies from 0 to 1]

$= 3\left[\frac{1}{3}x^3\right]_0^1 = 1.$

Again $\int_{C_2} F.dr = \int_{y=0}^{1} -14yz\,dy$

[$\because$ along C_2, dx = 0, dz = 0 and varies from 0 to 1]

$= \int_{y=0}^{1} -14y.0dy$ [$\because$ along C_2, z = 0]

$= 0.$

Finally $\int_{C_3} F.dr = \int_{z=0}^{1} 20xz^2dz$

[$\because$ along C_3, dx = 0, dy = 0 and z varies from 0 to 1]

$= 20\left[\frac{1}{3}z^3\right]_0^1 = \frac{20}{3}$

Hence $\int_{C} F.dr = 1 + 0 + \frac{20}{3} = \frac{23}{3}.$

45. *Integrate the function $F = x^2\,i - 2y\,j$ from the point (0, 0) to (1, 1) along parabola $y^2 = x$.*

Solution:

Here the parabola $y^2 = x$ lies in the xy-plane. It r is the position vector of any point (x, y) on this plane, then r = xi + yj so that

$$dr = dx\ i + dy\ j.$$

Let C be the curve $y^2 = x$ from (0, 0) to (1, 1). The parametric equations of $y^2 = x$ can be taken as $x = t^2$, y = t. At the point (0, 0) we have t = 0 and at the point (1, 1) we have t = 1.

$\therefore \int_{C} F.dr = \int_{C} (x^2i - xy\,j).(dx\,i + dy\,j)$

$= \int_{C} (x^2dx - xy\,dy)$, where $x = t^2$, y = t

$$= \int_{t=0}^{1}(t^4.2t\,dt - t^2.t\,dt) = \int_0^1(2t^5 - t^3)dt$$

$$= 2\left[\frac{1}{6}t^6\right]_0^1 - \left[\frac{1}{4}t^4\right]_0^1 = \frac{1}{3} - \frac{1}{4} = \frac{1}{12}.$$

46. *Evaluate* $\int$ F.dr *where* $F = (x^2 + y^2)\, i + xy\, j$ *and the curve C is the are of the parabola* $y = x^2$ *from (0, 0) to (3, 9) in the xy-plane.*

Solution: We have $\int_C F.dr$

$$= \int_C [(x^2 + y^2)i + xy\, j].(dx\, i + dy\, j) \qquad [\because r = ri + yj]$$

$$= \int_C [(x^2 + y^2)dx + xy\, dy] = \int_C (x^2 + y^2)dx + \int_C xy\, dy$$

$$= \int_{x=0}^{3}(x^2x^4)dx + \int_{y=0}^{9} y^{1/2}.y\, dy$$

[Θ $y^2 = r$ and for the curve C, x varies form 0 to 3 and y varies from 0 to 9]

$$= \left[\frac{1}{3}x^3 + \frac{1}{5}x^5\right]_0^3 + \frac{2}{5}\left[y^{5/2}\right]_0^9$$

$$= 9 + \frac{243}{5} + \frac{2}{5}.243 = \frac{1}{5}[45 + 243 + 486] = \frac{774}{4}.$$

47. *Evaluate* $\int$ F.dr *along the curve* $x^2 + y^2 = 1$, $z = 1$ *in the positive direction from (0, 1, 1) to (1, 0, 1),*

where $F = (2x + yz)\, i + xz\, j + (xy + 2z)\, k$.

Solution:

Let the given curve be denoted by C ane let A and B be points (0, 1, 1) and (1, 0, 1) respectively.

Along the given curve C, we have r = xi + yj + zk

$\therefore$ dr = dx i + dy j + dz k.

$$\therefore \int F.dr = \int_A^B [(2x + yz)i + xz\, j + (xy + 2z)k].(dx\, i + dy\, j + dz\, k)$$

$$= \int_A^B [(2x + yz)dx + xz\, dy + (xy + 2z)dz]. \qquad ...(1)$$

In moving from A to B, x varies from 0 to 1, y varies from 1 to 0 and z remains constant. W have z = 1 and so dz = 0.

Hence from (1)

$$\int_C F.dr = \int_0^1 (2x+y)dx + \int_1^0 x\,dy + 0$$

$$= \int_0^1 [2x+\sqrt{(1-x^2)}]dx - \int_0^1 \sqrt{(1-y^2)}dy = [x^2]_0^1 = 1,$$

The last two integrals cancel by a property of definite integrals.

48. *Find the work done when a force $F = (x^2 - y^2 + x)\, i - (2xy + y)\, j$ moves a particle in xy-plane from (0, 0) to (1, 1) along the parabola $y^2 = x$.*

Solution:

Let C denote the are of the parabola $y^2 = x$ from the point (0, 0) to the point (1, 1). The parametric equations of the parabola $y^2 = x$ can be taken as $x = t^2$, $y = t$. At the point (0, 0), t = 0 and at the point (1, 1), t = 1. The required work done

$$= \int_C F.dr = \int_C \{(x^2 - y^2 + x)i - (2xy + y)j\}.(dxi + dyj)$$

$$= \int_C [(x^2 - y^2 x)dx - (2xy + y)dy]$$

$$= \int_{t=0}^1 \left[(x^2 - y^2 + x)\frac{dx}{dt} - (2xy + y)\frac{dy}{dt}\right]dt$$

$$= \int_0^1 [(t^4 - t^2 + t^2).2t - (2t^3 + t).1]dt$$

$$= \int_0^1 [2t^5 - 2t^3 - t]dt = \left[2.\frac{t^6}{6} - 2.\frac{t^4}{4} - \frac{t^2}{2}\right]_0^1 = \frac{1}{3} - \frac{1}{2} - \frac{1}{2} = -\frac{2}{3}.$$

49. *Find the work done in moving a particle in a force field*

$$F = 3x^2 i + (2xz - y)\, j + zk$$

along the line joining (0, 0, 0) to (2, 1, 3).

Solution:

Let C be the straight line joining (0, 0, 0) to (2, 1, 3). The parametric equations of this straight line are

$$\frac{x-0}{2-0} = \frac{y-0}{1-0} = \frac{z-0}{3-0} = t \text{ say}$$

or $x = 2t,\ y = t,\ z = 3t.$

At the point (0, 0, 0), we have $t = 0$ and at the point (2, 1, 3), we have $t = 1$.

The required work done

$$= \int_C \mathbf{F}.d\mathbf{r} = \int_C [3x^2 i + (2xz - y) j + 3k].(dx\, i + dy\, j + dz\, k)$$

$$= \int_C [3x^2 dx + (2xz - y) dy + z dz]$$

$$= \int_C [3(2t)^2 2dt + \{2(2t)(3t) - t\} dt + (3t) 3 dt]$$

$$= \int_{t=0}^{1} (24t^2 + 12t^2 - t + 9t) dt = \int_0^1 (36t^2 + 8t) dt$$

$$= 36 \left[\frac{1}{3} t^3\right]_0^1 + 8 \left[\frac{1}{2} t^2\right]_0^1 = 12 + 4 = 16.$$

50. *If $\phi(x, y, z) = xy^2 z$ and $f = xz\, i - xy\, j + yz^2\, k$, show that $\frac{\partial^2}{\partial x^2 \partial z}(\phi f)$ at $(2, -1, 1)$ is $4i + 2j$.*

Solution: We have

$$\phi f = xy^2 z\,(xz\, i - xy\, j + yz^2 k)$$

$$= x^2 y^2 z^2\, i - x^3 y^3 z\, j + xy^3 z^3\, k.$$

$$\therefore \quad \frac{\partial}{\partial x}(\phi f) = 2xy^2 z^2\, i - 2xy^3 z\, j + y^3 z^3\, k.$$

$$\frac{\partial^2}{\partial x^2}(\phi f) = \frac{\partial}{\partial x}\left[\frac{\partial}{\partial x}(\phi f)\right] = \frac{\partial}{\partial x}[2xy^2 z^2 i - 2xy^3 z j + y^3 z^3 k]$$

$$= 2y^2 z^2\, i - 2y^3 z\, j + 0\, k = 2y^2 z^2\, i - 2y^3 z\, j$$

and $$\frac{\partial^3}{\partial x^2 \partial z}(\phi f) = \frac{\partial}{\partial z}\left[\frac{\partial^2}{\partial x^2}(\phi f)\right]$$

$$= \frac{\partial}{\partial z}(2y^2 z^2 i - 2y^3 - j) = 4y^2 z\, i - 2y^3\, j.$$

$\therefore$ at the point $(2, -1, 1)$, $\frac{\partial^3}{\partial x^2 \partial z}(\phi f)$

$= 4 \cdot (-1)^2 \cdot 1\, i - 2 \cdot (-1)^3\, j = 4\, i + 2\, j.$

51. *If $f = (2x^2y - x^4)\, i + (e^{xy} - y \sin x)\, j + x^2 \cos y\, k$, erify that*

$$\frac{\partial^2 f}{\partial y\, \partial x} = \frac{\partial^2 f}{\partial x\, \partial y}.$$

Solution: We have $\dfrac{\partial f}{\partial x}$

$$= \left[\frac{\partial}{\partial x}(2x^2y - x^4)\right] i + \left[\frac{\partial}{\partial x}(e^{xy} - y \sin x)\right] j + \left[\frac{\partial}{\partial x}(x^2 \cos y)\right] k$$

$$= (4xy - 4x^3)\, i + (ye^{xy} - y \cos x)\, j + 2x \cos y\, k.$$

$$\therefore \quad \frac{\partial^2 f}{\partial x\, \partial y} = \frac{\partial}{\partial y}\left(\frac{\partial f}{\partial x}\right)$$

$$= \frac{\partial}{\partial y}[(xxy - 4x^3) i + (ye^{xy} - y \cos x) j + 2x \cos y\, k]$$

$$= \left[\frac{\partial}{\partial y}(4xy - 4x^3)\right] i + \left[\frac{\partial}{\partial y}(ye^{xy} - y \cos x\right] j + \left[\frac{\partial}{\partial y}(2x \cos y)\right] k$$

$$= (4x)\, i + (e^{xy} + xy\, e^{xy} - \cos x)\, j + (-2x \sin y)\, k$$

$$= 4x\, i + (e^{xy} + xy\, e^{xy} - \cos x)\, j - 2x \sin y\, k. \qquad \text{...(i)}$$

Again $$\frac{\partial f}{\partial y} = \left[\frac{\partial}{\partial y}(2x^2y - x^4)\right] j + \left[\frac{\partial}{\partial y}(e^{xy} - y \sin x)\right]$$

$$j + \left[\frac{\partial}{\partial x}(x^2 \sin y)\right] k$$

$$= 2x^2\, i + (xe^{xy} - \sin x)\, j - x^2 \sin y\, k.$$

$$\therefore \quad \frac{\partial^2 f}{\partial y\, \partial x} = \frac{\partial}{\partial x}\left(\frac{\partial f}{\partial y}\right) = \left[\frac{\partial}{\partial x}(2x^2)\right] i + \left[\frac{\partial}{\partial x}(xe^{xy} - \sin x)\right]$$

$$j - \left[\frac{\partial}{\partial x}(x^2 \sin y)\right] k$$

$$= 4x\, i + (e^{xy} + xye^{xy} - \cos x)\, j - 2x \sin y\, k. \qquad \text{...(2)}$$

From (1) and (2), we have

$$\frac{\partial^2 f}{\partial x\, \partial y} = \frac{\partial^2 f}{\partial y\, \partial x}.$$

52. *If $f = xyz\, i + xz^2\, j - y^3\, k$ and $v = x^3\, i - xyz\, j + x^2 z\, k$, calculate $\dfrac{\partial^2 u}{\partial y^2} \times \dfrac{\partial^2 v}{\partial x^2}$ at the point (1, 1, 0).*

Solution: We have

$$\frac{\partial u}{\partial y} = \left[\frac{\partial}{\partial y}(xyz)\right]i+\left[\frac{\partial}{\partial y}(xz^2)\right]j-\left[\frac{\partial}{\partial y}(y^3)\right]k$$

$$= xz\ i + 0\ j - 3y^2\ k = xz\ i - 3y^2\ k.$$

and $$\frac{\partial^2 u}{\partial y^2} = \frac{\partial}{\partial y}\left(\frac{\partial u}{\partial y}\right) = 0\ i - 6y\ k = -6y\ k.$$

Again $$\frac{\partial v}{\partial x} = \left[\frac{\partial}{\partial x}(x^3)\right]i-\left[\frac{\partial}{\partial x}(xyz)\right]j+\left[\frac{\partial}{\partial x}(x^2z)\right]k$$

$$= 3x^2\ i - yz\ j + 2xz\ k$$

and $$\frac{\partial^2 v}{\partial x^2} = \frac{\partial}{\partial x}\left(\frac{\partial v}{\partial x}\right)$$

$$= 6x\ i - 0\ j + 2z\ k = 6x\ i + 2z\ k.$$

$\therefore$ $$\frac{\partial^2 u}{\partial y^2}\times\frac{\partial^2 v}{\partial x^2} = (-6y\ k) \times (6x\ i + 2z\ k)$$

$$= -36xy\ k \times i - 12yz\ k \times k$$

$\therefore$ at the point (1, 1, 0), we have [$\because$ $k \times i = j$ and $k \times k = 0$]

$$\frac{\partial^2 u}{\partial y^2}\times\frac{\partial^2 v}{\partial x^2} = -36\ j.$$

53. *If* $A = x^2yz\ i - 2xz^3\ j + xz^2\ k$, $B = 2z\ i + y\ j - x^2\ k$, *find the value of* $\frac{\partial^2}{\partial x\,\partial y}(A\times B)$ *at (1, 0 – 2).*

Solution: We have

$$A \times B = \begin{vmatrix} i & j & k \\ x^2yz & -2xz^3 & xz^2 \\ 2z & y & -x^2 \end{vmatrix}$$

$$= (2x^3z^3 - xyz^2)\ i + (2xz^3 + x^4yz)\ j + (x^2y^2z + 4xz^4)\ k.$$

$\therefore$ $$\frac{\partial}{\partial y}(A\times B) = -xz^2\ i + x^4z\ j + 2x^2yz\ k.$$

Again $$\frac{\partial^2}{\partial x\,\partial y}(A\times B) = \frac{\partial}{\partial x}\left\{\frac{\partial}{\partial y}(A\times B)\right\}$$

$= -z^2 j + 4x^3z j + 4xyz k.$ ...(1)

Putting x = 1, y = 0 and z = − 2 in (1), we get the required derivative at the point (1, 0, − 2) = − 4 i − 8j.

54. *If f (x, y, z) = $3x^2y - y^3z^2$, find grad f at the point (1, − 2, − 1).*

Solution: We have

$$\text{grad } f = \nabla f = \left(i\frac{\partial}{\partial x} + j\frac{\partial}{\partial y} + k\frac{\partial}{\partial z}\right)(3x^2y - y^3z^2)$$

$$= i\frac{\partial}{\partial x}(3x^2y - y^3z^2) + j\frac{\partial}{\partial y}(3x^2y - y^3z^2) + k\frac{\partial}{\partial z}(3x^2y - y^3z^2)$$

= i (6xy) + j(3x2 − 3y2z2) + k (−2y3z)

$= 6xy\, i + (3x^2 - 3y^2z^2)\, j - 2y^3z\, k.$

Putting x = 1, y = − 2, z = − 1, we get

$\nabla f = 6\,(1)\,(-2)\, i + \{3(1)^2 - 3\,(-2)^2\,(-1)^2\}\, j - 2\,(-2)^3\,(-1)\, k$

= − 121 − 9j − 16k.

55. *Show that*

(i) *grad (r.a) = a.*

(ii) *grad [r, a, b] = a × b,*

where a and be are constant vectors.

Solution:

(i) Let a = a_1 i + a_2j + a_3k. Then a_1, a_2, a_3 are constants.

Also r = xi + yj + zk.

∴ r.a = $a_1x + a_2y + a_3z$.

∴ grad (r.a) = ∇ (r.a) = ∇ ($a_1x + a_2y + a_3z$)

$$= i\frac{\partial}{\partial x}(a_1x + a_2y + a_3z) + j\frac{\partial}{\partial y}(a_1x + a_2y + a_3z) + k\frac{\partial}{\partial z}(a_1x + a_2y + a_3z)$$

$= a_1\, i + a_2\, j + a_3\, k = a.$

(ii) grad [r, a, b] = grad {r.(a × b)}, where a × b is a constant vector = a × b, as in part (i).

56. *If φ (x, y, z) = $x^2y + y^2x + z^2$, find ∇φ at the point (1, 1, 1).*

Solution: We have

$$\nabla\phi = \frac{\partial\phi}{\partial x}i + \frac{\partial\phi}{\partial y}j + \frac{\partial\phi}{\partial z}k$$

$= (2xy + y^2)\, i + (x^2 + 2xy)\, j + 2z\, k.$

Putting x = 1, y = 1, z = 1, we get

$\nabla\phi$ at the point (1, 1, 1) = 3i + 3j + 2k.

57. *Find grad f, where f is given by $f = x^3 - y^3 + xz^2$, at the point (1, –1, 2).*

Solution:

We have

$$\text{grad } f = \nabla f = \frac{\partial f}{\partial x} i + \frac{\partial f}{\partial y} j + \frac{\partial f}{\partial z} k$$

$$= \left[\frac{\partial f}{\partial x}(x^3 - y^3 + xz^2)\right] i + \left[\frac{\partial}{\partial y}(x^3 - y^3 + xz^2\right] j + \left[\frac{\partial}{\partial z}(x^3 - y^3 + xz^2)\right] k$$

$$= (3x^2 + z^2)\, i + (-3y^2)\, j + 2xz\, k.$$

Putting x = 1, y = – 1, z = 2, we get

∇f at the point (1,–1, 2) = 7i – 3j + 4k.

58. *Prove that* $A \,.\, \left(\nabla \frac{1}{r}\right) = -\frac{A.r}{r^3}$.

Solution: We have

$$\nabla\left(\frac{1}{r}\right) = i\frac{\partial}{\partial x}\left(\frac{1}{r}\right) + j\frac{\partial}{\partial y}\left(\frac{1}{r}\right) + k\frac{\partial}{\partial}\left(\frac{1}{r}\right)$$

$$= i\left(-\frac{1}{r^2}\frac{\partial r}{\partial x}\right) + j\left(-\frac{1}{r^2}\frac{\partial r}{\partial y}\right) + k\left(-\frac{\partial}{r^2}\frac{\partial r}{\partial z}\right)$$

$$= -\frac{1}{r^2}\left(\frac{\partial r}{\partial x} i + \frac{\partial r}{\partial y} j + \frac{\partial r}{\partial z} k\right) = -\frac{1}{r^2}\left(\frac{x}{r} i + \frac{y}{r} j + \frac{z}{r} k\right)$$

$$= -\frac{1}{r^3}(xi + yj + zk) = -\frac{1}{r^3} r.$$

$$\therefore\ A \,.\, \nabla\left(\frac{1}{r}\right) = A \,.\left(-\frac{r}{r^3}\right) = -\frac{A.r}{r^3}.$$

59. *Prove that* $\nabla r^{-3} = -3r^{-5}\, r$.

Solution: We have r = x i + y j + z k

and $\quad r = |r| = \sqrt{(x^2 + y^2 + z^2)}$

so that $r^2 = x^2 + y^2 + z^2$.

Now $\quad \nabla r^{-3} = \left(\frac{\partial}{\partial x} r^{-3}\right) i + \left(\frac{\partial}{\partial y} r^{-3}\right) j + \left(\frac{\partial}{\partial z} r^{-3}\right) k$

$$= -3r^{-4} \frac{\partial r}{\partial x} i - r^{-4} \frac{\partial r}{\partial y} - 3r^{-4} \frac{\partial r}{\partial z} k$$

$$= -3r^{-4} \left(\frac{\partial r}{\partial x} i + \frac{\partial r}{\partial y} j + \frac{\partial r}{\partial z} k\right). \quad \text{...(1)}$$

Differentiating both sides of

$$r^2 = x^2 + y^2 + z^2$$

partially w.r.t. w, we have

$$2r \frac{\partial r}{\partial x} = 2x \text{ or } \frac{\partial r}{\partial x} = \frac{x}{r}.$$

Similarly $\frac{\partial r}{\partial y} = \frac{y}{r}$ and $\frac{\partial r}{\partial z} = \frac{z}{r}$.

$\therefore$ from (1), we have

$$\nabla r^{-3} = -3r^{-4} \left(\frac{x}{r} i + \frac{y}{r} j + \frac{z}{r} k\right)$$

$$= -3r^{-5} (xi + yj + zk) = -3r^{-5} r.$$

60. *Prove that* $\nabla\phi \,.\, dr = d\phi$.

Solution: We have

$$\nabla\phi = \frac{\partial\phi}{\partial x} i + \frac{\partial\phi}{\partial y} j + \frac{\partial\phi}{\partial z} k. \quad \text{...(1)}$$

Also $\quad r = xi + yj + zk$

$\Rightarrow dr = dx\, i + dy\, j + dz\, k. \quad$ ...(2)

From (1) and (2), $\nabla\phi \,.\, dr$

$$= \left(\frac{\partial\phi}{\partial x} i + \frac{\partial\phi}{\partial y} j + \frac{\partial\phi}{\partial z} k\right).(dx\, i + dy\, j + dz\, k)$$

$$= \frac{\partial\phi}{\partial x} dx + \frac{\partial\phi}{\partial y} dy + \frac{\partial\phi}{\partial z} dz = d\phi.$$

61. *Show that* $\frac{d\phi}{ds} = \nabla\phi \,.\, \frac{dr}{ds}$, *where* $r = xi + yj + zk$ *and* ϕ *is a function of* x, y *and* z.

Solution: We have

$$\nabla\phi = \frac{\partial\phi}{\partial x}i+\frac{\partial\phi}{\partial y}j+\frac{\partial\phi}{\partial z}k \qquad ...(1)$$

Also $\quad r = xi + yj + zk$

$$\Rightarrow \frac{dr}{ds} = \frac{dx}{ds}i+\frac{dy}{ds}j+\frac{dz}{ds}k. \qquad ...(2)$$

From (1) and (2), we have

$$\nabla\phi \cdot \frac{dr}{ds} = \left(\frac{\partial\phi}{\partial x}i+\frac{\partial\phi}{\partial y}j+\frac{\partial\phi}{\partial z}k\right)\cdot\left(\frac{dx}{ds}i+\frac{dy}{ds}j+\frac{dz}{ds}k\right)$$

$$= \frac{\partial\phi}{\partial x}\frac{dx}{ds}+\frac{\partial\phi}{\partial y}\frac{dy}{ds}+\frac{\partial\phi}{\partial z}\frac{dz}{ds} = \frac{d\phi}{ds}.$$

62. *ρ and p are two scalar point functions such that ρ is a function of p: show that*

$$\nabla\rho = \frac{d\rho}{dp}\nabla p.$$

Solution:

We have

$$\nabla\rho = \frac{\partial\rho}{\partial x}i+\frac{\partial\rho}{\partial y}j+\frac{\partial\rho}{\partial z}k.$$

Since ρ is a function of p, therefore

$$\frac{\partial\rho}{\partial x} = \frac{d\rho}{dp}\frac{\partial p}{\partial x},\ \frac{\partial\rho}{\partial y} = \frac{d\rho}{dp}\frac{\partial p}{\partial y},\ \frac{\partial\rho}{\partial z} = \frac{d\rho}{dp}\frac{\partial p}{\partial z}.$$

$\therefore$ from (1), we have

$$\nabla\rho = \frac{d\rho}{dp}\frac{\partial p}{dx}i+\frac{\partial\rho}{dp}\frac{\partial p}{\partial y}j+\frac{d\rho}{dp}\frac{\partial p}{\partial z}k$$

$$= \frac{d\rho}{dp}\left(\frac{\partial p}{\partial x}i+\frac{\partial p}{\partial y}j+\frac{\partial p}{\partial z}k\right) = \frac{d\rho}{dp}\nabla p.$$

63. *If* $f = (3r^2 - 4r^{1/2} + 6r^{-1/3})$, show that $\nabla\phi = 2(3 - r^{-3/2} - r^{-7/3})$ r.

Solution:

We have $r = xi + yj + zk$

and $r = |r| = \sqrt{(x^2 + y^2 + z^2)}$

so that $\quad r^2 = x^2 + y^2 + z^2$.

Now ϕ is a function of r.

$$\therefore \quad \nabla\phi = \frac{d\phi}{dr}\nabla r \qquad [\because \nabla f(r) = f'(r)\,\nabla r]$$

$$= \left[6r - 4.\left(\frac{1}{2}\right)r^{-1/2} + 6.\left(-\frac{1}{3}\right)r^{-4/3}\right]\nabla r$$

$$= (6r - 2r^{-1/2} - 2r^{-4/3})\,\frac{1}{r}r \qquad \left[\because \nabla r = \frac{1}{r}r\right]$$

$$= (6 - 2r^{-3/2} - 2r^{-7/3})\,r$$

$$= 2\,(3 - r^{-3/2} - r^{-7/3})\,r.$$

64. *Evaluate* $\int_C F.dr$ *where* $F = c\,[-3a\sin^2 t\cos t\, i + a\,(2\sin t - 3\sin^3 t)\, j + b\sin 2t\, k]$ *and C is given by* $r = a\cos t\, i + a\sin t\, j + bt\, k$ *from* $t = \pi/4$ *to* $\pi/2$.

Solution:

We have $\frac{dr}{dt} = -a\sin t\, i + a\cos t\, j + b\, k.$

$$\therefore \int_C F.dr = \int_{t=\pi/4}^{\pi/2}\left(F.\frac{dr}{dt}\right)dt$$

$$= \int_{\pi/4}^{\pi/2} c[(-3a\sin^2 t\cos t)(-a\sin t) + a(2\sin t - 3\sin^3 t)(a\cos t) + (b\sin 2t)(b)]dt$$

$$= c\int_{\pi/4}^{\pi/2}[3a^2\sin^3 t\cos t + a^2(2\sin t\cos t - 3\sin^3 t\cos t) + b^2\sin 2t]dt$$

$$= c\int_{\pi/4}^{\pi/2}(a^2 + b^2)\sin 2t\,dt = c\,(c^2 + b^2)\left[-\frac{1}{2}\cos 2t\right]_{\pi/4}^{\pi/2}$$

$$= c\,(a^2 + b^2)\left[-\frac{1}{2}\{(-1) - 0\}\right] = \frac{1}{2}c(a^2 + b^2).$$

65. *If* $u = x + y + z$, $v = x^2 + y^2 + z^2$, $w = yz + zx + xy$, *prove that* $(grd\ u).[(grad\ v) \times (grad\ w)] = 0.$

Solution: We have

$$\text{grad } u = \frac{\partial u}{\partial x}i + \frac{\partial u}{\partial y}j + \frac{\partial u}{\partial z}k$$

$$= 1\, i + 1\, j + 1\, k = i + j + k,$$

$$\text{grad } v = \frac{\partial v}{\partial x}i + \frac{\partial v}{\partial y}j + \frac{\partial v}{\partial z}k = 2x\, i + 2y\, j + 2z\, k$$

$$\text{and grad } w = \frac{\partial w}{\partial x}i + \frac{\partial w}{\partial y}j + \frac{\partial w}{\partial z}k$$

$$= (y + z)\, i + (z + x)\, j + (x + y)\, k.$$

$\therefore$ grad u . [(grad v) × (grad w)] = scalar triple product of the vectors grad u. grad v and grad w

$$= \begin{vmatrix} 1 & 1 & 1 \\ 2x & 2y & 2z \\ y+z & z+x & x+y \end{vmatrix} = 2 \begin{vmatrix} 1 & 1 & 1 \\ x & y & z \\ y+z & z+x & x+y \end{vmatrix}$$

$$= 2 \begin{vmatrix} 1 & 1 & 1 \\ x+y+z & x+y+z & x+y+z \\ y+z & z+x & x+y \end{vmatrix}, \text{ by } R_2 + R_3$$

$$= 2\,(x + y + z) \begin{vmatrix} 1 & 1 & 1 \\ 1 & 1 & 1 \\ y+z & z+x & x+y \end{vmatrix}$$

$= 2\,(x+y+z), 0$, the first two rows of the determinant being identical

$= 0.$

EXERCISES

1. Evaluate $\int_C F.dr$, where F = xy i + yz j + zx k and curve C is r = t –i + t^2 j + t^3 k,t varying from – 1 to + 1.
2. Gauss's divergence theorem is used to transform:

   (a) a volume integral into surface integral

   (b) a line integral into surface integral.
3. Verify Stoke's theorem for $\vec{F} = (2x - y)\hat{i} - yz^2\hat{j} - y^2 z\hat{k}$ where s is the upper half surface of the sphere $x^2 + y^2 + z^2 = 1$ and C is its boundary.
4. What are Solonoidal and Lamellar or scalar potential fields ? Give examples.
5. Evaluate $\int_s$ F . n dS, where F = $x^2 + y^2 + z^2$ k and S is the surface of the solid cut off by the plane x + y + z = a from the first octant.

6. Verify Stokes' theorem for the function $F = x(ix + jy)$, integrated round the square, in the plane $z = 0$, whose sides are along the lines $x = 0$, $y = 0$, $x = a$, $y = a$.

7. If $F = (y^2 + z^2 + x^2)\, i + (z^2 + x^2 - y^2)\, j + (x^2 + y^2 - z^2)\, k$, evaluate $\int \text{curl } F.\, n\, dS$, integrated over the portion of the surface $x^2 + y^2 - 2ax + az = 0$ above the plane $z = 0$ and verify Stoke's theorem.

8. If $f(x, y, z) = 3x^2y - y^3z^2$, find grad f at the point $(1, -2, -1)$.

9. Apply Stoke's theorm to prove that

   $\int_c (y\, dx + z\, dy + x\, dz) = -2\sqrt{2}\, \pi\, a^2$,

   where C is te curve given by $x^2 + y^2 + z^2 - 2a\, x = 2a\, y = 4$, $x + y = 2a$ and begins at the point $(2a, 0, 0)$ and goes at first below the xy-plane.

10. State and prove Gauss's divergence theorem. What is its importance?

11. Find $\int_C t.dr$ where t is the unit tangent vector and C is the unit circle, in xy-plane, with centre at the origin.

12. ρ and p are two scalar point functions such that ρ is a function of p: show that

$$\nabla\rho = \frac{d\rho}{dp}\nabla p.$$

13. If $f = (2x^2y - x^4)\, i + (e^{xy} - y \sin x)\, j + x^2 \cos y\, k$, erify that

$$\frac{\partial^2 f}{\partial y\, \partial x} = \frac{\partial^2 f}{\partial x\, \partial y}.$$

14. Evaluate $\int_s F\, .\, n\, dS$ where S denotes the surface of the ellipsoid $(x^2/a^2) + (y^2/b^2) + (z^2/c^2) = 1$ and $F = x^2 i + y^2 j + z^2 k$.

    Find $\int F.\, dr$ over both the curvesx and explain why despite the fact that the beginning and end points of curves are the same, the two integrals are not equal.

15. Find the work done when a force $F = (x^2 - y^2 + x)\, i - (2xy + y)\, j$ moves a particle in xy-plane from (0, 0) to (1, 1) along the parabola $y^2 = x$.

16. Evaluate $\int_c (e^{-x} \sin y\, dx + e^{-x} \cos y\, dy)$ by Green's theorem when C is the rectangle whose vertices are

$$(0, 0), (\pi, 0)\left(\pi, \frac{1}{2}\pi\right) \text{ and } \left(0, \frac{1}{2}\pi\right).$$